SCHAUM'S OUTLINE OF

THEORY AND PROBLEMS

of

MECHANICAL
VIBRATIONS

S. GRAHAM KELLY, Ph.D.

Associate Professor of Mechanical Engineering and Assistant Provost
The University of Akron

SCHAUM'S OUTLINE SERIES
McGRAW-HILL

New York   St. Louis   San Francisco   Auckland   Bogotá   Caracas
Lisbon   London   Madrid   Mexico City   Milan   Montreal   New Delhi
San Juan   Singapore   Sydney   Tokyo   Toronto

S. GRAHAM KELLY is Associate Professor of Mechanical Engineering and Assistant Provost at The University of Akron. He has been on the faculty at Akron since 1982, serving before at the University of Notre Dame. He holds a B.S. in Engineering Science and Mechanics and an M.S. and a Ph.D. in Engineering Mechanics from Virginia Tech. He is also the author of *Fundamentals of Mechanical Vibrations* and the accompanying software *VIBES,* published by McGraw-Hill.

Credits

Selected material reprinted from S. Graham Kelly, *Fundamentals of Mechanical Vibrations,* © 1993 McGraw-Hill, Inc. Reprinted by permission of McGraw-Hill, Inc.

Maple-based material produced by permission of Waterloo Maple Inc. *Maple* and *Maple V* are registered trademarks of Waterloo Maple Inc.; *Maple* system itself copyrighted by Waterloo Maple Inc.

Mathcad material produced by permission of MathSoft, Inc.

Schaum's Outline of Theory and Problems of
MECHANICAL VIBRATIONS

5 6 7 8 9 10 11 12 13 14 15 16 17 18 19 20 PRS PRS 05 04 03 02

ISBN 0-07-034041-2

Sponsoring Editor: John Aliano
Production Supervisor: Suzanne Rapcavage
Project Supervision: The Total Book

Library of Congress Cataloging-in-Publication Data

Kelly, S. Graham.
  Schaum's outline of theory and problems of mechanical vibrations /
S. Graham Kelly.
    p.   cm. — (Schaum's outline series)
  Includes index.
  ISBN 0-07-034041-2
  1. Vibration—Problems, exercises, etc.   2. Vibration—Outlines,
syllabi, etc.   I. Title.
QA935.K383   1996
620.3′076—dc20   95-47554
                                                              CIP

McGraw-Hill

A Division of The McGraw-Hill Companies

To my son, Graham

Preface

A student of mechanical vibrations must draw upon knowledge of many areas of engineering science (statics, dynamics, mechanics of materials, and even fluid mechanics) as well as mathematics (calculus, differential equations, and linear algebra). The student must then synthesize this knowledge to formulate the solution of a mechanical vibrations problem.

Many mechanical systems require modeling before their vibrations can be analyzed. After appropriate assumptions are made, including the number of degrees of freedom necessary, basic conservation laws are applied to derive governing differential equations. Appropriate mathematical methods are applied to solve the differential equations. Often the modeling results in a differential equation whose solution is well known, in which case the existing solution is used. If this is the case the solution must be studied and written in a form which can be used in analysis and design applications.

A student of mechanical vibrations must learn how to use existing knowledge to do all of the above. The purpose of this book is to provide a supplement for a student studying mechanical vibrations that will guide the student through all aspects of vibration analysis. Each chapter has a short introduction of the theory used in the chapter, followed by a large number of solved problems. The solved problems mostly show how the theory is used in design and analysis applications. A few problems in each chapter examine the theory in more detail.

The coverage of the book is quite broad and includes free and forced vibrations of 1-degree-of-freedom, multi-degree-of-freedom, and continuous systems. Undamped systems and systems with viscous damping are considered. Systems with Coulomb damping and hysteretic damping are considered for 1-degree-of-freedom systems. There are several chapters of special note. Chapter 8 focuses on design of vibration control devices such as vibration isolators and vibration absorbers. Chapter 9 introduces the finite element method from an analytical viewpoint. The problems in Chapter 9 use the finite element method using only a few elements to analyze the vibrations of bars and beams. Chapter 10 focuses on nonlinear vibrations, mainly discussing the differences between linear and nonlinear systems including self-excited vibrations and chaotic motion. Chapter 11 shows how applications software can be used in vibration analysis and design.

The book can be used to supplement a course using any of the popular vibrations textbooks, or can be used as a textbook in a course where theoretical development is limited. In any case the book is a good source for studying the solutions of vibrations problems.

The author would like to thank the staff at McGraw-Hill, especially John Aliano, for making this book possible. He would also like to thank his wife and son, Seala and Graham, for patience during preparation of the manuscript and Gara Alderman and Peggy Duckworth for clerical help.

S. GRAHAM KELLY

v

Contents

Problems and Examples Also Found in the Companion SCHAUM'S ELECTRONIC TUTOR

Some of the problems and examples in this book have software components in the companion *Schaum's Electronic Tutor*. The Mathcad Engine, which "drives" the Electronic Tutor, allows every number, formula, and graph chosen to be completely live and interactive. To identify those items that are available in the Electronic Tutor software, please look for the Mathcad icons, ![Mathcad icon], placed under the problem number or adjacent to a numbered item. A complete list of these Mathcad entries follows below. For more information about the software, including the sample screens, see Appendix on page 333.

Problem 1.4	Problem 3.10	Problem 4.26	Problem 7.1	Problem 8.24
Problem 1.5	Problem 3.12	Problem 5.19	Problem 7.4	Problem 8.25
Problem 1.7	Problem 3.15	Problem 5.20	Problem 7.5	Problem 8.26
Problem 1.12	Problem 3.18	Problem 5.25	Problem 7.6	Problem 8.27
Problem 1.14	Problem 3.19	Problem 5.26	Problem 7.13	Problem 8.28
Problem 1.19	Problem 3.20	Problem 5.27	Problem 7.16	Problem 8.32
Problem 2.8	Problem 3.23	Problem 5.28	Problem 7.22	Problem 8.34
Problem 2.9	Problem 3.24	Problem 5.30	Problem 7.23	Problem 8.35
Problem 2.14	Problem 3.25	Problem 5.31	Problem 7.25	Problem 8.37
Problem 2.15	Problem 3.26	Problem 5.32	Problem 8.3	Problem 9.5
Problem 2.16	Problem 3.27	Problem 5.35	Problem 8.4	Problem 9.6
Problem 2.17	Problem 3.28	Problem 5.38	Problem 8.5	Problem 9.7
Problem 2.18	Problem 3.34	Problem 5.40	Problem 8.6	Problem 9.13
Problem 2.19	Problem 3.35	Problem 5.41	Problem 8.10	Problem 9.14
Problem 2.20	Problem 3.36	Problem 5.42	Problem 8.11	Problem 10.6
Problem 2.21	Problem 3.38	Problem 5.44	Problem 8.12	Problem 10.8
Problem 2.22	Problem 3.40	Problem 5.45	Problem 8.13	Problem 10.11
Problem 2.23	Problem 4.3	Problem 6.3	Problem 8.14	Problem 11.4
Problem 2.25	Problem 4.5	Problem 6.9	Problem 8.15	Problem 11.5
Problem 2.29	Problem 4.6	Problem 6.10	Problem 8.16	Problem 11.6
Problem 3.4	Problem 4.13	Problem 6.11	Problem 8.17	Problem 11.7
Problem 3.5	Problem 4.18	Problem 6.12	Problem 8.18	Problem 11.8
Problem 3.7	Problem 4.19	Problem 6.15	Problem 8.19	Problem 11.9
Problem 3.8	Problem 4.24	Problem 6.16		

Chapter 1

Mechanical System Analysis

1.1 DEGREES OF FREEDOM AND GENERALIZED COORDINATES

The number of *degrees of freedom* used in the analysis of a mechanical system is the number of kinematically independent coordinates necessary to completely describe the motion of every particle in the system. Any such set of coordinates is called a *set of generalized coordinates*. The choice of a set of generalized coordinates is not unique. Kinematic quantities such as displacements, velocities, and accelerations are written as functions of the generalized coordinates and their time derivatives. A system with a finite number of degrees of freedom is called a *discrete system,* while a system with an infinite number of degrees of freedom is called a *continuous system* or a *distributed parameter system.*

1.2 MECHANICAL SYSTEM COMPONENTS

A mechanical system comprises inertia components, stiffness components, and damping components. The inertia components have kinetic energy when the system is in motion. The kinetic energy of a rigid body undergoing planar motion is

$$T = \tfrac{1}{2}m\bar{v}^2 + \tfrac{1}{2}\bar{I}\omega^2 \qquad (1.1)$$

where $\bar{v}$ is the velocity of the body's mass center, ω is its angular velocity about an axis perpendicular to the plane of motion, m is the body's mass, and $\bar{I}$ is its mass moment of inertia about an axis parallel to the axis of rotation through the mass center.

A linear stiffness component (a linear spring) has a force displacement relation of the form

$$F = kx \qquad (1.2)$$

where F is applied force and x is the component's change in length from its unstretched length. The stiffness k has dimensions of force per length.

A *dashpot* is a mechanical device that adds viscous damping to a mechanical system. A linear viscous damping component has a force-velocity relation of the form

$$F = cv \qquad (1.3)$$

where c is the damping coefficient of dimensions mass per time.

1.3 EQUIVALENT SYSTEMS ANALYSIS

All linear 1-degree-of-freedom systems with viscous damping can be modeled by the simple mass-spring-dashpot system of Fig. 1-1. Let x be the chosen generalized coordinate. The kinetic energy of a linear system can be written in the form

$$T = \tfrac{1}{2}m_{eq}\dot{x}^2 \qquad (1.4)$$

1

The potential energy of a linear system can be written in the form

$$V = \tfrac{1}{2}k_{eq}x^2 \qquad (1.5)$$

The work done by the viscous damping force in a linear system between two arbitrary locations x_1 and x_2 can be written as

$$W = -\int_{x_1}^{x_2} c_{eq}\dot{x}\,dx \qquad (1.6)$$

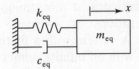

Fig. 1-1

1.4 TORSIONAL SYSTEMS

When an angular coordinate is used as a generalized coordinate for a linear system, the system can be modeled by the equivalent torsional system of Fig. 1-2. The moment applied to a linear torsional spring is proportional to its angular rotation while the moment applied to a linear torsional viscous damper is proportional to its angular velocity. The equivalent system coefficients for a torsional system are determined by calculating the total kinetic energy, potential energy, and work done by viscous damping forces for the original system in terms of the chosen generalized coordinate and setting them equal to

$$T = \tfrac{1}{2}I_{eq}\dot{\theta}^2 \qquad (1.7)$$

$$V = \tfrac{1}{2}k_{t_{eq}}\theta^2 \qquad (1.8)$$

$$W = -\int_{\theta_1}^{\theta_2} c_{t_{eq}}\dot{\theta}\,d\theta \qquad (1.9)$$

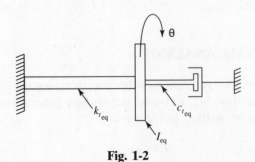

Fig. 1-2

1.5 STATIC EQUILIBRIUM POSITION

Systems, such as the one in Fig. 1-3, have elastic elements that are subject to force when the system is in equilibrium. The resulting deflection in the elastic element is called its *static deflection,* usually denoted by Δ_{st}. The static deflection of an elastic element in a linear system has no effect on the system's equivalent stiffness.

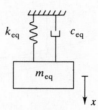

Fig. 1-3

Solved Problems

1.1 Determine the number of degrees of freedom to be used in the vibration analysis of the rigid bar of Fig. 1-4, and specify a set of generalized coordinates that can be used in its vibration analysis.

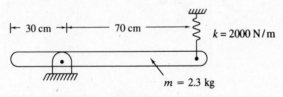

Fig. 1-4

Since the bar is rigid, the system has only 1 degree of freedom. One possible choice for the generalized coordinate is θ, the angular displacement of the bar measured positive clockwise from the system's equilibrium position.

1.2 Determine the number of degrees of freedom needed for the analysis of the mechanical system of Fig. 1-5, and specify a set of generalized coordinates that can be used in its vibration analysis.

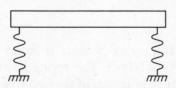

Fig. 1-5

Let x be the displacement of the mass center of the rigid bar, measured from the system's equilibrium position. Knowledge of x, by itself, is not sufficient to determine the displacement of any other particle on the bar. Thus the system has more than 1 degree of freedom.

Let θ be the clockwise angular rotation of the bar with respect to the axis of the bar in its equilibrium position. If θ is small, then the displacement of the right end of the bar is $x + (L/2)\theta$. Thus the system has 2 degrees of freedom, and x and θ are a possible set of generalized coordinates, as illustrated in Fig. 1-6.

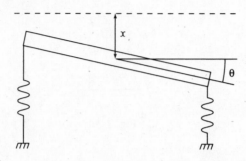

Fig. 1-6

1.3   Determine the number of degrees of freedom used in the analysis of the mechanical system of Fig. 1-7. Specify a set of generalized coordinates that can be used in the system's vibration analysis.

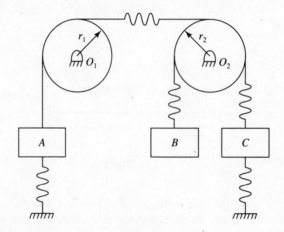

Fig. 1-7

The system of Fig. 1-7 has 4 degrees of freedom. A possible set of generalized coordinates are θ_1, the clockwise angular displacement from equilibrium of the disk whose center is at O_1; θ_2, the clockwise angular displacement from equilibrium of the disk whose center is at O_2; x_1, the downward displacement of block B; and x_2, the downward displacement of block C. Note that the upward displacement of block A is given by $r_1\theta_1$ and hence is not kinematically independent of the motion of the disk.

1.4   A tightly wound helical coil spring is made from an 18-mm-diameter bar of 0.2 percent hardened steel ($G = 80 \times 10^9$ N/m^2). The spring has 80 active coils with a coil diameter of 16 cm. What is the change in length of the spring when it hangs vertically with one end fixed and a 200-kg block attached to its other end?

The stiffness of a helical coil spring is

$$k = \frac{GD^4}{64Nr^3}$$

where D is the bar diameter, r is the coil radius, and N is the number of active turns. Substituting known values leads to

$$k = \frac{\left(80 \times 10^8 \frac{N}{m^2}\right)(0.018 \text{ m})^4}{64(80)(0.08 \text{ m})^3} = 3.20 \times 10^3 \frac{N}{m}$$

Using Eq. (1.2), the change in length of the spring is

$$x = \frac{F}{k} = \frac{mg}{k} = \frac{(200 \text{ kg})\left(9.81 \frac{m}{s^2}\right)}{3.20 \times 10^3 \frac{N}{m}} = 0.613 \text{ m}$$

1.5   Determine the longitudinal stiffness of the bar of Fig. 1-8.

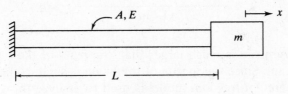

Fig. 1-8

The longitudinal motion of the block of Fig. 1-8 can be modeled by an undamped system of the form of Fig. 1-1. When a force F is applied to the end of the bar, its change in length is

$$\delta = \frac{FL}{AE}$$

or

$$F = \frac{AE}{L}\delta$$

which is in the form of Eq. (1.2). Thus

$$k_{eq} = \frac{AE}{L}$$

1.6   Determine the torsional stiffness of the shaft in the system of Fig. 1-9.

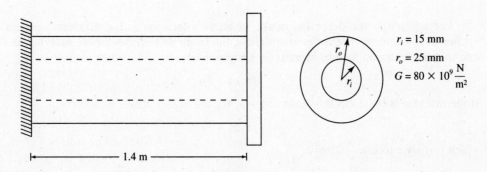

Fig. 1-9

If a moment M is applied to the end of the shaft, the angle of twist at the end of the shaft is determined from mechanics of materials as

$$\theta = \frac{ML}{JG}$$

where J is the polar moment of inertia of the shaft's cross section. Thus

$$M = \frac{JG}{L}\theta$$

and the shaft's equivalent torsional stiffness is

$$k_t = \frac{JG}{L}$$

For the shaft of Fig. 1-9

$$J = \frac{\pi}{2}(r_o^4 - r_i^4) = \frac{\pi}{2}[(0.025\ \mathrm{m})^4 - (0.015\ \mathrm{m})^4] = 5.34 \times 10^{-7}\ \mathrm{m}^4$$

Thus
$$k_t = \frac{(5.34 \times 10^{-7}\ \mathrm{m}^4)\left(80 \times 10^9\ \dfrac{\mathrm{N}}{\mathrm{m}^2}\right)}{1.4\ \mathrm{m}} = 3.05 \times 10^4\ \frac{\mathrm{N\text{-}m}}{\mathrm{rad}}$$

1.7 A machine whose mass is much larger than the mass of the beam shown in Fig. 1-10 is bolted to the beam. Since the inertia of the beam is small compared to the inertia of the machine, a 1-degree-of-freedom model is used to analyze the vibrations of the machine. The system is modeled by the system of Fig. 1-3. Determine the equivalent spring stiffness if the machine is bolted to the beam at

(*a*) $z = 1$ m

(*b*) $z = 1.5$ m

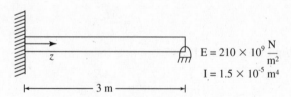

Fig. 1-10

Let $w(z; a)$ be the deflection of the beam at a location z due to a unit concentrated load applied at $z = a$. From mechanics of materials, the beam deflection is linear, and thus the deflection due to a concentrated load of magnitude F is given by

$$y(z; a) = Fw(z; a)$$

If the machine is bolted to the beam at $z = a$, the deflection at this location is

$$y(a; a) = Fw(a; a)$$

which is similar to Eq. (1.2) with

$$k = \frac{1}{w(a; a)} \tag{1.10}$$

From mechanics of materials, the deflection of a beam fixed at $z = 0$ and pinned at $z = L$ due to a unit concentrated load at $z = a$, for $z < a$ is

$$w(z;a) = \frac{1}{EI}\left\{\frac{z^3}{12}\left(1 - \frac{a}{L}\right)\left[\left(\frac{a}{L}\right)^2 - 2\frac{a}{L} - 2\right] + \frac{z^2 a}{4}\left(1 - \frac{a}{L}\right)\left(2 - \frac{a}{L}\right)\right\} \qquad (1.11)$$

(a) For $a = 1$ m, $a/L = \frac{1}{3}$. Then using Eqs. (1.10) and (1.11),

$$k_{eq} = \frac{1}{w(a;a)} = \frac{81EI}{11a^3}$$

$$= \frac{81\left(210 \times 10^9\ \frac{N}{m^2}\right)(1.5 \times 10^{-5}\ m^4)}{11(1\ m)^3} = 2.32 \times 10^7\ \frac{N}{m}$$

(b) For $a = 1.5$ m, $a/L = \frac{1}{2}$. Then using Eqs. (1.10) and (1.11),

$$k_{eq} = \frac{1}{w(a;a)} = \frac{96EI}{7a^3}$$

$$= \frac{96\left(210 \times 10^9\ \frac{N}{m^2}\right)(1.5 \times 10^{-5}\ m^4)}{7(1.5\ m)^3} = 1.28 \times 10^7\ \frac{N}{m}$$

1.8   A machine of mass m is attached to the midspan of a simply supported beam of length L, elastic modulus E, and cross-sectional moment of inertia I. The mass of the machine is much greater than the mass of the beam; thus the system can be modeled using 1 degree of freedom. What is the equivalent stiffness of the beam using the midspan deflection as the generalized coordinate?

The deflection of a simply supported beam at its midspan due to a concentrated load F applied at the midspan is

$$\delta = \frac{FL^3}{48EI}$$

The equivalent stiffness is the reciprocal of the midspan deflection due to a midspan concentrated unit load. Thus

$$k_{eq} = \frac{48EI}{L^3}$$

1.9   The springs of Fig. 1-11 are said to be in parallel. Derive an equation for the equivalent stiffness of the parallel combination of springs if the system of Fig. 1-11 is to be modeled by the equivalent system of Fig. 1-1.

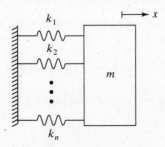

Fig. 1-11

If the block is subject to an arbitrary displacement x, the change in length of each spring in the parallel combination is x. The free body diagram of Fig. 1-12 shows that the total force acting on the block is

$$F = k_1 x + k_2 x + k_3 x + \cdots + k_n x = \left(\sum_{i=1}^{n} k_i \right) x \qquad (1.12)$$

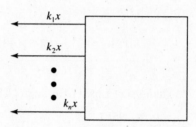

Fig. 1-12

The system of Fig. 1-1 can be used to model the system of Fig. 1-11 if the force acting on the block of Fig. 1-1 is equal to the force of Eq. (1.12) when the spring has a displacement x. If the spring of Fig. 1-1 has a displacement x, then the force acting on the block of Fig. 1-1 is

$$F = k_{eq} x \qquad (1.13)$$

Then for the forces from Eqs. (1.12) and (1.13) to be equal:

$$k_{eq} = \sum_{i=1}^{n} k_i$$

1.10   The springs in the system of Fig. 1-13 are said to be *in series*. Derive an equation for the series combination of springs if the system of Fig. 1-13 is to be modeled by the equivalent system of Fig. 1-1.

k_1 k_2 k_3 k_n

m

Fig. 1-13

Let x be the displacement of the block of Fig. 1-13 at an arbitrary instant. Let x_i be the change in length of the ith spring from the fixed support. If each spring is assumed massless, then the force developed at each end of the spring has the same magnitude but opposite in direction, as shown in Fig. 1-14. Thus the force is the same in each spring:

$$k_1 x_1 = k_2 x_2 = k_3 x_3 = \cdots = k_n x_n \qquad (1.14)$$

In addition,

$$x = x_1 + x_2 + x_3 + \cdots + x_n = \sum_{i=1}^{n} x_i \qquad (1.15)$$

Solving for x_i from Eq. (1.14) and substituting into Eq. (1.15) leads to

$$F = \frac{x}{\displaystyle\sum_{i=1}^{n} \frac{1}{k_i}} \qquad (1.16)$$

Noting that the force acting on the block of the system of Fig. 1-1 for an arbitrary x is $k_{eq}x$ and equating this to the force from Eq. (1.16) leads to

$$k_{eq} = \frac{1}{\sum\limits_{i=1}^{n} \frac{1}{k_i}}$$

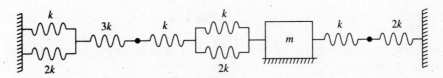

$$k_i x_i \qquad\qquad\qquad\qquad k_i x_i$$

Fig. 1-14

1.11 Model the system shown in Fig. 1-15 by a block attached to a single spring of an equivalent stiffness.

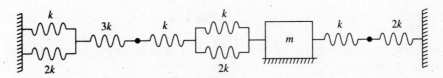

Fig. 1-15

The first step is to replace the parallel combinations by springs of equivalent stiffnesses using the results of Problem 1.9. The result is shown in Fig. 1-16a. The springs on the left of the block are in series with one another. The result of Problem 1.10 is used to replace these springs by a spring whose stiffness is calculated as

$$\frac{1}{\dfrac{1}{3k} + \dfrac{1}{3k} + \dfrac{1}{k} + \dfrac{1}{3k}} = \frac{k}{2}$$

The springs attached to the right of the block are in series and are replaced by a spring of stiffness

$$\frac{1}{\dfrac{1}{k} + \dfrac{1}{2k}} = \frac{2k}{3}$$

The result is the system of Fig. 1-16b. When the block has an arbitrary displacement x, the displacements in each of the springs of Fig. 1-16b are the same, and the total force acting on the block is the sum of the forces developed in the springs. Thus these springs behave as if they are in parallel and can be replaced by a spring of stiffness

$$\frac{k}{2} + \frac{2k}{3} = \frac{7k}{6}$$

as illustrated in Fig. 1-16c.

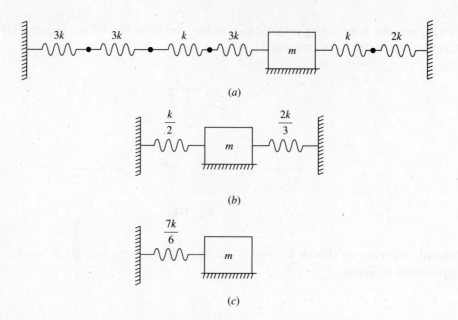

(a)

(b)

(c)

Fig. 1-16

1.12 Model the torsional system of Fig. 1-17 by a disk attached to a torsional spring of an equivalent stiffness.

Mathcad

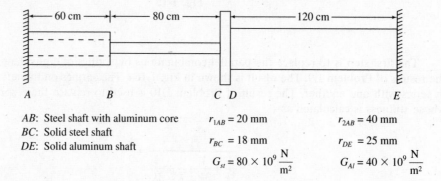

Fig. 1-17

AB: Steel shaft with aluminum core $r_{1AB} = 20$ mm $r_{2AB} = 40$ mm
BC: Solid steel shaft $r_{BC} = 18$ mm $r_{DE} = 25$ mm
DE: Solid aluminum shaft
$G_{st} = 80 \times 10^9 \dfrac{\text{N}}{\text{m}^2}$ $G_{Al} = 40 \times 10^9 \dfrac{\text{N}}{\text{m}^2}$

The stiffness of each of the shafts of Fig. 1-17 are calculated as

$$k_{AB_{st}} = \frac{J_{AB_{st}} G_{AB_{st}}}{L_{AB}} = \frac{\frac{\pi}{2}[(0.04 \text{ m})^4 - (0.02 \text{ m})^4]\left(80 \times 10^9 \dfrac{\text{N}}{\text{m}^2}\right)}{0.6 \text{ m}}$$

$$= 5.03 \times 10^5 \frac{\text{N-m}}{\text{rad}}$$

$$k_{AB_{Al}} = \frac{J_{AB_{Al}} G_{AB_{Al}}}{L_{AB}} = \frac{\frac{\pi}{2}(0.02 \text{ m})^4\left(40 \times 10^9 \dfrac{\text{N}}{\text{m}^2}\right)}{0.6 \text{ m}} = 1.68 \times 10^4 \frac{\text{N-m}}{\text{rad}}$$

$$k_{BC} = \frac{J_{BC} G_{BC}}{L_{BC}} = \frac{\frac{\pi}{2}(0.018 \text{ m})^4\left(80 \times 10^9 \dfrac{\text{N}}{\text{m}^2}\right)}{0.8 \text{ m}} = 1.65 \times 10^4 \frac{\text{N-m}}{\text{rad}}$$

$$k_{DE} = \frac{J_{DE} G_{DE}}{L_{DE}} = \frac{\frac{\pi}{2}(0.025 \text{ m})^4\left(40 \times 10^9 \dfrac{\text{N}}{\text{m}^2}\right)}{1.2 \text{ m}} = 2.05 \times 10^4 \frac{\text{N-m}}{\text{rad}}$$

The angle of twist of the end of aluminum core of shaft AB is the same as the angle of twist of the end of the steel shell of shaft AB. Also, the total torque on the end of shaft AB is the sum of the resisting torque in the aluminum core and the resisting torque in the steel shell. Hence the aluminum core and steel shell of shaft AB behave as torsional springs in parallel with an equivalent stiffness of

$$k_{AB} = k_{AB_{st}} + k_{AB_{Al}} = 5.03 \times 10^5 \ \frac{\text{N-m}}{\text{rad}} + 1.68 \times 10^4 \ \frac{\text{N-m}}{\text{rad}} = 5.20 \times 10^5 \ \frac{\text{N-m}}{\text{rad}}$$

The torques developed in shafts AB and BC are the same, and the angle of rotation of the disk is $\theta_{AB} + \theta_{BC}$. Thus shafts AB and BC behave as torsional springs in series whose combination acts in parallel with shaft DE. Hence the equivalent stiffness is

$$k_{eq} = \frac{1}{\dfrac{1}{k_{AB}} + \dfrac{1}{k_{BC}}} + k_{DE} = 3.65 \times 10^4 \ \frac{\text{N-m}}{\text{rad}}$$

1.13   Derive an expression for the equivalent stiffness of the system of Fig. 1-18 when the deflection of the machine is used as the generalized coordinate.

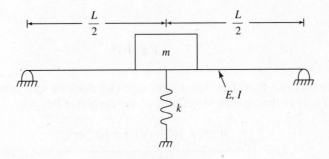

Fig. 1-18

Consider a concentrated downward load F_1 applied to the midspan of the simply supported beam leading to a midspan deflection x. A compressive force kx is developed in the spring. The total downward force acting on the beam at its midspan is $F_1 - kx$. As noted in Problem 1.8, the midspan deflection of a simply supported beam due to a concentrated load at its midspan is

$$x = \frac{FL^3}{48EI}$$

Thus for the beam of Fig. 1-18,

$$x = (F_1 - kx) \frac{L^3}{48EI}$$

which leads to

$$x = \frac{F_1}{k + \dfrac{48EI}{L^3}}$$

The equivalent stiffness is obtained by setting $F_1 = 1$, leading to

$$k_{eq} = k + \frac{48EI}{L^3}$$

Using the results of Problem 1.9, it is observed that the beam and the spring act as two springs in parallel.

1.14 What is the equivalent stiffness of the system of Fig. 1-19 using the displacement of the block as the generalized coordinate?

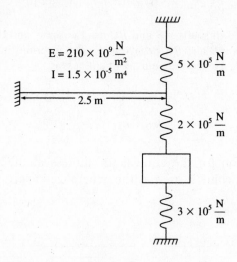

$E = 210 \times 10^9 \dfrac{N}{m^2}$

$I = 1.5 \times 10^{-5} \, m^4$

2.5 m

$5 \times 10^5 \dfrac{N}{m}$

$2 \times 10^5 \dfrac{N}{m}$

$3 \times 10^5 \dfrac{N}{m}$

Fig. 1-19

The deflection of a fixed-free beam at its free end due to a unit concentrated load at its free end is $L^3/(3EI)$. Thus the equivalent stiffness of the cantilever beam is

$$k_b = \frac{3EI}{L^3} = \frac{3\left(210 \times 10^9 \dfrac{N}{m^2}\right)(1.5 \times 10^{-5} \, m^4)}{(2.5 \, m)^3} = 6.05 \times 10^5 \frac{N}{m}$$

The analysis of Problem 1.13 suggests that the beam and the upper spring act in parallel. This parallel combination is in series with the spring placed between the beam and the block. This series combination is in parallel with the spring between the block and the fixed surface. Thus using the formulas for parallel and series combinations, the equivalent stiffness is calculated as

$$k_{eq} = \frac{1}{\dfrac{1}{6.05 \times 10^5 \dfrac{N}{m} + 5 \times 10^5 \dfrac{N}{m}} + \dfrac{1}{2 \times 10^5 \dfrac{N}{m}}} + 3 \times 10^5 \frac{N}{m} = 4.69 \times 10^5 \frac{N}{m}$$

1.15 The viscous damper shown in Fig. 1-20 contains a reservoir of a viscous fluid of viscosity μ and depth h. A plate slides over the surface of the reservoir with an area of contact A. What is the damping coefficient for this viscous damper?

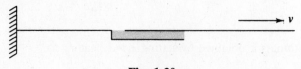

v

Fig. 1-20

Let y be a coordinate up into the fluid, measured from the bottom of the reservoir. If h is small and unsteady effects are neglected, the velocity profile $u(y)$ in the fluid is linear with $u(0) = 0$ and $u(h) = v$, as shown in Fig. 1-21 where v is the velocity of the plate. The mathematical form of the velocity profile is

$$u(y) = v\frac{y}{h}$$

The shear stress acting on the surface of the plate is calculated using Newton's viscosity law,

$$\tau = \mu\frac{du}{dy}$$

leading to

$$\tau = \frac{\mu v}{h}$$

The total viscous force is the resultant of the shear stress distribution

$$F = \tau A = \frac{\mu A}{h}v$$

The constant of proportionality between the force and the plate velocity is the damping coefficient

$$c = \frac{\mu A}{h}$$

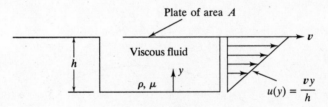

Fig. 1-21

1.16 The torsional viscous damper of Fig. 1-22 consists of a thin disk attached to a rotating shaft. The face of the disk has a radius R and rotates in a dish of fluid of depth h and viscosity μ. Determine the torsional viscous damping coefficient for this damper.

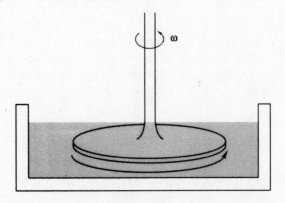

Fig. 1-22

Let τ be the shear stress acting on the differential area $dA = r\,dr\,d\theta$ on the surface of the disk, as illustrated in Fig. 1-23. The resultant moment about the axis of rotation due to the shear stress distribution is

$$M = \int_0^{2\pi} \int_0^R r\tau(r\,dr\,d\theta) \qquad (1.17)$$

If ω is the angular velocity of the shaft and disk, then the velocity of the differential element is $r\omega$. Let y be a coordinate, measured upward into the fluid from the bottom of the dish. Neglecting unsteady effects and assuming the depth of the fluid is small, the velocity distribution $u(r, y)$ in the fluid is approximately linear in y with $u(r, 0) = 0$ and $u(r, h) = r\omega$, leading to

$$u(r, y) = \frac{r\omega}{h} y$$

The shear stress acting on the fact of the disk is calculated from Newton's viscosity law,

$$\tau = \mu \frac{\partial u}{\partial y}(r, h) = \frac{r\mu\omega}{h}$$

which, when substituted into Eq. (1.17), leads to

$$M = \int_0^{2\pi} \int_0^R \frac{\mu r^3 \omega}{h}\,dr\,d\theta = \frac{\mu\pi R^4}{2h}\omega$$

The torsional damping coefficient is the constant of proportionality between the moment and the angular velocity,

$$c_t = \frac{\mu\pi R^4}{2h}$$

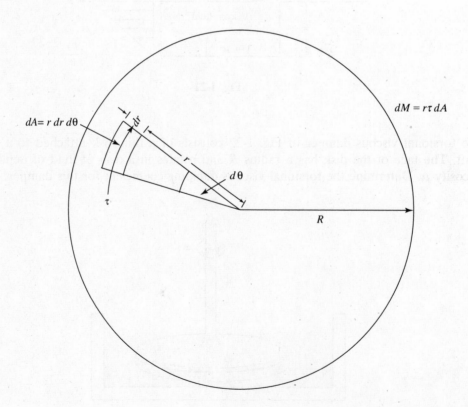

Fig. 1-23

1.17 Show that the inertia effects of a linear spring connecting a fixed support and a 1-degree-of-freedom system can be approximated by placing a particle of mass equal to one-third the mass of the spring at the system location where the spring is attached.

Let m_s be the mass of a uniform spacing of unstretched length ℓ that is connected between a fixed support and a particle in a 1-degree-of-freedom system whose displacement is given by $x(t)$. Let m_{eq} be the mass of a particle placed at the end of the spring. This particle can be used to approximate the inertia effects of the spring if the kinetic energy of the spring is

$$T = \tfrac{1}{2} m_{eq} \dot{x}^2$$

Let z be a coordinate along the axis of the spring in its unstretched position, $0 \le z \le \ell$, as illustrated in Fig. 1-24. Assume the displacement function $u(z, t)$ is linear along the length of the spring at any instant with $u(0) = 0$ and $u(\ell) = x$:

$$u(z, t) = \frac{x}{\ell} z$$

The kinetic energy of a differential spring element is

$$dT = \frac{1}{2} \left(\frac{\partial u}{\partial t}\right)^2 dm = \frac{1}{2} \left(\frac{\dot{x}}{\ell} z\right)^2 \frac{m_s}{\ell} dz$$

from which the total kinetic energy of the spring is calculated

$$T = \int dT = \frac{\dot{x}^2 m_s}{2\ell^3} \int_0^\ell z^2 \, dz = \frac{1}{2} \frac{m_s}{3} \dot{x}^2$$

Thus if a particle of mass $m_s/3$ is placed at the location on the system where the spring is attached, its kinetic energy is the same as that of the linear spring assuming a linear displacement function.

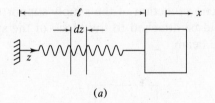

(a)

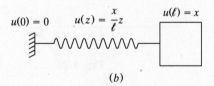

(b)

Fig. 1-24

1.18 What is the mass of a particle that should be added to the block of the system of Fig. 1-25 to approximate the inertia effects of the series combination of springs?

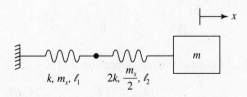

Fig. 1-25

Using the results of Problem 1.17, the inertia effects of the left spring can be approximated by placing a particle of mass $m_s/3$ at the junction between the two springs. Let x be the displacement of the block at an arbitrary instant. Let z_1 and z_2 be coordinates along the axes of the left and right springs, respectively. Let $u_1(z_1, t)$ and $u_2(z_2, t)$ be the displacement functions for these springs. It is known that $u_1(0, t) = 0$, $u_2(\ell_2, t) = x$. Also, $u_1(\ell_1, t) = u_2(0, t) = w$. Assuming linear displacement functions for each spring, this leads to

$$u_1(z_1, t) = \frac{w}{\ell_1} z_1$$

$$u_2(z_2, t) = \frac{x - w}{\ell_2} z_2 + w$$

Since the springs are in series, the forces in each spring are the same:

$$kw = 2k(x - w) \rightarrow w = \tfrac{2}{3}x$$

Thus the kinetic energy of the series combination is

$$T = \frac{1}{2}\left(\frac{m_s}{3}\right)\left(\frac{2}{3}\dot{x}\right)^2 + \frac{1}{2}\int_0^{\ell_2} \left(\frac{1}{3}\frac{z_2}{\ell_2} + \frac{2}{3}\right)^2 \dot{x}^2 \frac{m_s}{2\ell_2}\, dz_2$$

$$= \frac{1}{2}\left(\frac{1}{2}m_s\right)\dot{x}^2$$

which leads to an added particle mass of $m_s/2$.

1.19 Use the static deflection function of the simply supported beam to determine the mass of a particle that should be attached to the block of the system of Fig. 1-26 to approximate inertia effects of the beam.

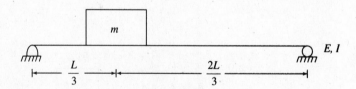

Fig. 1-26

The static deflection $y(z)$ of a simply supported beam due to a concentrated load F applied at $z = L/3$ is

$$y(z) = \frac{F}{EI}\begin{cases} \dfrac{5}{81}L^2 z - \dfrac{1}{9}z^3 & 0 \le z \le \dfrac{L}{3} \\[2mm] \dfrac{1}{6}\left(z - \dfrac{L}{3}\right)^3 + \dfrac{5}{81}L^2 z - \dfrac{1}{9}z^3 & \dfrac{L}{3} \le z \le L \end{cases}$$

The force required to cause a static deflection x at $z = L/3$ is calculated as

$$x = y\left(\frac{L}{3}\right) = \frac{4FL^3}{243EI} \rightarrow F = \frac{243EIx}{4L^3}$$

The kinetic energy of the beam is

$$T = \int_0^L \frac{1}{2}\dot{y}^2\rho A\, dz$$

$$= \frac{1}{2}\rho A\left(\frac{243}{4L^3}\right)^2\dot{x}^2\left\{\int_0^{L/3}\left(\frac{5}{81}L^2z - \frac{1}{9}z^3\right)^2 dz + \int_{L/3}^L\left[\frac{1}{6}\left(z - \frac{1}{3}L\right)^3 + \frac{5}{81}L^2z - \frac{1}{9}z^3\right]^2 dz\right\}$$

$$= 0.586\rho AL = 0.586m_b$$

Hence the inertia effects of the beam are approximated by adding a particle of mass $0.586m_b$ to the machine.

1.20 An approximation to the deflection of a fixed-fixed beam due to a concentrated load at its midspan is

$$y(z) = \frac{x}{2}\left(1 - \cos\frac{2\pi z}{L}\right)$$

where x is the midspan deflection. Use this approximation to determine the mass of a particle to be placed at the midspan of a beam to approximate the beam's inertia effects.

The kinetic energy of the beam is

$$T = \int_0^L \frac{1}{2}\dot{y}^2(z)\rho A\, dz$$

where ρ is the beam's mass density and A is its cross-sectional area. Substituting the suggested approximation,

$$T = \frac{1}{2}\rho A\frac{\dot{x}^2}{4}\int_0^L\left(1 - \cos\frac{2\pi z}{L}\right)^2 dz$$

$$= \frac{1}{2}\left(\frac{3}{8}\rho AL\right)\dot{x}^2 = \frac{1}{2}\left(\frac{3}{8}m_{\text{beam}}\right)\dot{x}^2$$

The inertia effects of the beam can be approximated by adding a particle of mass $3/8 m_{\text{beam}}$ at its midspan.

1.21 Let x be the displacement of the block of Fig. 1-27, measured positive downward from the system's equilibrium position. Show that the system's difference in potential energies between two arbitrary positions is independent of the mass of the block.

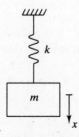

Fig. 1-27

Let x be the downward displacement of the block from the system's equilibrium position. When the system is in equilibrium, the spring has a static deflection $\Delta = mg/k$. If the datum for the potential energy due to gravity is taken as the system's equilibrium position, the potential energy of the system at an arbitrary instant is

$$V = \tfrac{1}{2}k(x + \Delta)^2 - mgx$$
$$= \tfrac{1}{2}kx^2 + (k\Delta - mg)x + \tfrac{1}{2}k\Delta^2$$
$$= \tfrac{1}{2}kx^2 + \tfrac{1}{2}k\Delta^2$$

Thus the difference in potential energies as the block moves between x_1 and x_2 is

$$V_2 - V_1 = \frac{1}{2}kx_2^2 + \frac{1}{k}\Delta^2 - \frac{1}{2}kx_1^2 - \frac{1}{2}k\Delta^2$$
$$= \frac{1}{2}k(x_2^2 - x_1^2)$$

which is independent of the mass of the block. The results of this problem are used to infer that the static deflection of a spring and the gravity force causing the static deflection cancel with one another in the potential energy difference.

1.22 Determine m_{eq} and k_{eq} for the system of Fig. 1-28 when x, the downward displacement of the block, measured from the system's equilibrium position, is used as the generalized coordinate.

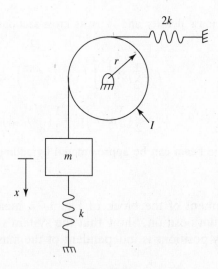

Fig. 1-28

From the results of Problem 1.21, it is evident that the effects of gravity and static deflections cancel in potential energy calculations and can thus be ignored. The potential energy of the system is

$$V = \tfrac{1}{2}kx^2 + \tfrac{1}{2}(2k)x^2 = \tfrac{1}{2}(3k)x^2 \;\rightarrow\; k_{eq} = 3k$$

The kinetic energy of the system is

$$T = \frac{1}{2}m\dot{x}^2 + \frac{1}{2}I\left(\frac{\dot{x}}{r}\right)^2 = \frac{1}{2}\left(m + \frac{I}{r^2}\right)\dot{x}^2 \;\rightarrow\; m_{eq} = m + \frac{I}{r^2}$$

1.23   Determine k_{eq} and m_{eq} for the system of Fig. 1-29 when x, the displacement of the center of the disk measured from equilibrium, is used as the generalized coordinate. Assume the disk is thin and rolls without slip.

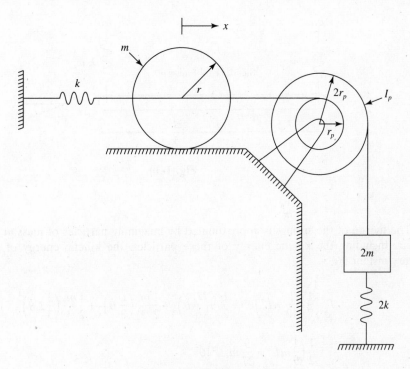

Fig. 1-29

If the disk rolls without slip, then the angular rotation of the pulley θ and the downward displacement of the block y are

$$\theta = \frac{x}{r_p} \qquad y = 2r_p\theta = 2x$$

Noting that the effects of gravity and static deflection cancel, potential energy calculations lead to

$$V = \tfrac{1}{2}kx^2 + \tfrac{1}{2}(2k)y^2 = \tfrac{1}{2}kx^2 + \tfrac{1}{2}2k(2x)^2 = \tfrac{1}{2}9kx^2 \rightarrow k_{eq} = 9k$$

The kinetic energy of the system is

$$T = \tfrac{1}{2}m\dot{x}^2 + \tfrac{1}{2}I_d\omega_d^2 + \tfrac{1}{2}I_p\dot{\theta}^2 + \tfrac{1}{2}(2m)\dot{y}^2$$

If the disk is thin $I_d = mr^2/2$ and if it rolls without slip, $\omega_d = \dot{x}/r$. Thus

$$T = \frac{1}{2}m\dot{x}^2 + \frac{1}{2}\left(\frac{1}{2}mr^2\right)\left(\frac{\dot{x}}{r}\right)^2 + \frac{1}{2}I_p\left(\frac{\dot{x}}{r_p}\right)^2 + \frac{1}{2}(2m)(2\dot{x})^2$$

$$= \frac{1}{2}\left(\frac{19}{2}m + \frac{I_p}{r_p^2}\right)\dot{x}^2 \rightarrow m_{eq} = \frac{19}{2}m + \frac{I_p}{r_p^2}$$

1.24   Calculate the parameters for an equivalent system model of the system of Fig. 1-30 when

θ, the clockwise angular displacement of the bar, measured from the system's equilibrium position, is used as the generalized coordinate. Include approximations for the inertia effects of the springs and assume small θ.

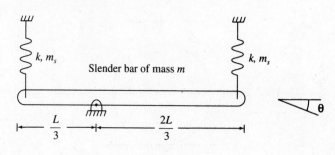

k, m_s Slender bar of mass m k, m_s

$\dfrac{L}{3}$ $\dfrac{2L}{3}$ θ

Fig. 1-30

The inertia of the springs is approximated by imagining particles of mass $m_s/3$ at each end of the bar. Including the kinetic energy of these particles, the kinetic energy of the system at an arbitrary instant is

$$T = \frac{1}{2}\left(\frac{1}{12}mL^2\right)\dot\theta^2 + \frac{1}{2}m\left(\frac{L}{6}\dot\theta\right)^2 + \frac{1}{2}\frac{m_s}{3}\left(\frac{L}{3}\dot\theta\right)^2 + \frac{1}{2}\frac{m_s}{3}\left(\frac{2}{3}L\dot\theta\right)^2$$

$$= \frac{1}{2}\left(\frac{1}{9}mL^2 + \frac{5}{27}m_sL^2\right)\dot\theta^2$$

If θ is small, then the displacement of the left end of the bar is approximated as $(L/3)\theta$ upward, and the displacement of the right end of the bar is approximated by $(2L/3)\theta$ downward. Assuming that the potential energy due to gravity cancels with potential energy due to static deflection, the potential energy at an arbitrary instant is

$$V = \frac{1}{2}k\left(\frac{L}{3}\theta\right)^2 + \frac{1}{2}k\left(\frac{2}{3}L\theta\right)^2$$

$$= \frac{1}{2}\left(\frac{5}{9}kL^2\right)\theta^2$$

From above

$$I_{eq} = \tfrac{1}{9}mL^2 + \tfrac{5}{27}m_sL^2 \qquad k_{t_{eq}} = \tfrac{5}{9}kL^2$$

1.25   Determine the parameters when an equivalent system is used to model the system of Fig. 1-31 when θ, the clockwise angular displacement of bar AB, measured from the system's equilibrium position, is used as the generalized coordinate. Assume small θ.

Let ϕ be the counterclockwise displacement of bar CD, measured from the system's equilibrium position. Since the ends of bars AB and CD are connected by a rigid link, their displacements must be the same. Thus assuming small θ and ϕ,

$$\tfrac{2}{3}L\theta = L\phi \;\rightarrow\; \phi = \tfrac{2}{3}\theta$$

The kinetic energy of the system at an arbitrary instant is

$$T = \tfrac{1}{2}I_{AB}\dot{\theta}^2 + \tfrac{1}{2}m_{AB}\bar{v}_{AB}^{\ 2} + \tfrac{1}{2}I_{CD}\dot{\phi}^2 + \tfrac{1}{2}m_{CD}\bar{v}_{CD}^{\ 2}$$
$$= \tfrac{1}{2}(\tfrac{1}{12}mL^2)\dot{\theta}^2 + \tfrac{1}{2}m(\tfrac{1}{6}L\dot{\theta})^2 + \tfrac{1}{2}(\tfrac{1}{12}mL^2)(\tfrac{2}{3}\dot{\theta})^2 + \tfrac{1}{2}m(\tfrac{1}{2}L\tfrac{2}{3}\dot{\theta})^2$$
$$= \tfrac{1}{2}(\tfrac{7}{27}mL^2)\dot{\theta}^2$$

The potential energy of the system at an arbitrary instant is

$$V = \frac{1}{2}3k\left(\frac{L}{3}\theta\right)^2 + \frac{1}{2}k\left(L\frac{2}{3}\phi\right)^2 = \frac{1}{2}3k\left(\frac{L}{3}\theta\right)^2 + \frac{1}{2}k\left(\frac{4}{9}L\theta\right)^2$$
$$= \frac{1}{2}\left(\frac{43}{81}kL^2\right)\theta^2$$

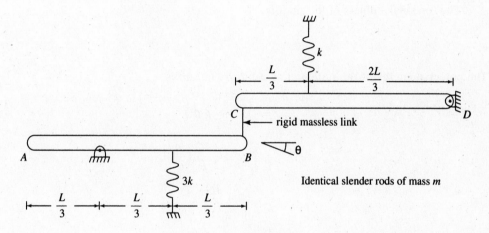

Fig. 1-31

Since θ is used as the generalized coordinate, the appropriate equivalent systems model is the torsional system. Thus from the above the appropriate equivalent system parameters are

$$I_{eq} = \frac{7}{27}mL^2 \qquad k_{t_{eq}} = \frac{43}{81}kL^2$$

1.26 The disk in the system of Fig. 1-32 rolls without slip on the plate. Determine the parameters for an equivalent system model of the system using x, the displacement of the plate from the system's equilibrium position, as the generalized coordinate.

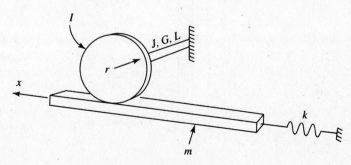

Fig. 1-32

Let θ be the clockwise angular displacement of the disk, measured from equilibrium. Since the center of the disk is fixed and it rolls without slip on the plate,

$$x = r\theta \;\rightarrow\; \theta = \frac{x}{r}$$

The kinetic energy of the system at an arbitrary instant is

$$T = \frac{1}{2}m\dot{x} + \frac{1}{2}I\dot{\theta}^2 = \frac{1}{2}m\dot{x}^2 + \frac{1}{2}I\left(\frac{\dot{x}}{r}\right)^2$$

$$= \frac{1}{2}\left(m + \frac{I}{r^2}\right)\dot{x}^2$$

The torsional stiffness of the shaft is

$$k_t = \frac{JG}{L}$$

The potential energy of the system at an arbitrary instant is

$$V = \frac{1}{2}kx^2 + \frac{1}{2}k_t\theta^2$$

$$= \frac{1}{2}kx^2 + \frac{1}{2}\frac{JG}{L}\left(\frac{x}{r}\right)^2$$

$$= \frac{1}{2}\left(k + \frac{JG}{Lr^2}\right)x^2$$

Hence the parameters for an equivalent system model are

$$m_{\text{eq}} = m + \frac{I}{r^2} \qquad k_{\text{eq}} = k + \frac{JG}{Lr^2}$$

1.27 Determine parameters for an equivalent system analysis of the system of Fig. 1-33, using θ the clockwise angular displacement of the bar from the system's equilibrium position, as the generalized coordinate. Assume small θ.

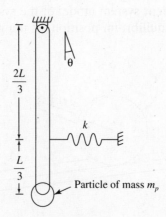

$\dfrac{2L}{3}$

$\dfrac{L}{3}$

k

Particle of mass m_p

Fig. 1-33

The kinetic energy of the system at an arbitrary instant is

$$T = \tfrac{1}{2}m\bar{I}\dot{\theta} + \tfrac{1}{2}m\bar{v}^2 + \tfrac{1}{2}m_p v_p^2$$

$$= \tfrac{1}{2}(\tfrac{1}{12}mL^2)\dot{\theta}^2 + \tfrac{1}{2}m(\tfrac{1}{2}L\dot{\theta})^2 + \tfrac{1}{2}m_p(L\dot{\theta})^2$$

$$= \tfrac{1}{2}(\tfrac{1}{3}mL^2 + m_p L^2)\dot{\theta}^2$$

Using a horizontal plane through the pin support as the datum for potential energy calculations, the difference in potential energies between an arbitrary system position and the system's initial position is

$$V = \frac{1}{2}k\left(\frac{2}{3}L\theta\right)^2 + mg\frac{L}{2}(1 - \cos\theta) + m_p gL(1 - \cos\theta)$$

For small θ

$$\cos\theta \approx 1 - \tfrac{1}{2}\theta^2$$

Hence

$$V = \frac{1}{2}\left(\frac{4}{9}kL^2\right)\theta^2 + mg\frac{L}{2}\left(\frac{1}{2}\theta^2\right) + m_p gL\left(\frac{1}{2}\theta^2\right)$$

$$= \frac{1}{2}\left(\frac{4}{9}kL^2 + mg\frac{L}{2} + m_p gL\right)\theta^2$$

Since an angular coordinate was chosen as the generalized coordinate, the appropriate equivalent system is the torsional system with

$$I_{\text{eq}} = \frac{1}{3}mL^2 + m_p L^2 \qquad k_{t_{\text{eq}}} = \frac{4}{9}kL^2 + mg\frac{L}{2} + m_p gL$$

1.28 Determine the parameters for an equivalent systems model for the system of Fig. 1-34, using x, the downward displacement of the block from the system's equilibrium position, as the generalized coordinate.

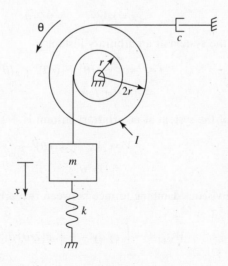

Fig. 1-34

The angular displacement θ of the disk and the displacement y of the particle on the cable connected to the viscous damper at an arbitrary instant are

$$\theta = \frac{x}{r}, \qquad y = 2r\theta = 2x$$

The system's kinetic energy at an arbitrary instant is

$$T = \frac{1}{2}m\dot{x}^2 + \frac{1}{2}I\dot{\theta}^2 = \frac{1}{2}m\dot{x}^2 + \frac{1}{2}I\left(\frac{\dot{x}}{r}\right)^2$$

$$= \frac{1}{2}\left(m + \frac{I}{r^2}\right)\dot{x}^2$$

Noting that the potential energy change due to gravity balances with the potential energy change due to static deflections, the potential energy of the system at an arbitrary instant is

$$V = \tfrac{1}{2}kx^2$$

The work done by the viscous damping force between two arbitrary instants is

$$W_{1\to 2} = -\int_{x_1}^{x_2} c\dot{y}\,dy = -\int_{x_1}^{x_2} c(2\dot{x})\,d(2x)$$

$$= -\int_{x_1}^{x_2} 4c\dot{x}\,dx$$

Thus from the above

$$m_{eq} = m + \frac{I}{r^2}, \qquad k_{eq} = k, \qquad c_{eq} = 4c$$

1.29 Repeat Problem 1.28 using θ, the angular displacement of the disk measured counterclockwise from the system's equilibrium position, as the generalized coordinate.

The downward displacement x of the block and the displacement y of a particle on the cable connected to the viscous damper at an arbitrary instant are

$$x = r\theta, \qquad y = 2r\theta$$

The kinetic energy of the system at an arbitrary instant is

$$T = \tfrac{1}{2}m\dot{x}^2 + \tfrac{1}{2}I\dot{\theta}^2 = \tfrac{1}{2}(r\dot{\theta})^2 + \tfrac{1}{2}I\dot{\theta}^2$$

$$= \tfrac{1}{2}(mr^2 + I)\dot{\theta}^2$$

The potential energy of the system at an arbitrary instant is

$$V = \tfrac{1}{2}kx^2 = \tfrac{1}{2}k(r\theta)^2$$

$$= \tfrac{1}{2}kr^2\theta^2$$

The work done by the viscous damping formce between two arbitrary instants is

$$W_{1\to 2} = -\int_{\theta_1}^{\theta_2} c\dot{y}\,dy = -\int_{\theta_1}^{\theta_2} c(2r\dot{\theta})\,d(2r\theta)$$

$$= -\int_{\theta_1}^{\theta_2} 4cr^2\dot{\theta}\,d\theta$$

Since and angular coordinate is used as the generalized coordinate, the appropriate equivalent systems model is the torsional system. Thus from the above

$$I_{eq} = I + mr^2, \qquad c_{t_{eq}} = 4cr^2, \qquad k_{t_{eq}} = kr^2$$

1.30 Determine the parameters in an equivalent system model of the system of Fig. 1-35 when θ, the clockwise angular displacement of the bar from the system's equilibrium position, is used as the generalized coordinate. Assume small θ.

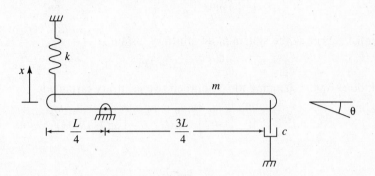

Fig. 1-35

Assuming small θ, the downward displacements of the bar's left end x_ℓ, right end x_r, and the mass center $\bar{x}$ are

$$x_\ell = -\frac{L}{4}\theta, \qquad x_r = \frac{3}{4}L\theta, \qquad \bar{x} = \frac{L}{4}\theta$$

The kinetic energy of the system at an arbitrary instant is

$$\begin{aligned}
T &= \tfrac{1}{2}\bar{I}\dot{\theta}^2 + \tfrac{1}{2}m\bar{v}^2 \\
&= \tfrac{1}{2}\tfrac{1}{12}mL^2\dot{\theta}^2 + \tfrac{1}{2}m(\tfrac{1}{4}L\dot{\theta})^2 \\
&= \tfrac{1}{2}(\tfrac{7}{48}mL^2)\dot{\theta}^2
\end{aligned}$$

The potential energy of the system at an arbitrary instant is

$$\begin{aligned}
V &= \tfrac{1}{2}kx_\ell^2 = \tfrac{1}{2}k(-\tfrac{1}{4}L\theta)^2 \\
&= \tfrac{1}{2}(\tfrac{1}{16}kL^2)\theta^2
\end{aligned}$$

The work done by the damping force between two arbitrary instants is

$$\begin{aligned}
W_{1\to 2} &= -\int_{x_{r1}}^{x_{r2}} c\dot{x}_r \, dx_r = -\int_{\theta_1}^{\theta_2} c\left(\frac{3}{4}L\dot{\theta}\right) d\left(\frac{3}{4}L\theta\right) \\
&= -\int_{\theta_1}^{\theta_2} \frac{9}{16}cL^2\dot{\theta}\, d\theta
\end{aligned}$$

Since the generalized coordinate is an angular displacement, the appropriate equivalent model is the torsional system. From the above,

$$I_{eq} = \tfrac{7}{48}mL^2, \qquad c_{t_{eq}} = \tfrac{9}{16}cL^2, \qquad k_{t_{eq}} = \tfrac{1}{16}kL^2$$

1.31 Repeat Problem 1.30 using x, the upward displacement of the left end of the bar, measured from equilibrium, as the generalized coordinate. Assume small x.

Assuming small x, the downward displacement of the right end x_r, mass center $\bar{x}$, and the clockwise angular rotation of the bar θ are

$$x_r = 3x, \qquad \bar{x} = x, \qquad \theta = \frac{4}{L}x$$

The kinetic energy of the system at an arbitrary instant is

$$T = \frac{1}{2}m\bar{v}^2 + \frac{1}{2}\bar{I}\dot{\theta}^2 = \frac{1}{2}m\dot{x}^2 + \frac{1}{2}\frac{1}{12}mL^2\left(\frac{4}{L}\dot{x}\right)^2$$

$$= \frac{1}{2}\left(\frac{7}{3}m\right)\dot{x}^2$$

The potential energy of the system at an arbitrary instant is

$$V = \tfrac{1}{2}kx^2$$

The work done by the damping force between two arbitrary instants is

$$W_{1\to2} = -\int_{x_{r_1}}^{x_{r_2}} c\dot{x}_r\, dx_r = -\int_{x_1}^{x_2} c(3\dot{x})\, d(3x)$$

$$= -\int_{x_1}^{x_2} 9c\dot{x}\, dx$$

Thus the equivalent system parameters are

$$m_{eq} = \tfrac{7}{3}m, \qquad c_{eq} = 9c, \qquad k_{eq} = k$$

1.32 Determine the parameters for an equivalent system model of the system of Fig. 1-36 when x, the displacement of the mass center of the disk measured from the system's equilibrium position, is used as the generalized coordinate. Assume the disk rolls without slip.

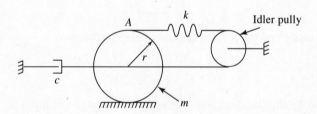

Fig. 1-36

If the disk rolls without slip, then the friction force does no work and the angular displacement of the disk is related to x by

$$\theta = \frac{x}{r}$$

The kinetic energy of the system at an arbitrary instant is

$$T = \frac{1}{2}m\dot{x}^2 + \frac{1}{2}\bar{I}\dot{\theta}^2 = \frac{1}{2}m\dot{x}^2 + \frac{1}{2}\left(\frac{1}{2}mr^2\right)\left(\frac{\dot{x}}{r}\right)^2$$

$$= \frac{1}{2}\frac{3}{2}m\dot{x}^2$$

The spring is attached to the disk at A. If the center of the disk moves a distance x to the left, then A moves relative to the disk a distance $r\theta$. The change in length of the spring is

$$\delta = x + r\theta + x = 2x + r\theta = 2x + x = 3x$$

Thus the potential energy of the system at an arbitrary instant is

$$V = \tfrac{1}{2}(3x)^2 = \tfrac{1}{2}9kx^2$$

The work done by the viscous damping force between two arbitrary instants is

$$W_{1\to 2} = -\int_{x_1}^{x_2} c\dot{x}\, dx$$

Hence the coefficients for an equivalent systems model is

$$m_{eq} = \tfrac{3}{2}m, \qquad c_{eq} = c, \qquad k_{eq} = 9k$$

Supplementary Problems

1.33   Determine the number of degrees of freedom necessary for the analysis of the system of Fig. 1-37.

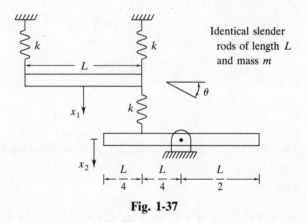

Fig. 1-37

Ans.   3

1.34   Determine the number of degrees of freedom necessary for the analysis of the system of Fig. 1-38.

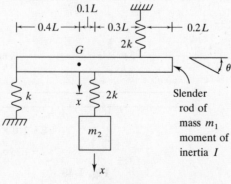

Fig. 1-38

Ans.   3

1.35 Determine the number of degrees of freedom necessary for the analysis of the system of Fig. 1-39.

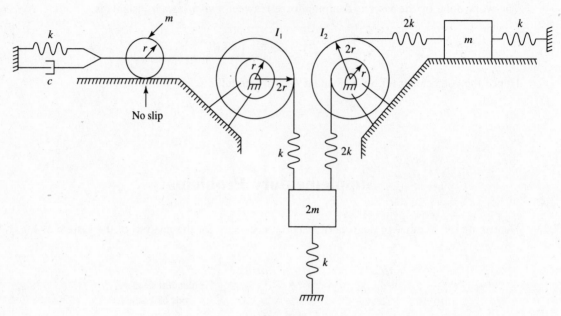

Fig. 1-39

Ans.   4

1.36 Determine the longitudinal stiffness of a rectangular, 30×50 mm steel bar ($E = 210 \times 10^9$ N/m^2) of length 2.1 m.

Ans. 1.5×10^8 N/m

1.37 Determine the torsional stiffness of a 60-cm-long annular aluminum shaft ($G = 40 \times 10^9$ N/m^2) of inner radius 25 mm and outer radius 35 mm.

Ans. 1.16×10^5 N-m/rad

1.38 A 200-kg machine is placed at the end of the beam of Fig. 1-40. Determine the stiffness of the beam for use in a 1-degree-of-freedom model of the system.

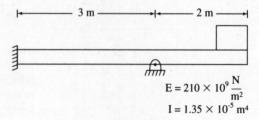

$$E = 210 \times 10^9 \frac{N}{m^2}$$
$$I = 1.35 \times 10^{-5} \text{ m}^4$$

Fig. 1-40

Ans. 5.00×10^5 N/m

1.39   Determine the equivalent stiffness of the beam of Fig. 1-41 at the location where the machine is placed.

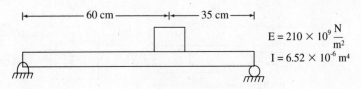

$$E = 210 \times 10^9 \frac{N}{m^2}$$
$$I = 6.52 \times 10^{-6} \ m^4$$

Fig. 1-41

Ans. 8.85×10^7 N/m

1.40   A helical coil spring is made from a steel $(G = 80 \times 10^9$ N/m$^2)$ bar of radius 6 mm. The spring has a coil diameter of 6 cm and has 46 active turns. What is the stiffness of the spring?

Ans. 2.09×10^4 N/m

1.41   What is the static deflection of the spring of Problem 1.40 when it is used in the system of Fig. 1.42?

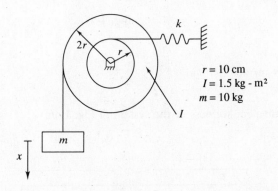

$r = 10$ cm
$I = 1.5$ kg - m^2
$m = 10$ kg

Fig. 1-42

Ans. 9.39×10^{-3} m

1.42   Determine the equivalent stiffness of the system of Fig. 1-43.

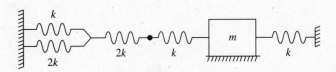

Fig. 1-43

Ans. $17k/11$

1.43   Determine the equivalent stiffness of the system of Fig. 1-44.

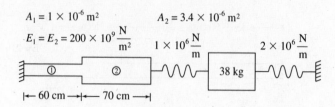

Fig. 1-44

Ans. 2.20×10^6 N/m

1.44   Determine the equivalent torsional stiffness of the system of Fig. 1-45.

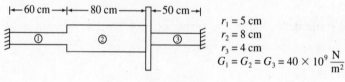

Fig. 1-45

Ans. 8.66×10^5 N-m/rad

1.45   Determine the equivalent stiffness of the system of Fig. 1-46.

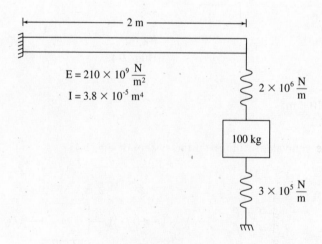

Fig. 1-46

Ans. 1.5×10^6 N/m

1.46 Determine the equivalent stiffness of the system of Fig. 1-47.

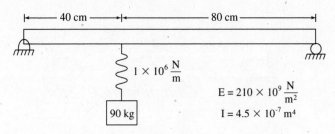

Fig. 1-47

Ans. 7.69×10^5 N/m

1.47 Determine the equivalent stiffness of the system of Fig. 1-48.

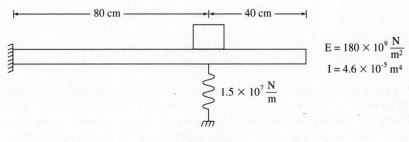

Fig. 1-48

Ans. 6.35×10^7 N/m

1.48 The torsional viscous damper of Fig. 1-49 consists of a cylinder of radius r that rotates inside a fixed cylinder. The cylinders are concentric with a clearance h. The gap between the cylinders is filled with a fluid of viscosity μ. The length of cylinder in contact with the fluid is ℓ. Determine the torsional viscous damping coefficient for this damper.

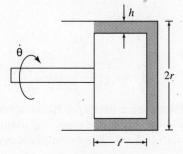

Fig. 1-49

Ans.

$$c_t = \frac{2\tau\mu r^3 \ell}{h}$$

1.49   Determine the kinetic energy of the system of Fig. 1-50 at an arbitrary instant in terms of $\dot{x}$, including the inertia effects of the springs

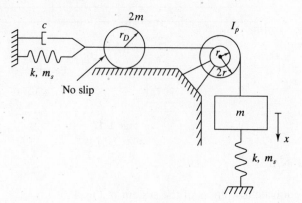

Fig. 1-50

Ans.

$$\frac{1}{2}\left(\frac{7}{4}m + \frac{5}{4}m_s + \frac{I_p}{4r^2}\right)\dot{x}^2$$

1.50   Let $\theta(t)$ represent the angular displacement of a thin disk attached at one end of circular shaft, fixed at its other end. The shaft has a mass moment of inertia I_s about its longitudinal centroidal axis. Using a linear displacement approximation, determine the equivalent moment of inertia of a disk to be added to the end of the shaft to approximate the inertia effects of the shaft.

Ans.

$$I_{\text{eq}} = \tfrac{1}{3}I_s$$

1.51   The static deflection for a fixed-free beam of length L, cross-sectional moment of inertia I, and elastic modulus E with a concentrated load F at its free end is

$$y(z) = \frac{Fz^2}{6EI}(3L - z)$$

Use this equation to develop the equivalent mass of the beam if it has a cross-sectional area A and a mass density ρ.

Ans. $0.236\rho AL$

1.52   The trigonometric function

$$y(z) = x\left[1 - \cos\left(\frac{\pi z}{2L}\right)\right]$$

satisfies all boundary conditions for a fixed-free beam of length L where x is the deflection at the free end. Use this function to determine the mass of a particle that can be placed at the end of the beam to approximate its inertia effects. The beam has a mass density ρ and a cross-sectional area A.

Ans. $0.227\rho AL$

1.53   Use a trigonometric function similar to that of Problem 1.20 to determine the mass of a particle to be placed $\frac{2}{3}$ along the span of a fixed-fixed beam to approximate the beam's inertia effects.

Ans. $\frac{2}{3}\rho AL$

1.54 Determine the equivalent moment of inertia of the gearing system of Fig. 1-51.

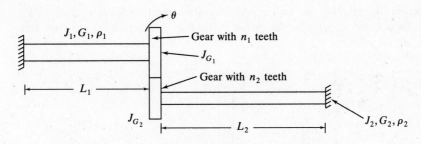

Fig. 1-51

Ans.

$$J_{G_1} + J_{G_2}\left(\frac{n_1}{n_2}\right)^2$$

1.55 Repeat Problem 1.54, including the inertia effects of the shafts.

Ans.

$$\frac{1}{3}J_1 + J_{G_1} + \left(J_{G_2} + \frac{1}{3}J_2\right)\left(\frac{n_1}{n_2}\right)^2$$

1.56 Determine m_{eq} and k_{eq} for an equivalent system model of the system of Fig. 1-42 using x as the generalized coordinate. Note that k is determined in Problem 1.40 as 2.09×10^4 N/m.

Ans.

$$m_{eq} = 47.5 \text{ kg}, \qquad k_{eq} = 5.23 \times 10^3 \frac{\text{N}}{\text{m}}$$

1.57 Determine I_{eq} and $k_{t_{eq}}$ for an equivalent system model of the system of Fig. 1-52 using θ, the clockwise displacement of the pulley, as the generalized coordinate.

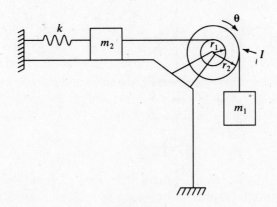

Fig. 1-52

Ans.

$$I_{eq} = I + m_1 r_2^2 + m_2 r_1^2, \qquad k_{t_{eq}} = kr_1^2$$

1.58   Determine I_{eq} and $k_{t_{eq}}$ for an equivalent system model of the system of Fig. 1-53 using θ, the counterclockwise angular displacement of the bar, as the generalized coordinate.

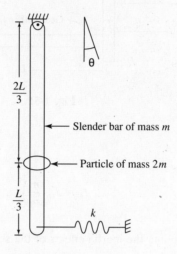

Fig. 1-53

Ans.

$$I_{eq} = \tfrac{11}{9} mL^2, \qquad k_{t_{eq}} = kL^2 + \tfrac{11}{6} mgL$$

1.59   Determine m_{eq}, k_{eq}, and c_{eq} for an equivalent system model of the system of Fig. 1-54 using x as the generalized coordinate.

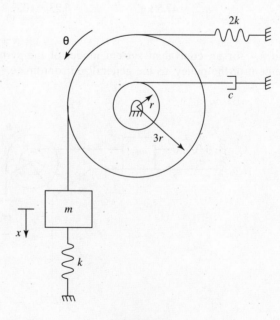

Fig. 1-54

Ans.

$$m_{eq} = m + \frac{I}{9r^2}, \qquad c_{eq} = \frac{c}{9}, \qquad k_{eq} = 3k$$

1.60 Determine I_{eq}, $k_{t_{eq}}$, and $c_{t_{eq}}$ for an equivalent system model of the system of Fig. 1-54 using θ as the generalized coordinate.

Ans.

$$I_{eq} = I + 9mr^2, \qquad c_{t_{eq}} = cr^2, \qquad k_{t_{eq}} = 27kr^2$$

1.61 Determine I_{eq}, $k_{t_{eq}}$, and $c_{t_{eq}}$ for an equivalent system model of the system of Fig. 1-55 using θ as the generalized coordinate.

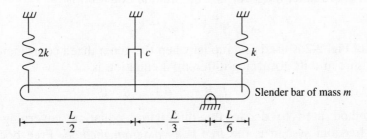

Fig. 1-55

Ans.

$$I_{eq} = \tfrac{7}{36}mL^2, \qquad c_{t_{eq}} = \tfrac{1}{9}cL^2, \qquad k_{t_{eq}} = \tfrac{17}{12}kL^2$$

1.62 Use an energy method to derive the equivalent stiffness of two identical springs in series.

Ans. $k/2$

Chapter 2

Free Vibrations of 1-Degree-of-Freedom Systems

2.1 DERIVATION OF DIFFERENTIAL EQUATIONS

All linear 1-degree-of-freedom systems can be modeled using either the system of Fig. 2-1 or the system of Fig. 2-2. The *equivalent system method*, or the *energy method*, uses the system's equivalent inertia, stiffness, and damping properties as described in Chap. 1. The system of Fig. 2-1 is used as a model when the generalized coordinate is a linear displacement coordinate. Its governing differential equation is

$$m_{eq}\ddot{x} + c_{eq}\dot{x} + k_{eq}x = 0 \tag{2.1}$$

The system of Fig. 2-2 is used as a model when the generalized coordinate is a coordinate of angular measure and its governing differential equation is

$$I_{eq}\ddot{\theta} + c_{t_{eq}}\dot{\theta} + k_{t_{eq}}\theta = 0 \tag{2.2}$$

Another method used to derive the differential equation governing the motion of a 1-degree-of-freedom system is the *free body diagram method*. Free body diagrams of the system components are drawn at an arbitrary instant. The external forces due to elastic elements and viscous damping are labeled in terms of the chosen generalized coordinate, with their directions drawn consistent with the chosen positive sense of the generalized coordinate. The basic laws of newtonian mechanics are applied to the free body diagrams, leading to the governing differential equation. For a rigid body undergoing planar motion, these equations are

$$\sum \vec{F} = m\vec{a} \tag{2.3}$$

and

$$\sum M_G = \bar{I}\alpha \tag{2.4}$$

where a quantity with an overbar is referenced to G, the body's mass center.

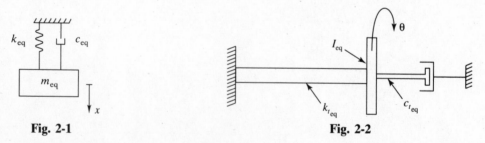

Fig. 2-1   **Fig. 2-2**

A version of the free body diagram method for rigid bodies undergoing planar motion uses a variation of D'Alembert's principle. In addition to the free body diagram showing external forces at an arbitrary instant, a second free body diagram is drawn at the same instant showing the system's effective forces. The effective forces for a rigid body are defined as force equal to $m\vec{a}$, acting at the mass center, and a couple equal to $\bar{I}\alpha$. Equations (2.3) and (2.4) are applied in the form

$$\left(\sum \vec{F}\right)_{ext} = \left(\sum \vec{F}\right)_{eff} \tag{2.5}$$

and
$$\left(\sum M_A\right)_{\text{ext}} = \left(\sum M_A\right)_{\text{eff}} \tag{2.6}$$

for any point A.

2.2 STANDARD FORM OF DIFFERENTIAL EQUATIONS

Equation (2.1) can be rewritten as
$$\ddot{x} + 2\zeta\omega_n\dot{x} + \omega_n^2 x = 0 \tag{2.7}$$

where
$$\omega_n = \sqrt{\frac{k_{\text{eq}}}{m_{\text{eq}}}} \tag{2.8}$$

is called the system's *undamped natural frequency* and
$$\zeta = \frac{c_{\text{eq}}}{2m_{\text{eq}}\omega_n} = \frac{c_{\text{eq}}}{2\sqrt{m_{\text{eq}}k_{\text{eq}}}} \tag{2.9}$$

is called the *damping ratio*. Equation (2.7) is subject to initial conditions of the form
$$x(0) = x_0 \tag{2.10}$$

and
$$\dot{x}(0) = \dot{x}_0 \tag{2.11}$$

2.3 UNDAMPED RESPONSE

For $\zeta = 0$, the solution of Eq. (2.7) subject to Eqs. (2.10) and (2.11) is
$$x(t) = A \sin(\omega_n t + \phi) \tag{2.12}$$

where A, the *amplitude*, is the maximum displacement from equilibrium and is given by
$$A = \sqrt{x_0^2 + \left(\frac{\dot{x}_0}{\omega_n}\right)^2} \tag{2.13}$$

and ϕ, the *phase angle*, is given by
$$\phi = \tan^{-1}\left(\frac{\omega_n x_0}{\dot{x}_0}\right) \tag{2.14}$$

Equation (2.12) is illustrated in Fig. 2-3.

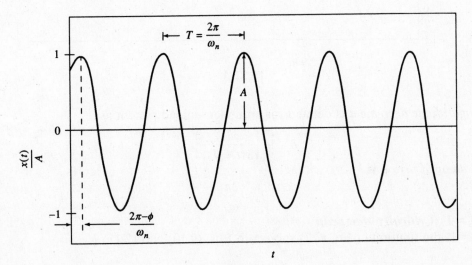

Fig. 2-3

2.4 DAMPED RESPONSE

The mathematical form of the free vibration response of a 1-degree-of-freedom system with viscous damping is dependent on the value of ζ.

Case 1: $\zeta < 1$ (Underdamped):

$$x(t) = Ae^{-\zeta\omega_n t} \sin(\omega_d t + \phi_d) \tag{2.15}$$

where

$$A = \sqrt{x_0^2 + \left(\frac{\dot{x}_0 + \zeta\omega_n x_0}{\omega_d}\right)^2} \tag{2.16}$$

$$\phi_d = \tan^{-1}\left(\frac{x_0\omega_d}{\dot{x}_0 + \zeta\omega_n x_0}\right) \tag{2.17}$$

and the *damped natural frequency* is

$$\omega_d = \omega_n\sqrt{1 - \zeta^2} \tag{2.18}$$

Equation (2.15) is illustrated in Fig. 2-4 for $x_0 \neq 0$.

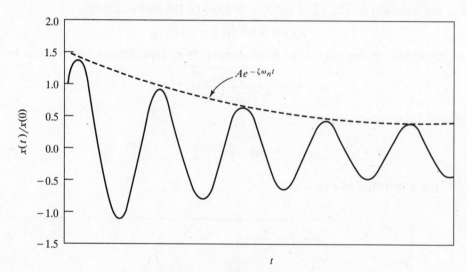

Fig. 2-4

The *logarithmic decrement* is defined for an underdamped system as

$$\delta = \ln\left[\frac{x(t)}{x(t + T_d)}\right] \tag{2.19}$$

where the damped period is

$$T_d = \frac{2\pi}{\omega_d} \tag{2.20}$$

Case 2: $\zeta = 1$ (Critically damped):

For $\zeta = 1$, the solution of Eq. (2.7) subject to Eqs. (2.10) and (2.11) is

$$x(t) = e^{-\omega_n t}[x_0 + (\dot{x}_0 + \omega_n x_0)t] \tag{2.21}$$

Equation (2.21) is illustrated in Fig. 2-5.

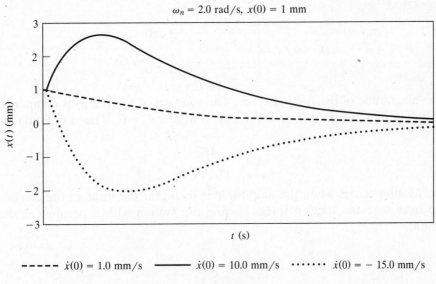

Fig. 2-5

Case 3: $\zeta > 1$ (Overdamped):

For $\zeta > 1$, the solution of Eq. (2.7) subject to Eqs. (2.10) and (2.11) is

$$x(t) = \frac{e^{-\zeta\omega_n t}}{2\sqrt{\zeta^2 - 1}}\left\{\left[\frac{\dot{x}_0}{\omega_n} + x_0(\zeta + \sqrt{\zeta^2 - 1})\right]e^{\omega_n\sqrt{\zeta^2 - 1}\,t}\right.$$
$$\left. + \left[-\frac{\dot{x}_0}{\omega_n} + x_0(-\zeta + \sqrt{\zeta^2 - 1})\right]e^{-\omega_n\sqrt{\zeta^2 - 1}\,t}\right\} \tag{2.22}$$

Equation (2.22) is illustrated in Fig. 2-6.

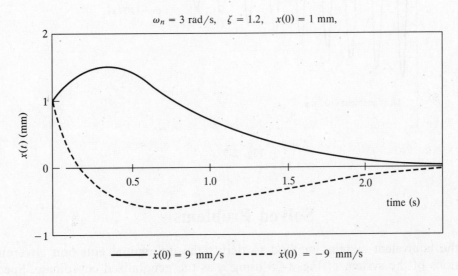

Fig. 2-6

2.5   FREE VIBRATION RESPONSE FOR SYSTEMS SUBJECT TO COULOMB DAMPING

The differential equation governing the motion of a system with Coulomb damping can be written in general as

$$\dot{x} + \omega_n{}^2 x = \begin{cases} -\dfrac{F_f}{m_{eq}} & \dot{x} > 0 \\[2mm] \dfrac{F_f}{m_{eq}} & \dot{x} < 0 \end{cases} \qquad (2.23)$$

where F_f is the magnitude of the friction force. The solution of Eq. (2.23) is complicated but can be attained and is illustrated in Fig. 2-7 for $x_0 = \delta$ and $\dot{x}_0 = 0$. The amplitude of response decreases by

$$\Delta A = \frac{4F_f}{m_{eq}\omega_n{}^2} \qquad (2.24)$$

on each cycle. Motion ceases when the amplitude is such that the force in the elastic element is insufficient to overcome the friction force, leaving the system with a permanent displacement from equilibrium.

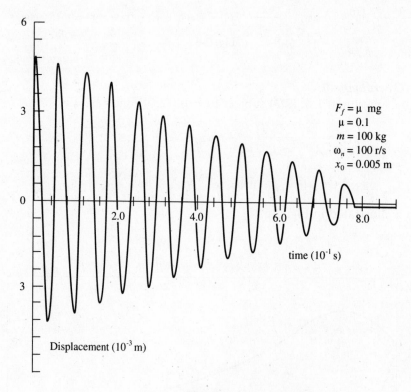

Fig. 2-7

Solved Problems

2.1   Use the equivalent systems method to derive the differential equation governing free vibrations of the system of Fig. 1-28 using x as the generalized coordinate. Specify the system's natural frequency.

The system of Fig. 1-28 can modeled by an undamped system of the form of Fig. 2-1. The equivalent system parameters are determined in Problem 1.22 as

$$m_{eq} = m + \frac{I}{r^2}, \qquad k_{eq} = 3k, \qquad c_{eq} = 0$$

Equation (2.1) is used to write the governing differential equation as

$$\left(m + \frac{I}{r^2}\right)\ddot{x} + 3kx = 0$$

$$\ddot{x} + \frac{3kr^2}{I + mr^2}x = 0$$

from which the natural frequency is determined as

$$\omega_n = \sqrt{\frac{3kr^2}{I + mr^2}}$$

2.2 Use the equivalent systems method to derive the differential equation governing the motion of the system of Fig. 1-31 using θ as the generalized coordinate. Specify the system's natural frequency.

From the results of Problem 1.25, the system of Fig. 1-31 is modeled by an undamped system of the form of Fig. 2-2 with

$$I_{eq} = \tfrac{7}{27}mL^2, \qquad k_{t_{eq}} = \tfrac{43}{81}kL^2$$

Then, using Eq. (2.2), the governing differential equation is

$$\tfrac{7}{27}mL^2\ddot{\theta} + \tfrac{43}{81}kL^2\theta = 0$$

$$\ddot{\theta} + \frac{43}{21}\frac{k}{m}\theta = 0$$

from which the natural frequency is determined as

$$\omega_n = \sqrt{\frac{43k}{21m}}$$

2.3 Use the equivalent system method to derive the differential equation governing the motion of the system of Fig. 1-35, using θ as the generalized coordinate. Specify the system's natural frequency.

From the results of Problem 1.30, the system of Fig. 1-35 can be modeled by the system of Fig. 2-2 with

$$I_{eq} = \tfrac{7}{48}mL^2, \qquad c_{t_{eq}} = \tfrac{9}{16}cL^2, \qquad k_{t_{eq}} = \tfrac{1}{16}kL^2$$

Then, using Eq. (2.2), the governing differential equation is

$$\tfrac{7}{48}mL^2\ddot{\theta} + \tfrac{9}{16}cL^2\dot{\theta} + \tfrac{1}{16}kL^2\theta = 0$$

$$\ddot{\theta} + \frac{27}{7}\frac{c}{m}\dot{\theta} + \frac{3}{7}\frac{k}{m}\theta = 0$$

from which the natural frequency is determined as

$$\omega_n = \sqrt{\frac{3k}{7m}}$$

2.4   Use the free body diagram method to derive the differential equation governing the motion of the system of Fig. 1-28 using x as the generalized coordinate.

Free body diagrams of the system, which includes the block and disk, at an arbitrary instant are shown in Fig. 2-8. It is noted that since gravity leads to static deflection in the springs, their effects cancel in the differential equation. Summing moments about the center of the disk

$$\left(\sum M_O\right)_{ext} = \left(\sum M_O\right)_{eff}$$

leads to

$$-(kx)r - (2kx)r = (m\ddot{x})(r) + I\left(\frac{\ddot{x}}{r}\right)$$

$$\left(mr + \frac{I}{r}\right)\ddot{x} + 3krx = 0$$

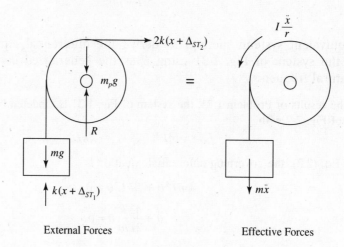

External Forces   Effective Forces

Fig. 2-8

2.5   Use the free body digram method to derive the differential equation governing the motion of the system of Fig. 1-31, using θ as the generalized coordinate.

Let ϕ represent the clockwise angular rotation of bar CD. Since the displacements of the particles where the rigid link is connected must be the same, assuming small θ and ϕ

$$L\phi = \tfrac{2}{3}L\theta \rightarrow \phi = \tfrac{2}{3}\theta$$

Free body diagrams of each of the bars are shown in Figs. 2-9 and 2-10, assuming small θ. It is noted that since gravity leads to static deflections, their effects cancel in the differential equations. Summing moments on bar AB

$$\left(\sum M_O\right)_{ext} = \left(\sum M_O\right)_{eff}$$

leads to

$$-3k\left(\frac{L}{3}\theta\right)\frac{L}{3} - F_{BC}\left(\frac{2}{3}L\right) = \frac{L}{6}m\ddot{\theta}\left(\frac{L}{6}\right) + \frac{1}{12}mL^2\ddot{\theta}$$

$$F_{BC} = -\tfrac{1}{6}mL\ddot{\theta} - \tfrac{1}{2}kL\theta$$

Summing moments on bar CD

$$\left(\sum M_Q\right)_{\text{ext}} = \left(\sum M_Q\right)_{\text{eff}}$$

leads to

$$F_{BC}L - \frac{4}{9}kL\theta\left(\frac{2}{3}L\right) = \frac{1}{2}mL\left(\frac{2}{3}\ddot{\theta}\right)\left(\frac{L}{2}\right) + \frac{1}{12}mL^2\left(\frac{2}{3}\ddot{\theta}\right)$$

Substitution for F_{BC} leads to

$$\tfrac{7}{18}mL^2\ddot{\theta} + \tfrac{43}{54}kL^2\theta = 0$$

$$\ddot{\theta} + \frac{43}{21}\frac{k}{m}\theta = 0$$

Bar AB

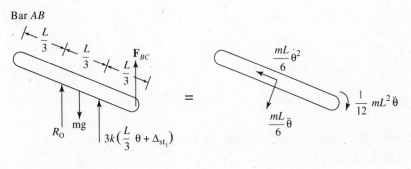

External forces   Effective forces

Fig. 2-9

Bar CD

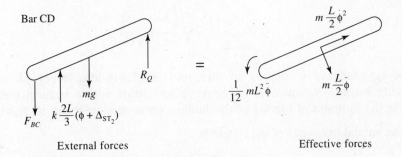

External forces   Effective forces

Fig. 2-10

2.6   Use the free body diagram method to derive the differential equation governing the motion of the system of Fig. 1-35 using θ as the generalized coordinate.

Free body diagrams of the bar at an arbitrary instant assuming small θ are shown in Fig. 2-11. It is noted that since gravity causes static deflection in the spring, terms containing these quantities cancel in the governing differential equation. Summing moments about the pin support

$$\left(\sum M_A\right)_{\text{ext}} = \left(\sum M_A\right)_{\text{eff}}$$

leads to

$$-k\frac{L}{4}\theta\left(\frac{L}{4}\right) - \frac{3}{4}cL\dot{\theta}\left(\frac{3}{4}L\right) = \frac{L}{4}m\ddot{\theta}\left(\frac{L}{4}\right) + \frac{1}{12}mL^2\ddot{\theta}$$

$$\tfrac{7}{48}mL^2\ddot{\theta} + \tfrac{9}{16}cL^2\dot{\theta} + \tfrac{1}{16}kL^2\theta = 0$$

$$\ddot{\theta} + \frac{27}{7}\frac{c}{m}\dot{\theta} + \frac{3}{7}\frac{k}{m}\theta = 0$$

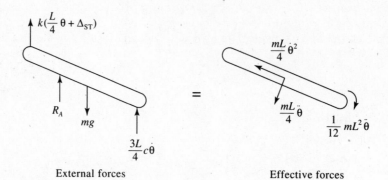

External forces   =   Effective forces

Fig. 2-11

2.7 Determine the natural frequency of the system of Fig. 1-15.

From Problem 1.11 the system of Fig. 1-15 can be modeled by the system of Fig. 2-1 with

$$k_{eq} = \frac{7}{6}k, \qquad m_{eq} = m$$

Thus from Eq. (2.8),

$$\omega_n = \sqrt{\frac{\frac{7}{6}k}{m}} = \sqrt{\frac{7k}{6m}}$$

2.8 A 200-kg machine is placed at the end of 1.8-m-long steel $(E = 210 \times 10^9 \text{ N/m}^2)$ cantilever beam. The machine is observed to vibrate with a natural frequency of 21 Hz. What is the moment of inertia of the beam's cross section about its neutral axis?

The natural frequency of the system is

$$\omega_n = 21 \text{ Hz} = \left(21 \,\frac{\text{cycle}}{\text{s}}\right)\left(2\pi \,\frac{\text{rad}}{\text{cycle}}\right) = 131.9 \,\frac{\text{rad}}{\text{s}}$$

and is related to the system properties by

$$\omega_n = \sqrt{\frac{k_{eq}}{m}} \;\rightarrow\; k_{eq} = m\omega_n^2 = (200 \text{ kg})\left(131.9 \,\frac{\text{rad}}{\text{s}}\right)^2 = 3.48 \times 10^6 \,\frac{\text{N}}{\text{m}}$$

The equivalent stiffness of a cantilever beam for a mass at its end is

$$k_{eq} = \frac{3EI}{L^3}$$

Thus $$I = \frac{k_{eq}L^3}{3E} = \frac{\left(3.48 \times 10^6 \,\frac{\text{N}}{\text{m}}\right)(1.8 \text{ m})^3}{3\left(210 \times 10^9 \,\frac{\text{N}}{\text{m}^2}\right)} = 3.22 \times 10^{-5} \text{ m}^4$$

2.9   A 2.5-kg slender bar of length 40 cm is pinned at one end. A 3-kg particle is to be attached to the bar. How far from the pin support should the particle be placed such that the period of the bar's oscillation is 1 s?

Mathcad

If the period of oscillation is 1 s, then

$$\omega_n = \frac{2\pi}{T} = \frac{2\pi}{1\,\text{s}} = 6.28\,\frac{\text{rad}}{\text{s}}$$

Let ℓ be the distance of the particle from the pin support. Let θ be the counterclockwise angular displacement of the system from the vertical equilibrium position. Free body diagrams of the bar and particle at an arbitrary instant are shown in Fig. 2-12. Summing moments about the point of support

$$\left(\sum M_O\right)_{ext} = \left(\sum M_O\right)_{eff}$$

and assuming small θ leads to

$$-m_b g \frac{L}{2}\theta - m_p g \ell\theta = m_p \ell\ddot{\theta}(\ell) + m_b \frac{L}{2}\ddot{\theta}\left(\frac{L}{2}\right) + \frac{1}{12}mL^2\ddot{\theta}$$

$$\left(m_b \frac{L^2}{3} + m_p \ell^2\right)\ddot{\theta} + \left(m_b \frac{L}{2} + m_p \ell\right)g\theta = 0$$

The natural frequency is determined from the differential equation as

$$\omega_n = \sqrt{\frac{\left(m_b \dfrac{L}{2} + m_p \ell\right)g}{m_b \dfrac{L^2}{3} + m_p \ell^2}}$$

which can be rearranged as

$$m_p \omega_n^2 \ell^2 - m_p g\ell + m_b \frac{L^2}{3}\omega_n^2 - m_b \frac{L}{2}g = 0$$

Substituting given and calculated values leads to

$$118.3\ell^2 - 29.43\ell + 0.36 = 0$$

whose positive solution is $\ell = 0.235$ m.

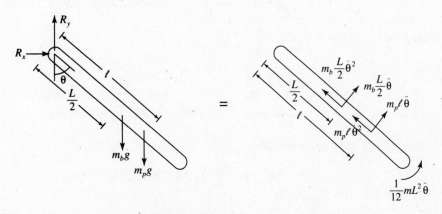

External forces   Effective forces

Fig. 2-12

2.10 A 5-kg wheel is mounted on a 1-kg plate whose center is attached to a 40-mm-diameter, 75-cm-long ($G = 80 \times 10^9$ N/m^2) steel bar, which is fixed at its other end. The centroidal polar moment of inertia of the plate is 1.4 kg-m^2. The period of torsional oscillation of this assembly is 0.15 s. What is the polar moment of inertia of the wheel?

The torsional oscillations are modeled by the system of Fig. 2-2 with $c_{t_{eq}} = 0$. The torsional stiffness of the bar is

$$k_t = \frac{JG}{L} = \frac{\frac{\pi}{2}(0.02 \text{ m})^4 \left(80 \times 10^9 \frac{\text{N}}{\text{m}^2}\right)}{0.75 \text{ m}} = 2.68 \times 10^4 \frac{\text{N-m}}{\text{rad}}$$

The observed natural frequency is

$$\omega_n = \frac{2\pi}{T} = \frac{2\pi}{0.15 \text{ s}} = 41.9 \frac{\text{rad}}{\text{s}}$$

The natural frequency of the torsional system is

$$\omega_n = \sqrt{\frac{k_{eq}}{I}} \rightarrow I = \frac{k_{eq}}{\omega_n^2} = \frac{2.68 \times 10^4 \frac{\text{N-m}}{\text{rad}}}{\left(41.9 \frac{\text{rad}}{\text{s}}\right)^2} = 15.3 \text{ kg-m}^2$$

Hence the moment of inertia of the wheel is

$$I_w = I - I_p = 15.3 \text{ kg-m}^2 - 1.4 \text{ kg-m}^2 = 13.9 \text{ kg-m}^2$$

2.11 A 60-kg drum of diameter 40 cm containing waste material of mass density 1100 kg/m^3 is being hoisted by a 30-mm-diameter steel ($E = 210 \times 10^9$ N/m^2) cable. When the drum is to be hoisted 10 m, the system's natural frequency is measured as 40 Hz. Determine the volume of waste in the drum.

The system is modeled as a mass attached to the end of an elastic bar. The measured frequency is the frequency of longitudinal vibrations. The equivalent stiffness of the cable is

$$k_{eq} = \frac{AE}{L} = \frac{\pi(0.015 \text{ m})^2 \left(210 \times 10^9 \frac{\text{N}}{\text{m}^2}\right)}{10 \text{ m}} = 1.48 \times 10^7 \frac{\text{N}}{\text{m}}$$

The total mass is calculated from

$$\omega_n = \sqrt{\frac{k_{eq}}{m}} \rightarrow m = \frac{k_{eq}}{\omega_n^2} = \frac{1.48 \times 10^7 \frac{\text{N}}{\text{m}}}{\left(40 \frac{\text{cycle}}{\text{s}} 2\pi \frac{\text{rad}}{\text{cycle}}\right)^2} = 234.3 \text{ kg}$$

Thus the mass of the waste material is

$$m_w = m - m_d = 234.3 \text{ kg} - 60 \text{ kg} = 174.3 \text{ kg}$$

and its volume is

$$V = \frac{m_w}{\rho} = \frac{174.3 \text{ kg}}{1100 \frac{\text{kg}}{\text{m}^3}} = 0.158 \text{ m}^3$$

2.12 A sway pole is used by aerialists for acrobatic tricks. A sway pole consists of a long thin pole fixed at one end designed such that an aerialist can sway and perform tricks at the end of the pole. What is the natural frequency, in hertz, of a 120-lb aerialist at the end of a 25-ft steel ($E = 29 \times 10^6$ psi) pole of 4 in diameter?

The aerialist at the end of the sway pole is modeled as a mass at the end of a cantilever beam. The equivalent stiffness of the pole is

$$k_{eq} = \frac{3EI}{L^3} = \frac{3\left(29 \times 10^6 \, \frac{\text{lb}}{\text{in}^2}\right)\left(12 \, \frac{\text{in}}{\text{ft}}\right)^2 \frac{\pi}{4}\left(\frac{1}{6} \, \text{ft}\right)^4}{(25 \, \text{ft})^3} = 4.86 \times 10^2 \, \frac{\text{lb}}{\text{ft}}$$

Thus the aerialist's natural frequency is

$$\omega_n = \sqrt{\frac{k_{eq}}{m}} = \sqrt{\frac{4.86 \times 10^2 \, \frac{\text{lb}}{\text{ft}}}{\frac{120 \, \text{lb}}{32.2 \, \frac{\text{ft}}{\text{s}^2}}}} = 3.30 \, \frac{\text{rad}}{\text{s}}\left(\frac{1 \, \text{cycle}}{2\pi \, \text{rad}}\right) = 1.82 \, \text{Hz}$$

2.13 Determine the natural frequency of the system of Fig. 2-13.

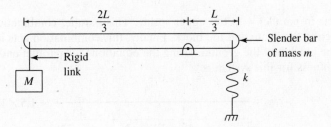

Fig. 2-13

Let θ be the clockwise angular displacement of the bar from the system's equilibrium position. Assuming small θ, the potential energy of the system at an arbitrary instant is

$$V = \frac{1}{2}k\left(\frac{L}{3}\theta\right)^2 = \frac{1}{2}\left(\frac{1}{9}kL^2\right)\theta^2 \quad \rightarrow \quad k_{t_{eq}} = \frac{1}{9}kL^2$$

The kinetic energy of the system at an arbitrary instant is

$$T = \frac{1}{2}\frac{1}{12}mL^2\dot{\theta}^2 + \frac{1}{2}m\left(\frac{1}{6}L\dot{\theta}\right)^2 + \frac{1}{2}M\left(\frac{2}{3}L\dot{\theta}\right)^2 = \frac{1}{2}\left(\frac{1}{9}mL^2 + \frac{4}{9}ML^2\right)\dot{\theta}^2 \quad \rightarrow \quad I_{eq} = \frac{1}{9}mL^2 + \frac{4}{9}ML^2$$

The governing differential equation is

$$\left(\tfrac{1}{9}mL^2 + \tfrac{4}{9}ML^2\right)\ddot{\theta} + \tfrac{1}{9}kL^2\theta = 0$$

$$\ddot{\theta} + \frac{k}{m+4M}\theta = 0$$

from which the natural frequency is determined as

$$\omega_n = \sqrt{\frac{k}{m+4M}}$$

2.14 What is the natural frequency of the 200-kg block of Fig. 2-14?

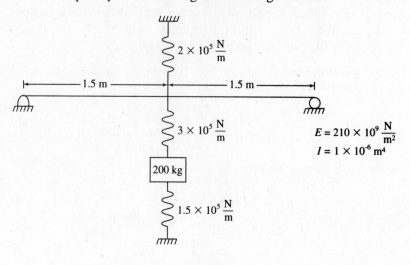

Fig. 2-14

The equivalent stiffness of the simply supported beam at its midpoint is

$$k_b = \frac{48EI}{L^3} = \frac{48\left(210 \times 10^9 \, \frac{N}{m^2}\right)\left(1 \times 10^{-6} \, m^4\right)}{(3 \, m)^3} = 3.73 \times 10^5 \, \frac{N}{m}$$

The beam acts in parallel with the upper spring. This parallel combination acts in series with the spring connecting the beam and the block. Finally, this combination acts in parallel with the spring connecting the block and the ground. Using the equations for parallel and series combinations, the equivalent stiffness for this system is

$$k_{eq} = \frac{1}{\dfrac{1}{3.73 \times 10^5 \, \frac{N}{m} + 2 \times 10^5 \, \frac{N}{m}} + \dfrac{1}{3 \times 10^5 \, \frac{N}{m}}} + 1.5 \times 10^5 \, \frac{N}{m}$$

$$= 3.47 \times 10^5 \, \frac{N}{m}$$

The natural frequency is

$$\omega_n = \sqrt{\frac{k_{eq}}{m}} = \sqrt{\frac{3.47 \times 10^5 \, \frac{N}{m}}{200 \, kg}} = 41.6 \, \frac{rad}{s}$$

2.15 A 500-kg vehicle is mounted on springs such that its static deflection is 1.5 mm. What is the damping coefficient of a viscous damper to be added to the system in parallel with the springs, such that the system is critically damped?

The static deflection is related to the natural frequency by

$$\omega_n = \sqrt{\frac{g}{\Delta_{ST}}} = \sqrt{\frac{9.81 \, \frac{m}{s^2}}{0.0015 \, m}} = 80.9 \, \frac{rad}{s}$$

The addition of a viscous damper of damping coefficient c leads to a damping ratio of

$$\zeta = \frac{c}{2m\omega_n} \quad \rightarrow \quad c = 2\zeta m\omega_n$$

The system is critically damped when the damping ratio is 1, requiring a damping coefficient of

$$c = 2(1)(500 \text{ kg})\left(80.9 \ \frac{\text{rad}}{\text{s}}\right) = 8.09 \times 10^4 \ \frac{\text{N-s}}{\text{m}}$$

2.16 For what value of c is the damping ratio of the system of Fig. 2-15 equal to 1.25?

Mathcad

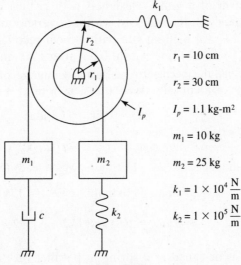

$r_1 = 10$ cm

$r_2 = 30$ cm

$I_p = 1.1$ kg-m²

$m_1 = 10$ kg

$m_2 = 25$ kg

$k_1 = 1 \times 10^4 \ \dfrac{\text{N}}{\text{m}}$

$k_2 = 1 \times 10^5 \ \dfrac{\text{N}}{\text{m}}$

Fig. 2-15

The equivalent systems method is used to derive the governing differential equation. Let θ be the counterclockwise angular displacement of the disk. The kinetic energy of the system at an arbitrary instant is

$$T = \tfrac{1}{2}I_p\dot{\theta}^2 + \tfrac{1}{2}m_1(r_2\dot{\theta})^2 + \tfrac{1}{2}m_2(r_1\dot{\theta})^2 = \tfrac{1}{2}(I_p + m_1r_2^2 + m_2r_1^2)\dot{\theta}^2$$

$$\rightarrow \ I_{eq} = I_p + m_1r_2^2 + m_2r_1^2$$

$$= 1.1 \text{ kg-m}^2 + (10 \text{ kg})(0.3 \text{ m})^2 + (25 \text{ kg})(0.1 \text{ m})^2 = 2.25 \text{ kg-m}^2$$

The potential energy of the system at an arbitrary instant is

$$V = \frac{1}{2}k_1(r_2\theta)^2 + \frac{1}{2}k_2(r_1\theta)^2 = \frac{1}{2}(k_1r_2^2 + k_2r_1^2)\theta^2$$

$$\rightarrow \ k_{t_{eq}} = k_1r_2^2 + k_2r_1^2 = \left(1 \times 10^4 \ \frac{\text{N}}{\text{m}}\right)(0.3 \text{ m})^2 + \left(1 \times 10^5 \ \frac{\text{N}}{\text{m}}\right)(0.1 \text{ m})^2$$

$$= 1900 \ \frac{\text{N-m}}{\text{rad}}$$

The work done by the damping force between two arbitrary instants is

$$W = -\int_{\theta_1}^{\theta_2} cr_2\dot{\theta} \, d(r_2\theta) = -\int_{\theta_1}^{\theta_2} cr_2^2\dot{\theta} \, d\theta$$

$$\rightarrow \ c_{t_{eq}} = cr_2^2 = (0.3 \text{ m})^2 c = 0.09c$$

Hence the governing differential equation is

$$2.25\ddot{\theta} + 0.09c\dot{\theta} + 1900\theta = 0$$

$$\ddot{\theta} + 0.04c\dot{\theta} + 844.4\theta = 0$$

From the governing differential equation,

$$\omega_n = \sqrt{844.4} = 29.1 \ \frac{\text{rad}}{\text{s}}$$

$$2\zeta\omega_n = 0.04c$$

For a damping ratio of 1.25,

$$c = \frac{2(1.25)\left(29.1 \ \frac{\text{rad}}{\text{s}}\right)}{0.04 \ \frac{\text{m}^2}{\text{kg-m}^2}} = 1820 \ \frac{\text{N-s}}{\text{m}}$$

2.17 The recoil mechanism of a gun is designed with critical damping such that the system returns to its firing position the quickest without overshooting. Design a recoil mechanism (by specifying c and k) for a 10-kg gun with a 5-cm recoil such that the firing mechanism returns to within 0.5 mm of firing within 0.5 s after maximum recoil.

Let $t = 0$ occur when the mechanism reaches maximum recoil. The response of the mechanism from this time is that of a critically damped system with $x(0) = 0.05$ m and $\dot{x}(0) = 0$. From Eq. (2.21)

$$x(t) = 0.05e^{-\omega_n t}(1 + \omega_n t)$$

Requiring that $x(0.5) = 0.0005$ m leads to

$$0.0005 = 0.05e^{-0.5\omega_n}(1 + 0.5\omega_n)$$

A trial-and-error solution leads to $\omega_n = 13.2$ rad/s, which leads to

$$k = m\omega_n^2 = (10 \text{ kg})\left(13.2 \ \frac{\text{rad}}{\text{s}}\right)^2 = 17400 \ \frac{\text{N}}{\text{m}}$$

$$c = 2m\omega_n = 2(10 \text{ kg})\left(13.2 \ \frac{\text{rad}}{\text{s}}\right) = 264 \ \frac{\text{N-s}}{\text{m}}$$

2.18 A railroad bumper is designed as a spring in parallel with a viscous damper. What is the bumper's damping coefficient such that the system has a damping ratio of 1.25 when the bumper is engaged by a 20,000-kg railroad car and has a stiffness of 2×10^5 N/m?

The damping coefficient is calculated from the damping ratio by

$$c = 2\zeta\sqrt{mk} = 2(1.25)\sqrt{(20,000 \text{ kg})\left(200,000 \ \frac{\text{N}}{\text{m}}\right)} = 1.58 \times 10^5 \ \frac{\text{N-s}}{\text{m}}$$

2.19 The railroad car of Problem 2.18 is traveling at a speed of 20 m/s when it engages the bumper. What is the maximum deflection of the bumper?

The natural frequency of the system of Problem 2.18 is

$$\omega_n = \sqrt{\frac{k}{m}} = \sqrt{\frac{200,000 \ \frac{\text{N}}{\text{m}}}{20,000 \text{ kg}}} = 3.16 \ \frac{\text{rad}}{\text{s}}$$

Let $t = 0$ occur when the car engages the bumper. Since the system is overdamped with $x(0) = 0$ and $\dot{x}(0) = 20$ m/s, application of Eq. (2.22) leads to

$$x(t) = \frac{20 \ \frac{\text{m}}{\text{s}}}{3.16 \ \frac{\text{rad}}{\text{s}}} \frac{e^{-(1.25)(3.16)t}}{2\sqrt{(1.25)^2 - 1}}$$

$$\times \left[e^{(3.16)\sqrt{(1.25)^2 - 1}\,t} - e^{-(3.16)\sqrt{(1.25)^2 - 1}\,t}\right]$$

$$x(t) = 4.22(e^{-1.58t} - e^{-6.32t}) \text{ m}$$

The time at which the maximum deflection occurs is obtained by setting

$$\frac{dx}{dt} = 0 = 4.22(-1.58e^{-1.58t} + 6.32e^{-6.32t})$$

$$\frac{6.32}{1.58} = \frac{e^{-1.58t}}{e^{-6.32t}} = e^{4.74t}$$

$$t = \frac{1}{4.74} \ln\left(\frac{6.32}{1.58}\right) = 0.292 \text{ s}$$

Thus the maximum bumper deflection is

$$x_{\max} = 4.22(e^{-1.58(0.292)} - e^{-6.32(0.292)}) = 1.99 \text{ m}$$

2.20 An empty railroad car has a mass of only 4500 kg. What are the natural frequency and damping ratio of a system with the bumper of Problem 2.18 when engaged by an empty railroad car?

The natural frequency and damping ratio are calculated as

$$\omega_n = \sqrt{\frac{k}{m}} = \sqrt{\frac{200{,}000 \ \frac{N}{m}}{4500 \ kg}} = 6.67 \ \frac{rad}{sec}$$

$$\zeta = \frac{c}{2\sqrt{mk}} = \frac{1.58 \times 10^5 \ \frac{N\text{-}s}{m}}{2\sqrt{(4500 \ kg)\left(200{,}000 \ \frac{N}{m}\right)}} = 2.63$$

2.21 *Overshoot for an underdamped system* is defined as the maximum displacement of the system at the end of its first half cycle. What is the minimum damping ratio for a system such that it is subject to no more than 5 percent overshoot?

The general response of an underdamped system is given by

$$x(t) = Ae^{-\zeta \omega_n t} \sin (\omega_d t + \phi_d)$$

Let Δ be the initial displacement of an underdamped system. Then

$$x(0) = \Delta = A \sin \phi_d$$

For 5 percent overshoot

$$x\left(\frac{T_d}{2}\right) = -0.05 \ \Delta$$

$$= Ae^{-\zeta \omega_n (T_d/2)} \sin \left(\frac{\omega_d T_d}{2} + \phi_d\right) = -A \sin \phi_d e^{-\zeta \omega_n (T_d/2)} = -\Delta e^{-\zeta \omega_n (T_d/2)}$$

The damped natural period is

$$T_d = \frac{2\pi}{\omega_n \sqrt{1 - \zeta^2}}$$

Thus

$$0.05 = e^{-(\zeta \pi / \sqrt{1 - \zeta^2})}$$

$$\zeta = \sqrt{\frac{[\ln (0.05)]^2}{\pi^2 + [\ln (0.05)]^2}} = 0.690$$

2.22 A suspension system is being designed for a 2000-kg vehicle (empty weight). It is estimated that the maximum added mass from passengers and cargo is 1000 kg. When the vehicle is empty, its static deflection is to be 3.1 mm. What is the minimum value of the damping coefficient such that the vehicle is subject to no more than 5 percent overshoot, empty or full?

The required suspension stiffness is

$$k = \frac{mg}{\Delta_{ST}} = \frac{(2000 \ kg)\left(9.81 \ \frac{m}{s^2}\right)}{0.0031 \ m} = 6.33 \times 10^6 \ \frac{N}{m}$$

The results of Problem 2.21 show that the damping ratio must be no smaller than 0.69 to limit the overshoot to 5 percent. The damping ratio of a system whose stiffness is fixed but whose mass can vary is smaller for a larger mass. Thus the damping coefficient must be limited to 0.69 when the vehicle is fully loaded:

$$c = 2\zeta \sqrt{mk} = (0.69)\sqrt{(3000 \ kg)\left(6.33 \times 10^6 \ \frac{N}{m}\right)} = 1.90 \times 10^5 \ \frac{N\text{-}s}{m}$$

2.23 The vehicle with the suspension designed in Problem 2.22 encounters a bump of height 5.5 cm. What is the vehicle's overshoot if it is carrying 25 kg of fuel, one 80-kg passenger, and 110 kg of cargo?

The mass of the vehicle is 2215 kg, leading to

$$\omega_n = \sqrt{\frac{k}{m}} = \sqrt{\frac{6.33 \times 10^6 \frac{N}{m}}{2215 \text{ kg}}} = 53.5 \frac{\text{rad}}{\text{sec}}$$

$$\zeta = \frac{c}{2\sqrt{mk}} = \frac{1.90 \times 10^5 \frac{N\text{-s}}{m}}{2\sqrt{(2215 \text{ kg})\left(6.33 \times 10^6 \frac{N}{m}\right)}} = 0.802$$

From Problem 2.20 the overshoot is calculated by

$$-x\left(\frac{T_d}{2}\right) = \Delta e^{-(\zeta\pi/\sqrt{1-\zeta^2})} = (0.055 \text{ m})e^{-(0.802\pi/\sqrt{1-(0.802)^2}]} = 0.081 \text{ cm}$$

2.24 A free vibrations test is run to determine the stiffness and damping properties of an elastic element. A 20-kg block is attached to the element. The block is displaced 1 cm and released. The resulting oscillations are monitored with the results shown in Fig. 2-16. Determine k and c for this element.

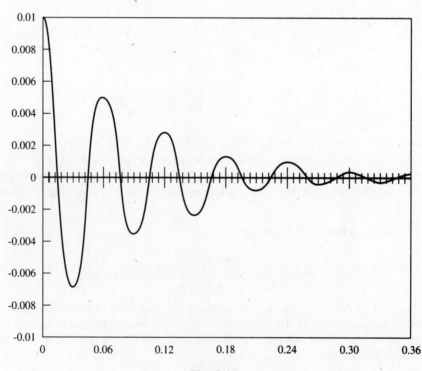

Fig. 2-16

From Fig. 2-16 the displacement of the block at the end of the first cycle is 0.005 m. The logarithmic decrement is calculated as

$$\delta = \ln\left(\frac{0.01 \text{ m}}{0.005 \text{ m}}\right) = 0.693$$

from which the damping ratio is determined

$$\zeta = \frac{\delta}{\sqrt{4\pi^2 + \delta^2}} = \frac{0.693}{\sqrt{4\pi^2 + (0.693)^2}} = 0.11$$

The damped natural period is determined from Fig. 2-16 as 0.06 s. The damped natural frequency is

$$\omega_d = \frac{2\pi}{T_d} = \frac{2\pi}{0.06 \text{ s}} = 104.7 \frac{\text{rad}}{\text{s}}$$

The natural frequency is calculated from

$$\omega_n = \frac{\omega_d}{\sqrt{1 - \zeta^2}} = \frac{104.7 \frac{\text{rad}}{\text{s}}}{\sqrt{1 - (0.11)^2}} = 105.3 \frac{\text{rad}}{\text{s}}$$

Thus the stiffness and damping coefficient are calculated

$$k = m\omega_n^2 = (20 \text{ kg})\left(105.3 \frac{\text{rad}}{\text{s}}\right)^2 = 2.22 \times 10^5 \frac{\text{N}}{\text{m}}$$

$$c = 2\zeta m\omega_n = 2(0.11)(20 \text{ kg})\left(105.3 \frac{\text{rad}}{\text{s}}\right) = 4.63 \times 10^2 \frac{\text{N-s}}{\text{m}}$$

2.25 The 25-kg block of Fig. 2-15 is displaced 20 mm and released. If $c = 100$ N-s/m, how many cycles will be executed before the amplitude is reduced to 1 mm?

Mathcad

Using the results of Problem 2.16, the damping ratio is calculated as

$$\zeta = \frac{0.04c}{2\omega_n} = \frac{0.04\left(100 \frac{\text{N-s}}{\text{m}}\right)}{2\left(29.1 \frac{\text{rad}}{\text{s}}\right)} = 0.069$$

Thus the logarithmic decrement is calculated as

$$\delta = \frac{2\pi\zeta}{\sqrt{1 - \zeta^2}} = \frac{2\pi(0.069)}{\sqrt{1 - (0.069)^2}} = 0.435$$

The concept of logarithmic decrement can be successively applied between cycles leading to an alternate form of Eq. (2.19)

$$\delta = \frac{1}{n} \ln\left(\frac{x(t)}{x(t + nT_d)}\right)$$

where n is an integer. Thus, for the amplitude to be reduced to 1 mm in n cycles from an initial amplitude of 20 mm,

$$0.435 = \frac{1}{n} \ln\left(\frac{20 \text{ mm}}{1 \text{ mm}}\right) \quad \rightarrow \quad n = 6.89$$

However, since n must be an integer the amplitude will be reduced to less than 1 mm after the 7th cycle.

2.26 List three differences between the free vibration response of a system with Coulomb damping and the free vibration response for a system with viscous damping whose free vibrations are underdamped.

Three differences between the systems are:

(a) The magnitude of the Coulomb damping has no effect on the frequency or period of motion while the magnitude of the viscous damping does affect the frequency ω_d and period T_d. An increase in viscous damping leads to a decrease in ω_d and an increase in T_d.

(b) The amplitude of vibration for a system with Coulomb damping decreases by a constant amount per cycle (linear decrease) while the amplitude of vibration for a system with viscous damping decreases exponentially.

(c) Motion ceases for a system with Coulomb damping when the amplitude becomes small enough such that the force in the elastic member is insufficient to overcome static friction, leading to a permanent displacement from equilibrium. Motion continues indefinitely for a system with viscous damping.

2.27 Use the work-energy method to determine the change in amplitude per cycle of motion for a block of mass m attached to a spring of stiffness k, sliding on a surface with a coefficient of friction μ.

Let X_1 be the amplitude of motion at the beginning of a cycle, defined as a time where the velocity is zero. Let X_2 be the amplitude at the end of the next half cycle, when the velocity is again zero. The principle of work and energy applied between these times is

$$T_1 + V_1 + W_{1\to2} = T_2 + V_2$$

where

$$T_1 = T_2 = 0$$

$$V_1 = \tfrac{1}{2}kX_1^2 \qquad V_2 = \tfrac{1}{2}kX_2^2$$

and the work done by the friction is

$$W_{1\to2} = -\mu mg(X_1 + X_2)$$

Thus

$$\tfrac{1}{2}kX_1^2 - \mu mg(X_1 + X_2) = \tfrac{1}{2}kX_2^2$$

$$\tfrac{1}{2}kX_2^2 + \mu mgX_2 + (\mu mgX_1 - \tfrac{1}{2}kX_1^2) = 0$$

The quadratic formula is used to solve for X_2 in terms of X_1:

$$X_2 = \frac{1}{2}\left[-\mu mg \pm \sqrt{(\mu mg)^2 - 2k\left(\mu mgX_1 - \frac{1}{2}kX_1^2\right)}\right]$$

$$X_2 = -X_1, \qquad X_1 - \frac{2\mu mg}{k}$$

However, X_2 must be positive; thus the change in amplitude over one half cycle is

$$X_1 - X_2 = \frac{2\mu mg}{k}$$

Since the change in amplitude is independent of the amplitude, it is constant over each half cycle. Thus the change in amplitude over 1 cycle of motion is

$$\Delta A = \frac{4\mu mg}{k}$$

2.28 The block in the system of Fig. 2-17 is displaced 10 mm and released. How many cycles of motion will be executed?

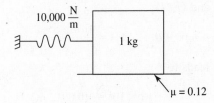

Fig. 2-17

The differential equation governing the motion of the system of Fig. 2-17 can be shown to be

$$m\ddot{x} + kx = \begin{cases} -\mu mg, & \dot{x} > 0 \\ \mu mg, & \dot{x} < 0 \end{cases}$$

Comparing the above equation to Eq. (2.23),

$$F_f = \mu mg$$

Thus the decrease in amplitude per cycle is given by

$$\Delta A = \frac{4\mu mg}{k} = \frac{4(0.12)(1\text{ kg})\left(9.81\ \dfrac{\text{m}}{\text{s}^2}\right)}{10,000\ \dfrac{\text{N}}{\text{m}}} = 0.47 \text{ mm}$$

Motion will cease when the amplitude is such that the spring force cannot overcome the friction force. The resulting permanent displacement is given by

$$kx_f = \mu mg \;\rightarrow\; x_f = \frac{\mu mg}{k} = 0.118 \text{ mm}$$

Hence the number of cycles is

$$x_0 - n\Delta A < x_f$$

$$n > \frac{x_0 - x_f}{\Delta A} = \frac{10 \text{ mm} - 0.118 \text{ mm}}{0.47 \text{ mm}} = 21.02 \text{ cycles}$$

Hence 22 cycles will be executed.

2.29 The block of Fig. 2-18 is displaced 25 mm and released. It is observed that the amplitude decreases 1.2 mm each cycle. What is the coefficient of friction between the block and the surface?

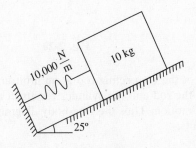

Fig. 2-18

Application of Newton's law to the free body diagrams of Fig. 2-19 leads to

$$m\ddot{x} + kx = \begin{cases} -\mu mg \cos \theta, & \dot{x} > 0 \\ \mu mg \cos \theta, & \dot{x} < 0 \end{cases}$$

Hence, using Eqs. (2.23) and (2.24),

$$F_f = \mu mg \cos \theta, \qquad \Delta A = \frac{4\mu mg \cos \theta}{k} \; \rightarrow \; \mu = \frac{k \, \Delta A}{4mg \cos \theta}$$

Substituting given values leads to

$$\mu = \frac{\left(1 \times 10^4 \, \dfrac{\text{N}}{\text{m}}\right)(0.0012 \text{ m})}{4(10 \text{ kg})\left(9.81 \, \dfrac{\text{m}}{\text{s}^2}\right) \cos (25°)} = 0.034$$

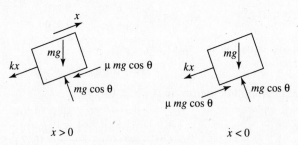

$$\dot{x} > 0 \qquad\qquad\qquad \dot{x} < 0$$

Fig. 2-19

2.30 The connecting rod of Fig. 2-20 is fitted around a cylinder of diameter 5 cm. The coefficient of friction between the rod and the cylinder is 0.08. If the rod is rotated 13° and released, what is the decrease in angle on each cycle?

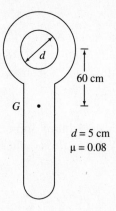

Fig. 2-20

Free body diagrams of the connecting rod at an arbitrary instant are shown in Fig. 2-21. The effect of friction between the rod and the cylinder results in a moment $\mu mgd/2$ acting on the rod, opposing the direction of motion. Thus the free body diagrams are shown for a clockwise angular velocity. Summing moments,

$$\left(\sum M_0\right)_{\text{eff}} = \left(\sum M_0\right)_{\text{ext}}$$

and assuming small θ leads to

$$(I + m\ell^2)\ddot{\theta} + mg\ell\theta = \begin{cases} -\mu mg\dfrac{d}{2}, & \dot{\theta} < 0 \\[2mm] \mu mg\dfrac{d}{2}, & \dot{\theta} > 0 \end{cases}$$

Comparing the above to Eq. (2.23) leads to

$$\omega_n = \sqrt{\frac{mg\ell}{I + m\ell^2}}, \qquad F_f = \frac{1}{2}\mu mgd$$

which from Eq. (2.24) leads to

$$\Delta\Theta = \frac{4(\frac{1}{2}\mu mgd)}{mg\ell} \approx \frac{2\mu d}{\ell}$$

Substituting given values leads to

$$\Delta\Theta = \frac{2(0.08)(0.05 \text{ m})}{0.6 \text{ m}} = 0.0133 \text{ rad} = 0.76°$$

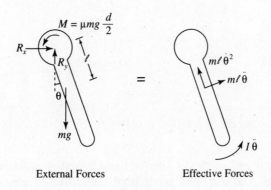

Fig. 2-21

Supplementary Problems

2.31 Use the equivalent systems method to derive the differential equation governing the motion of the system of Fig. 2-22. Use x as the generalized coordinate. Determine the system's natural frequency.

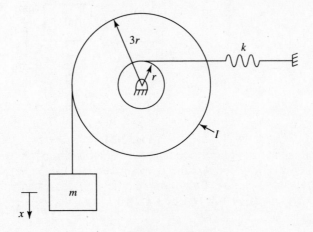

Fig. 2-22

Ans.

$$\left(m + \frac{I}{9r^2}\right)\ddot{x} + \frac{k}{9}x = 0, \qquad \omega_n = \sqrt{\frac{kr^2}{I + 9mr^2}}$$

2.32   Use the equivalent system method to derive the differential equation governing the motion of the system of Fig. 2-23. Use θ as the generalized coordinate assuming small θ. Determine the system's natural frequency.

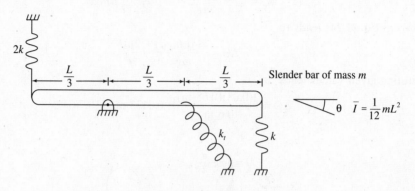

Fig. 2-23

Ans.

$$\frac{1}{9}mL^2\ddot{\theta} + \left(\frac{2}{3}kL^2 + k_t\right)\theta = 0, \qquad \omega_n = \sqrt{\frac{6kL^2 + 9k_t}{mL^2}}$$

2.33   Use the equivalent system method to derive the differential equation governing the motion of the system of Fig. 2-24. Use x as the generalized coordinate. Assume small x and determine the system's natural frequency.

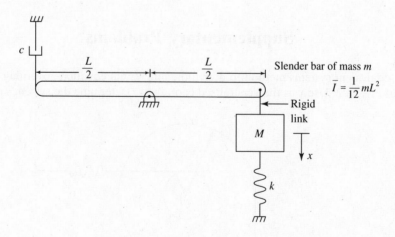

Fig. 2-24

Ans.

$$\left(\frac{1}{3}m + M\right)\ddot{x} + c\dot{x} + kx = 0, \qquad \omega_n = \sqrt{\frac{3k}{m + 3M}}$$

2.34   Use the free body diagram method to derive the differential equation governing the motion of the system of Fig. 2-25. Use θ as the generalized coordinate, assuming small θ. Assume the structure is composed of two slender rods welded together.

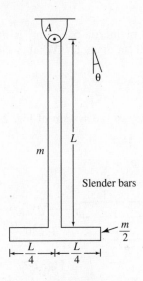

Fig. 2-25

Ans.

$$\tfrac{81}{96}mL^2\ddot{\theta} + mgL\theta = 0$$

2.35   Use the free body diagram method to derive the differential equation governing the motion of the system of Fig. 2-26. Use θ as the generalized coordinate, assuming small θ.

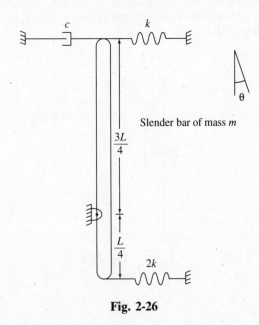

Fig. 2-26

Ans.

$$\frac{7}{48}mL^2\ddot{\theta} + \frac{9}{16}cL^2\dot{\theta} + \left(\frac{11}{16}kL^2 - mg\frac{L}{4}\right)\theta = 0$$

2.36   A 300-kg block is attached to four identical springs, each of stiffness 2.3×10^5 N/m, placed in parallel. Determine the system's natural frequency in hertz.

Ans.   8.81 Hz

2.37   A thin disk of mass moment of inertia 5.8 kg-m² is attached to the end of a 2.5-m aluminum ($G = 40 \times 10^9$ N/m²) shaft of 10 cm diameter. What is the natural frequency of torsional oscillation of the disk?

Ans.   164.6 rad/s

2.38   Determine the natural frequency of the system of Fig. 2-27.

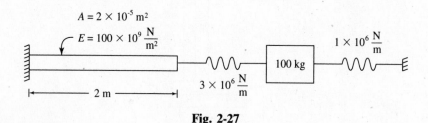

$A = 2 \times 10^{-5}$ m²
$E = 100 \times 10^9 \dfrac{N}{m^2}$
$3 \times 10^6 \dfrac{N}{m}$
100 kg
$1 \times 10^6 \dfrac{N}{m}$
2 m

Fig. 2-27

Ans.   132.3 rad/s

2.39   When empty the static deflection of a 2000-lb vehicle is 0.8 in. What is the vehicle's natural frequency when it is carrying a 200-lb passenger and 250 lb of cargo?

Ans.   19.9 rad/s

2.40   The location of the center of mass and the mass moment of inertia of the connecting rod of Fig. 2-28 are unknown. When the rod is pinned at *A*, its natural frequency is observed as 20 rad/s. When a 250-g mass is added to the free end, the system's natural frequency is observed as 10 rad/s. Determine the location of the center of mass ℓ.

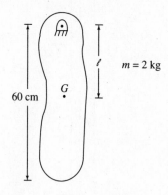

60 cm
ℓ
$m = 2$ kg
G

Fig. 2-28

Ans.   0.512 m

2.41   A rotor of mass moment of inertia 2.5 kg-m² is to be attached at the end of a 60-cm circular steel ($G = 80 \times 10^9$ N/m²) shaft. What is the range of shaft diameters such that the torsional natural frequency of the system is between 100 and 200 Hz?

Ans.   9.32 cm $< D <$ 13.2 cm

2.42 A uniform 45-kg flywheel of inner radius 80 cm and outer radius 100 cm is swung as a pendulum about a knife edge support on its inner rim. Its period is observed as 2.1 s. Determine the flywheel's centroidal moment of inertia.

Ans. $I = 10.65$ kg-m^2

2.43 A particle of mass m is attached to the midpoint of a taut string of length L and tension T, as shown in Fig. 2-29. Determine the particle's natural frequency of vertical vibration.

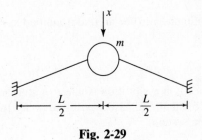

Fig. 2-29

Ans.

$$\omega_n = \sqrt{\frac{4T}{mL}}$$

2.44 The disk in the system of Fig. 2-30 rolls without slip. Determine the value of c such that the system has a damping ratio of 0.2.

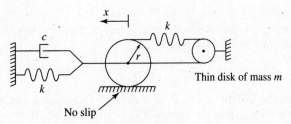

Thin disk of mass m

No slip

Fig. 2-30

Ans.

$$c = 1.55\sqrt{mk}$$

2.45 What is the value of c such that the system of Fig. 2-31 is critically damped if $m = 20$ kg and $k = 10,000$ N/m?

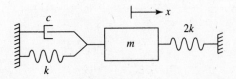

Fig. 2-31

Ans. $c = 1.55 \times 10^3$ N-s/m

2.46 A 200-kg block is attached to a spring of stiffness 50,000 N/m in parallel with a viscous damper. The period of free vibration of this system is observed as 0.417 s. What is the value of the damping coefficient?

Ans. $c = 1.91 \times 10^3$ N-s/m

2.47 For the recoil mechanism designed in the solution of Problem 2.17, what is the initial velocity of the recoil mechanism that leads to a recoil of 5 cm?

Ans.   1.79 m/s

2.48 What is the minimum damping ratio for an underdamped system such that its overshoot is limited to 10 percent.

Ans. $\zeta = 0.591$

2.49 A 1000-kg machine is placed on a vibration isolator of stiffness 1×10^6 N/m. The machine is given an initial displacement of 5 cm and released. After 10 cycles the machine's amplitude is 1 cm. What is the damping ratio of the system?

Ans. $\zeta = 0.026$

2.50 A 100-kg block is attached to a spring of stiffness 1.5×10^6 N/m in parallel with a viscous damper of damping coefficient 4900 N-s/m. The block is given an initial velocity of 5 m/s. What is its maximum displacement?

Ans.   30.9 mm

2.51 Solve Problem 2.50 if $c = 29,000$ N-s/m.

Ans.   13.4 mm

2.52 How long after being given the initial velocity will it take the system of Problem 2.51 to return permanently to within 1 mm of equilibrium.

Ans.   0.0515 s

2.53 The slender bar in the system of Fig. 2-32 is rotated 5° from equilibrium and released. Determine the time dependent response of the system if $m = 2$ kg, $L = 80$ cm, $r = 10$ cm, $k = 20,000$ N/m, and $c = 300$ N-s/m.

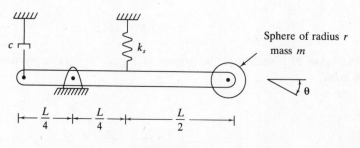

Fig. 2-32

Ans.

$$\theta(t) = 0.0895e^{-6.61t} \sin (28.9t + 1.35)$$

2.54 Solve Problem 2.53 if $c = 1500$ N-s/m.

Ans.

$$\theta(t) = 0.144e^{-18.6t} - 0.055e^{-47.6t}$$

2.55   A spring-dashpot mechanism is designed such that a system is critically damped when the system has a mass m. What is the damping ratio of a system using this mechanism with a mass (*a*) $3m/4$, (*b*) $4m/3$?

Ans.   (a) 1.15, (b) 0.866

2.56   If the initial conditions for the motion of a critically damped system are of opposite sign, overshoot is possible. Derive a relationship that the initial conditions x_0 and $\dot{x}_0$ must satisfy in order for overshoot to occur.

Ans.

$$\frac{x_0}{\dot{x}_0 + \omega_n x_0} < 0$$

2.57   A 35-kg block is connected to a spring of stiffness 1.7×10^5 N/m. The coefficient of friction between the block and the surface on which it slides is 0.11. The block is displaced 10 mm from equilibrium and released. (*a*) What is the amplitude of motion at the end of the first cycle? (*b*) How many cycles of motion occur?

Ans.   (*a*) 9.11 mm, (*b*) 11

2.58   A 50-kg block is attached to a spring of stiffness 200,000 N/m and slides on a surface that makes an angle of 34° with the horizontal. For what values of μ, the coefficient of friction between the block and the surface, will motion cease during the 10th cycle when the block is displaced 1 cm from equilibrium and released?

Ans. $0.120 < \mu < 0.133$

Chapter 3

Harmonic Excitation of 1-Degree-of-Freedom Systems

3.1 DERIVATION OF DIFFERENTIAL EQUATIONS

Differential equations governing the forced vibrations of a 1-degree-of-freedom system can be derived using the free body diagram method as discussed in Chap. 3. Time-dependent forces and moments are illustrated on the free body diagram showing external forces.

Any linear 1-degree-of-freedom system can be modeled by one of the systems of Figs. 3-1 or 3-2 using the equivalent system method. The system of Fig. 3-1 is appropriate if the chosen generalized coordinate represents a linear displacement while the system of Fig. 3-2 is appropriate if the generalized coordinate is an angular displacement. The equivalent inertia, stiffness, and damping properties are determined as in Chaps. 1 and 2. The equivalent external force F_{eq} or moment M_{eq} is determined using the *method of virtual work*. Let δx be a *variation* in the generalized coordinate (a virtual displacement). Let δW be the work done by the external forces as the system displaces from x to $x + \delta x$. The equivalent force is determined from

$$\delta W = F_{eq}\,\delta x \tag{3.1}$$

If the generalized coordinate is an angular displacement,

$$\delta W = M_{eq}\,\delta\theta \tag{3.2}$$

The general form of the differential equation governing the forced vibrations of a 1-degree-of-freedom system is

$$m_{eq}\ddot{x} + c_{eq}\dot{x} + k_{eq}x = F_{eq}(t) \tag{3.3}$$

or dividing by m_{eq},

$$\ddot{x} + 2\zeta\omega_n\dot{x} + \omega_n^2 x = \frac{1}{m_{eq}}F_{eq}(t) \tag{3.4}$$

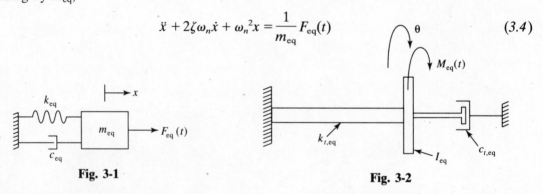

Fig. 3-1   **Fig. 3-2**

3.2 HARMONIC EXCITATION

A single-frequency harmonic excitation is of the form

$$F_{eq}(t) = F_0 \sin(\omega t + \psi) \tag{3.5}$$

where F_0 is the *amplitude of excitation*, ω is the *frequency of excitation*, and ψ is the *phase of the excitation*.

3.3 UNDAMPED SYSTEM RESPONSE

For $\zeta = 0$ and $\omega \neq \omega_n$, the solution of Eq. (3.4) with $F_{eq}(t)$ given by Eq. (3.5) and subject to initial conditions $x(0) = x_0$ and $\dot{x}(0) = \dot{x}_0$, is

$$x(t) = \left[x_0 - \frac{F_0 \sin \psi}{m_{eq}(\omega_n^2 - \omega^2)} \right] \cos \omega_n t$$

$$+ \frac{1}{\omega_n} \left[\dot{x}_0 - \frac{F_0 \omega \cos \psi}{m_{eq}(\omega_n^2 - \omega^2)} \right] \sin \omega_n t$$

$$+ \frac{F_0}{m_{eq}(\omega_n^2 - \omega^2)} \sin (\omega t + \psi) \qquad (3.6)$$

This response is illustrated in Fig. 3-3.

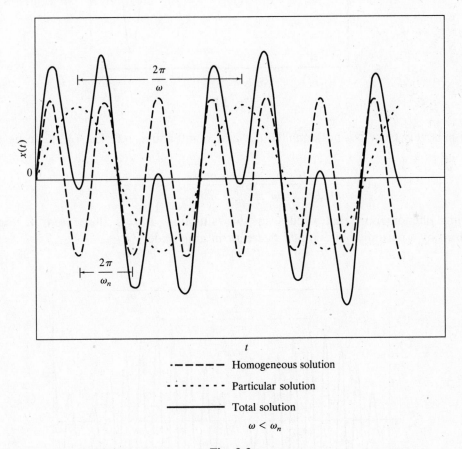

- - - - - Homogeneous solution

· · · · · · · Particular solution

———— Total solution

$\omega < \omega_n$

Fig. 3-3

When $\omega = \omega_n$, the solution of Eq. (3.4) is

$$x(t) = x_0 \cos \omega_n t + \left(\frac{\dot{x}_0}{\omega_n} + \frac{F_0 \cos \psi}{2m_{eq}\omega_n^2} \right) \sin \omega_n t$$

$$- \frac{F_0}{2m_{eq}\omega_n} t \cos (\omega_n t + \psi) \qquad (3.7)$$

Figure 3-4 illustrates the condition of *resonance*, the unbounded growth of the response when the excitation frequency coincides with the natural frequency.

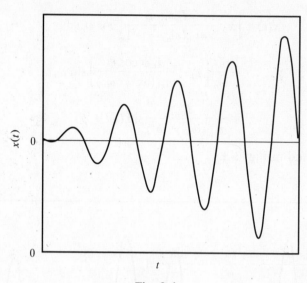

Fig. 3-4

When ω is very close but not equal to ω_n, and $x_n = 0$ and $\dot{x}_0 = 0$, Eq. (3.7) can be rewritten as

$$x(t) = \frac{2F_0}{m_{\text{eq}}(\omega_n^2 - \omega^2)} \sin\left(\frac{\omega - \omega_n}{2}\right)t \cos\left(\frac{\omega + \omega_n}{2}\right)t \tag{3.8}$$

The *beating* phenomenon that occurs in this situation and is illustrated in Fig. 3-5 is characterized by a periodic buildup and decrease in amplitude.

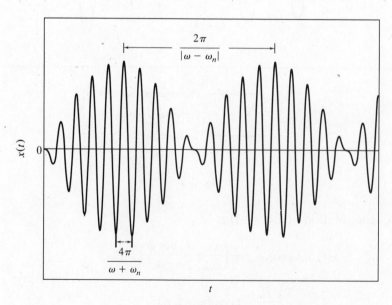

Fig. 3-5

3.4 DAMPED SYSTEM RESPONSE

For $\zeta \neq 0$, the homogeneous solution of Eq. (3.5) decays with increasing time, and eventually only the particular solution is important in the response. This condition is called *steady state,* and the corresponding particular solution, the *steady-state response,* is given by

$$x_{ss}(t) = X \sin(\omega t + \psi - \phi) \tag{3.9}$$

where X is called the *steady-state amplitude* and ϕ is called the *phase difference.* The steady-state amplitude is calculated from

$$\frac{m_{eq}\omega_n^2 X}{F_0} = M(r, \zeta) \tag{3.10}$$

where

$$r = \frac{\omega}{\omega_n} \tag{3.11}$$

is called the *frequency ratio* and

$$M(r, \zeta) = \frac{1}{\sqrt{(1 - r^2)^2 + (2\zeta r)^2}} \tag{3.12}$$

is called the *magnification factor.* The phase difference between the response and the excitation is

$$\phi = \tan^{-1}\left(\frac{2\zeta r}{1 - r^2}\right) \tag{3.13}$$

Figures 3-6 and 3-7 illustrate the nondimensional magnification factor and phase difference as functions of r for several values of ζ. It is noted that for a fixed $\zeta < 1/\sqrt{2}$, the maximum value of M is

$$M_{max} = \frac{1}{2\zeta\sqrt{1 - \zeta^2}} \tag{3.14}$$

and occurs for a frequency ratio of

$$r_{m_M} = \sqrt{1 - 2\zeta^2} \tag{3.15}$$

3.5 FREQUENCY SQUARED EXCITATIONS

A common form of harmonic excitation is one whose amplitude is proportional to the square of its frequency. That is

$$F_0 = A\omega^2 \tag{3.16}$$

where A is a constant of proportionality. For a frequency squared excitation, Eq. (3.10) can be rewritten as

$$\frac{m_{eq}X}{A} = \Lambda(r, \zeta) \tag{3.17}$$

where

$$\Lambda(r, \zeta) = r^2 M(r, \zeta) = \frac{r^2}{\sqrt{(1 - r^2)^2 + (2\zeta r)^2}} \tag{3.18}$$

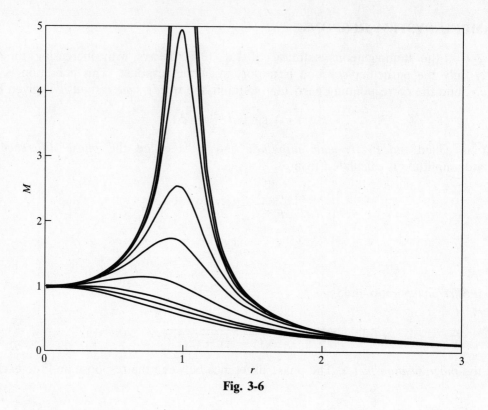

Fig. 3-6

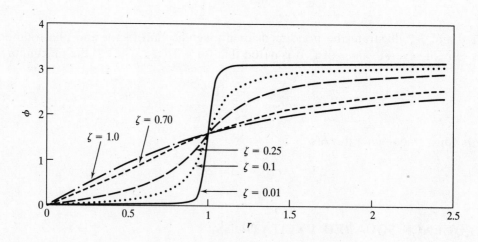

Fig. 3-7

The nondimensioal function Λ is illustrated in Fig. 3-8 as a function of r for several values of ζ. It is noted that for a fixed $\zeta < 1/\sqrt{2}$, the maximum value of Λ is

$$\Lambda_{\max} = \frac{1}{2\zeta\sqrt{1 - \zeta^2}} \qquad (3.19)$$

and occurs for a frequency ratio of

$$r_{m_\Lambda} = \frac{1}{\sqrt{1 - 2\zeta^2}} \qquad (3.20)$$

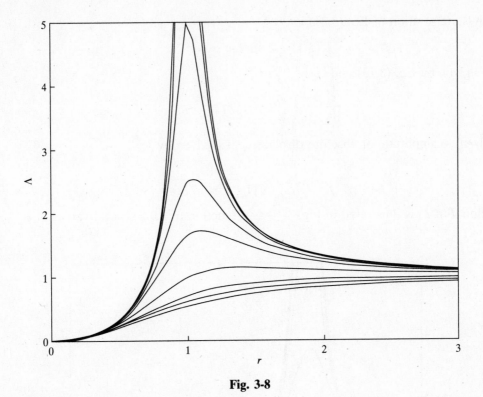

Fig. 3-8

3.6 HARMONIC SUPPORT EXCITATION

The block in the system of Fig. 3-9 is connected through a spring in parallel with a viscous damper to a movable support. The support is given a harmonic motion of the form

$$y(t) = Y \sin \omega t \qquad (3.21)$$

The absolute displacement of the block x is governed by

$$\ddot{x} + 2\zeta\omega_n\dot{x} + \omega_n^2 x = \omega_n^2 Y \sin \omega t + 2\zeta\omega_n Y \cos \omega t \qquad (3.22)$$

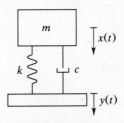

Fig. 3-9

Let

$$z(t) = x(t) - y(t) \qquad (3.23)$$

be the displacement of the block relative to its support. The differential equation governing $z(t)$ is

$$\ddot{z} + 2\zeta\omega_n\dot{z} + \omega_n^2 = \omega^2 Y \sin \omega t \qquad (3.24)$$

The steady-state solution of Eq. (3.24) is

$$z(t) = Z \sin(\omega t - \phi) \qquad (3.25)$$

where ϕ is given by Eq. (3.13) and

$$\frac{Z}{Y} = \Lambda(r, \zeta) \qquad (3.26)$$

The steady-state amplitude of absolute displacement is given by

$$\frac{X}{Y} = T(r, \zeta) = \sqrt{\frac{1 + (2\zeta r)^2}{(1 - r^2)^2 + (2\zeta r)^2}} \qquad (3.27)$$

The function $T(r, \zeta)$ is illustrated in Fig. 3-10 as a function of r for several values of ζ.

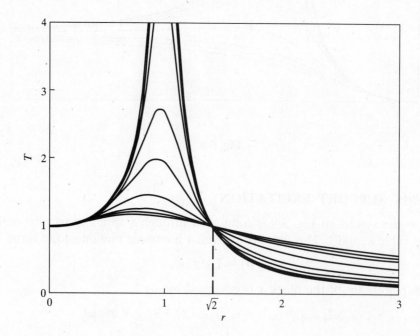

Fig. 3-10

3.7 MULTIFREQUENCY EXCITATIONS

The steady-state solution of Eq. (3.4) with

$$F(t) = \sum_{i=1}^{n} F_i \sin(\omega_i t + \psi_i) \qquad (3.28)$$

is obtained using the principle of linear superposition

$$x(t) = \sum_{i=1}^{n} F_i M(r_i, t) \sin(\omega_i t + \psi_i - \phi_i) \qquad (3.29)$$

where

$$r_i = \frac{\omega_i}{\omega_n} \qquad (3.30)$$

and
$$\phi_i = \tan^{-1}\left(\frac{2\zeta r_i}{1 - r_i^2}\right) \tag{3.31}$$

3.8 GENERAL PERIODIC EXCITATIONS: FOURIER SERIES

If $F(t)$ is a periodic function of period T, then $F(t)$ has the *Fourier series representation*

$$F(t) = \frac{a_0}{2} + \sum_{i=1}^{\infty} (a_i \cos \omega_i t + b_i \sin \omega_i t) \tag{3.32}$$

where
$$\omega_i = \frac{2\pi i}{T} \tag{3.33}$$

$$a_i = \frac{2}{T}\int_0^T F(t) \cos \omega_i t \, dt \tag{3.24}$$

$$b_i = \frac{2}{T}\int_0^T F(t) \sin \omega_i t \, dt \tag{3.35}$$

The series of Eq. (3.32) converges to $F(t)$ at all t where F is continuous. If $F(t)$ has a jump discontinuity at t, then the series of Eq. (3.32) converges to the average value of F as t is approached from the left and right.

If $F(t)$ is an even function, then $F(-t) = F(t)$ for all t and $b_i = 0$, $i = 1, 2, \ldots$. If $F(t)$ is an odd function, then $F(-t) = -F(t)$ for all t, and $a_i = 0$, $i = 0, 1, 2, \ldots$.

An alternate representation of Eq. (3.32) is

$$F(t) = \frac{a_0}{2} + \sum_{i=1}^{\infty} c_i \sin (\omega_i t - \kappa_i) \tag{3.36}$$

where
$$c_i = \sqrt{a_i^2 + b_i^2} \tag{3.37}$$

and
$$\kappa_i = \tan^{-1}\left(\frac{a_i}{b_i}\right) \tag{3.38}$$

The response of a 1-degree-of-freedom system subject to a periodic excitation is

$$x(t) = \frac{1}{m_{eq}\omega_n^2}\left[\frac{a_0}{2} + \sum_{i=1}^{\infty} c_i M(r_i, \zeta) \sin (\omega_i t + \kappa_i - \phi_i)\right] \tag{3.39}$$

An upper bound for the maximum steady-state displacement is

$$x_{max} \leq \frac{1}{m_{eq}\omega_n^2}\left[\frac{a_0}{2} + \sum_{i=1}^{\infty} c_i M(r_i, \zeta)\right] \tag{3.40}$$

3.9 COULOMB DAMPING

An approximation of the response of a system with Coulomb damping subject to a single-frequency harmonic excitation is obtained by modeling the system using viscous damping with an equivalent viscous damping ratio, ζ_{eq}, calculated such that the work done over 1 cycle of motion by the system with Coulomb damping is the same as the work done by the system with viscous damping with the equivalent damping coefficient. To this end,

$$\zeta_{eq} = \frac{2\iota}{\pi r M} \tag{3.41}$$

where
$$\iota = \frac{F_f}{F_0}$$
(3.42)

where F_f is the magnitude of the Coulomb damping force, F_0 is the amplitude of the excitation, and M is the magnification factor. Substituting into Eq. (3.12) leads to

$$M = \sqrt{\frac{1 - \left(\frac{4\iota}{\pi}\right)^2}{(1 - r^2)^2}}$$
(3.43)

Equation (3.43) provides an approximation to the magnification factor for $\iota < \pi/4$.

3.10 HYSTERETIC DAMPING

Empirical evidence indicates that the energy dissipated over 1 cycle of motion due to hysteretic damping is independent of frequency but proportional to the square of amplitude. The free vibration response of a system with hysteretic damping is similar to that of a system with viscous damping. A dimensionless hysteretic damping coefficient h is determined from the logarithmic decrement δ as

$$h = \frac{\delta}{\pi}$$
(3.44)

For forced vibration, the equivalent viscous damping ratio is

$$\zeta_{eq} = \frac{h}{2r}$$
(3.45)

which leads to a magnification factor of

$$M = \frac{1}{\sqrt{(1 - r^2)^2 + h^2}}$$
(3.46)

Solved Problems

3.1 Use the free body diagram method to derive the differential equation governing the motion of the system of Fig. 3-11 using θ as the generalized coordinate.

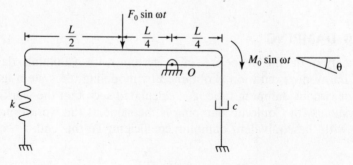

Fig. 3-11

Free body diagrams of the system at an arbitrary instant assuming small θ are shown in Fig. 3-12. Summing moments about the pin support leads to

$$\left(\sum M_O\right)_{\text{ext}} = \left(\sum M_O\right)_{\text{eff}}$$

$$-\tfrac{3}{4}kL\theta(\tfrac{3}{4}L) - F_0\sin\omega t(\tfrac{1}{4}L) - \tfrac{1}{4}cL\dot\theta(\tfrac{1}{4}L) + M_0\sin\omega t = \tfrac{1}{12}mL^2\ddot\theta + \tfrac{1}{4}mL\ddot\theta(\tfrac{1}{4}L)$$

$$\tfrac{7}{48}mL^2\ddot\theta + \tfrac{1}{16}cL^2\dot\theta + \tfrac{9}{16}kL^2\theta = M_0\sin\omega t - \tfrac{1}{4}F_0L\sin\omega t$$

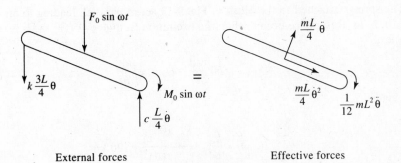

External forces   Effective forces

Fig. 3-12

3.2   Use the equivalent system method to derive the differential equation governing the motion of the system of Fig. 3-11 using θ as the generalized coordinate.

The kinetic energy of the system at an arbitrary instant is

$$T = \tfrac{1}{2}m(\tfrac{1}{4}L\dot\theta)^2 + \tfrac{1}{2}\tfrac{1}{12}mL^2\dot\theta^2 = \tfrac{1}{2}\tfrac{7}{48}mL^2\dot\theta^2$$

The potential energy of the system at an arbitrary instant is

$$V = \tfrac{1}{2}k(\tfrac{3}{4}L\theta)^2 = \tfrac{1}{2}\tfrac{9}{16}kL^2\theta^2$$

The work done by the damping force between two arbitrary instants is

$$W = -\int_{\theta_1}^{\theta_2} \tfrac{1}{4}cL\dot\theta\, d(\tfrac{1}{4}L\theta) = -\int_{\theta_1}^{\theta_2} \tfrac{1}{16}cL^2\dot\theta\, d\theta$$

The work done by the external forces as the system moves through a variation $\delta\theta$ is

$$\delta W = -F_0(\sin\omega t)\,\delta(\tfrac{1}{4}L\theta) + M_0(\sin\omega t)\,\delta\theta$$

$$= (-\tfrac{1}{4}F_0L + M_0)(\sin\omega t)\,\sigma\theta$$

Hence the governing differential equation is

$$\tfrac{7}{48}mL^2\ddot\theta + \tfrac{1}{16}cL^2\dot\theta + \tfrac{9}{16}kL^2\theta = (-\tfrac{1}{4}F_0L + M_0)\sin\omega t$$

3.3   For what value of m will resonance occur for the system of Fig. 3-13?

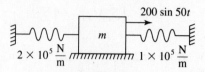

Fig. 3-13

The springs attached to the block in Fig. 3-13 act in parallel, leading to an equivalent stiffness of 3×10^5 N/m. Resonance occurs when the excitation frequency of 50 rad/s is equal to the natural frequency,

$$50 \, \frac{\text{rad}}{\text{s}} = \omega_n = \sqrt{\frac{k_{eq}}{m}}$$

which leads to

$$m = \frac{k_{eq}}{\omega^2} = \frac{3 \times 10^5 \, \dfrac{\text{N}}{\text{m}}}{\left(50 \, \dfrac{\text{rad}}{\text{s}}\right)^2} = 120 \text{ kg}$$

3.4   A 45-kg machine is placed at the end of a 1.6-m cantilever beam of elastic modulus of 200×10^9 N/m² and cross-sectional moment of inertia 1.6×10^{-5} m⁴. As it operates, the machine produces a harmonic force of magnitude 125 N. At what operating speeds will the machine's steady-state amplitude be less than 0.2 mm?

The equivalent stiffness of the beam is

$$k_{eq} = \frac{3EI}{L^3} = \frac{3\left(200 \times 10^9 \, \dfrac{\text{N}}{\text{m}^2}\right)(1.6 \times 10^{-5} \text{ m}^4)}{(1.6 \text{ m})^3} = 2.34 \times 10^6 \, \frac{\text{N}}{\text{m}}$$

The system's natural frequency is

$$\omega_n = \sqrt{\frac{k_{eq}}{m}} = \sqrt{\frac{2.34 \times 10^6 \, \dfrac{\text{N}}{\text{m}}}{45 \text{ kg}}} = 228.0 \, \frac{\text{rad}}{\text{s}}$$

In order to limit the steady-state amplitude to 0.2 mm, the allowable value of the magnification factor is

$$M = \frac{m\omega_n^2 X}{F_0} = \frac{(45 \text{ kg})\left(228.0 \, \dfrac{\text{rad}}{\text{s}}\right)^2 (0.0002 \text{ m})}{125 \text{ N}} = 3.74$$

For an undamped system, Eq. (3.12) becomes

$$M = \frac{1}{|1 - r^2|}$$

For $r < 1$, requiring $M < 3.74$ leads to

$$r < \sqrt{1 - \frac{1}{3.74}} \doteq 0.856$$

For $r > 1$, requiring $M < 3.74$ leads to

$$r > \sqrt{1 + \frac{1}{M}} = \sqrt{1 + \frac{1}{3.74}} = 1.126$$

Thus the allowable ranges of frequencies are

$$\omega < 0.856\omega_n = 195.2 \,\frac{\text{rad}}{\text{s}} \quad \text{and} \quad \omega > 1.125\omega_n = 256.5 \,\frac{\text{rad}}{\text{s}}$$

3.5 A thin disk of mass 0.8 kg and radius 60 mm is attached to the end of a 1.2-m steel ($G = 80 \times 10^9$ N/m^2, $\rho = 7500$ kg/m^3) shaft of diameter 20 mm. The disk is subject to a harmonic torque of amplitude 12.5 N-m at a frequency of 700 rad/s. What is the steady-state amplitude of angular oscillations of the disk?

The torsional stiffness of the shaft is

$$k_t = \frac{JG}{L} = \frac{\frac{\pi}{2}(0.01 \text{ m})^4\left(80 \times 10^9 \,\frac{\text{N}}{\text{m}^2}\right)}{1.2 \text{ m}} \doteq 1.05 \times 10^3 \,\frac{\text{N-m}}{\text{rad}}$$

The mass moment of inertia of the shaft is

$$I_s = \frac{1}{2}\rho\pi L r_s^4 = \frac{1}{2}\left(7500 \,\frac{\text{kg}}{\text{m}^3}\right)\pi(1.2 \text{ m})(0.01 \text{ m})^4$$

$$= 1.41 \times 10^{-4} \text{ kg-m}^2$$

The inertia effects of the shaft are included in a 1-degree-of-freedom model by

$$I_{eq} = I_d + \tfrac{1}{3}I_s = \tfrac{1}{2}m_d r_d^2 + \tfrac{1}{3}I_s$$

$$= \tfrac{1}{2}(0.8 \text{ kg})(0.06 \text{ m})^2 + \tfrac{1}{3}(1.41 \times 10^{-4} \text{ kg-m}^2) = 1.49 \times 10^{-3} \text{ kg-m}^2$$

The natural frequency of the system is

$$\omega_n = \sqrt{\frac{k_t}{I_{eq}}} = \sqrt{\frac{1.05 \times 10^3 \,\dfrac{\text{N-m}}{\text{rad}}}{1.49 \times 10^{-3} \text{ kg-m}^2}} = 839.5 \,\frac{\text{rad}}{\text{s}}$$

The frequency ratio is

$$r = \frac{\omega}{\omega_n} = \frac{700 \,\dfrac{\text{rad}}{\text{s}}}{839.5 \,\dfrac{\text{rad}}{\text{s}}} = 0.834$$

which leads to a magnification factor of

$$M = \frac{1}{1 - r^2} = \frac{1}{1 - (0.834)^2} = 3.28$$

Let Θ be the steady-state amplitude of torsional oscillation. The torsional oscillation equivalent of Eq. (3.10) is

$$\frac{I_{eq}\omega_n^2\Theta}{T_0} = M$$

$$\Theta = \frac{MT_0}{I_{eq}\omega_n^2} = \frac{3.28(12.5 \text{ N-m})}{(1.49 \times 10^{-3} \text{ kg-m}^2)\left(839.5 \frac{\text{rad}}{\text{s}}\right)^2}$$

$$= 0.0390 \text{ rad} = 2.24°$$

3.6 A 45-kg machine is mounted on four parallel springs each of stiffness 2×10^5 N/m. When the machine operates at 32 Hz, the machine's steady-state amplitude is measured as 1.5 mm. What is the magnitude of the excitation provided to the machine at this speed?

The system's natural frequency is

$$\omega_n = \sqrt{\frac{k_{eq}}{m}} = \sqrt{\frac{4\left(2 \times 10^5 \frac{\text{N}}{\text{m}}\right)}{45 \text{ kg}}} = 133.3 \frac{\text{rad}}{\text{s}}$$

The system's frequency ratio is

$$r = \frac{\omega}{\omega_n} = \frac{\left(32 \frac{\text{cycle}}{\text{s}}\right)\left(2\pi \frac{\text{rad}}{\text{cycle}}\right)}{133.3 \frac{\text{rad}}{\text{s}}} = 1.51$$

The magnification factor for an undamped system with a frequency ratio greater than 1 is

$$M = \frac{1}{r^2 - 1} = \frac{1}{(1.51)^2 - 1} = 0.781$$

Equation (3.10) is rearranged to solve for the excitation force as

$$F_0 = \frac{m\omega_n^2 X}{M} = \frac{(45 \text{ kg})\left(133.3 \frac{\text{rad}}{\text{s}}\right)^2 (0.0015 \text{ m})}{0.781} = 1.54 \times 10^3 \text{ N}$$

3.7 A system that exhibits beating has a period of oscillation of 0.05 s and a beating period of 1.0 s. Determine the system's natural frequency and its excitation frequency if the excitation frequency is greater than the natural frequency.

From Fig. 3-5 it is observed that the period of oscillation is

$$T = \frac{4\pi}{\omega + \omega_n} = 0.05 \text{ s}$$

and the period of beating is

$$T_b = \frac{2\pi}{|\omega - \omega_n|} = 1 \text{ s}$$

These equations are rearranged to

$$\omega + \omega_n = 80\pi$$

$$\omega - \omega_n = 2\pi$$

which are solved simultaneously yielding

$$\omega = 41\pi \frac{\text{rad}}{\text{s}}, \qquad \omega_n = 39\pi \frac{\text{rad}}{\text{s}}$$

3.8   Repeat Problem 3.4 as if the beam had a damping ratio of 0.08.

From the solution of Problem 3.4, the system's natural frequency is 228.0 rad/s, and the maximum allowable value of the magnification factor is 3.74. Thus in order to limit the magnification factor to 3.74,

$$3.74 > \frac{1}{\sqrt{(1 - r^2)^2 + [2(0.08)r]^2}}$$

$$r^4 - 1.9744r^2 + 1 > 0.07149$$

$$r^4 - 1.9744r^2 + 0.9285 > 0$$

The above is quadratic in r^2. Application of the quadratic formula leads to positive solutions of $r = 0.879$ and $r = 1.096$. The magnification factor is less than 3.74 if $\omega < 0.879(228.0 \text{ rad/s}) = 200.4$ rad/s or $\omega > 1.096(228.0 \text{ rad/s}) = 249.9$ rad/s.

3.9   A 110-kg machine is mounted on an elastic foundation of stiffness 2×10^6 N/m. When operating at 150 rad/s, the machine is subject to a harmonic force of magnitude 1500 N. The steady-state amplitude of the machine is measured as 1.9 mm. What is the damping ratio of the foundation?

The natural frequency of the system is

$$\omega_n = \sqrt{\frac{k}{m}} = \sqrt{\frac{2 \times 10^6 \frac{\text{N}}{\text{m}}}{110 \text{ kg}}} = 134.8 \frac{\text{rad}}{\text{s}}$$

The magnification factor during operation is

$$M = \frac{m\omega_n^2 X}{F_0} = \frac{(110 \text{ kg})\left(134.8 \frac{\text{rad}}{\text{s}}\right)^2 (0.0019 \text{ m})}{1500 \text{ N}} = 2.53$$

The frequency ratio for operation at 150 rad/s is

$$r = \frac{\omega}{\omega_n} = \frac{150 \frac{\text{rad}}{\text{s}}}{134.8 \frac{\text{rad}}{\text{s}}} = 1.113$$

Equation (3.12) can be rearranged to solve for the damping ratio as

$$\zeta = \frac{1}{2r} \sqrt{\frac{1}{M^2} - (1 - r^2)^2}$$

which for this problem leads to

$$\zeta = \frac{1}{2(1.113)}\sqrt{\frac{1}{(2.53)^2} - [1 - (1.113)^2]^2} = 0.142$$

3.10 The differential equation governing the motion of the system of Fig. 3-14 is

$$\left(m + \frac{I}{r^2}\right)\ddot{x} + c\dot{x} + 5kx = \frac{M_0}{r}\sin \omega t$$

Using the given values, determine the steady-state amplitude of the block.

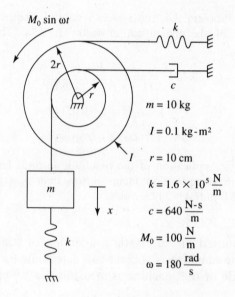

$M_0 \sin \omega t$

$2r$

r

$m = 10 \text{ kg}$

$I = 0.1 \text{ kg-m}^2$

$r = 10 \text{ cm}$

$k = 1.6 \times 10^5 \dfrac{\text{N}}{\text{m}}$

$c = 640 \dfrac{\text{N-s}}{\text{m}}$

$M_0 = 100 \dfrac{\text{N}}{\text{m}}$

$\omega = 180 \dfrac{\text{rad}}{\text{s}}$

Fig. 3-14

The system's natural frequency and damping ratio are

$$\omega_n = \sqrt{\frac{5k}{m + \frac{I}{r^2}}} = \sqrt{\frac{5\left(1.6 \times 10^5 \frac{\text{N}}{\text{m}}\right)}{10 \text{ kg} + \frac{0.1 \text{ kg-m}^2}{(0.1 \text{ m})^2}}} = 200 \frac{\text{rad}}{\text{s}}$$

$$\zeta = \frac{c}{2\omega_n\left(m + \frac{I}{r^2}\right)} = \frac{640 \frac{\text{N-s}}{\text{m}}}{2\left(200 \frac{\text{rad}}{\text{s}}\right)\left(10 \text{ kg} + \frac{0.1 \text{ kg-m}^2}{(0.1 \text{ m})^2}\right)} = 0.08$$

The frequency ratio is

$$r = \frac{\omega}{\omega_n} = \frac{180 \frac{\text{rad}}{\text{s}}}{200 \frac{\text{rad}}{\text{s}}} = 0.9$$

The magnification factor for the system is

$$M = M(0.9, 0.08) = \frac{1}{\sqrt{[1 - (0.9)^2]^2 + [2(0.08)(0.9)]^2}} = 4.19$$

The steady-state amplitude is determined using Eq. (3.10) with

$$F_0 = \frac{M_0}{r} = \frac{100 \text{ N-m}}{0.1 \text{ m}} = 1000 \text{ N}$$

and

$$m_{eq} = m + \frac{I}{r^2} = 20 \text{ kg}$$

Thus

$$X = \frac{F_0 M(0.9, 0.08)}{m_{eq}\omega_n^2} = \frac{(1000 \text{ N})(4.19)}{(20 \text{ kg})\left(200 \dfrac{\text{rad}}{\text{s}}\right)^2} = 5.24 \text{ mm}$$

3.11 Derive Eq. (3.14) from Eq. (3.12).

For a fixed ζ, the value of r for which the maximum of $M(r, \zeta)$ occurs is obtained by finding the value of r such that $\partial M / \partial r = 0$. To this end,

$$\frac{\partial M}{\partial r} = -\frac{1}{2}[(1 - r^2)^2 + (2\zeta r)^2]^{-3/2}[2(1 - r^2)(-2r) + 2(2\zeta r)(2\zeta)]$$

$$\frac{\partial M}{\partial r} = 0 \rightarrow (1 - r^2) + 2\zeta^2 = 0$$

$$r = \sqrt{1 - 2\zeta^2}$$

Substituting this value of r into Eq. (3.12) leads to

$$M_{max} = \frac{1}{\sqrt{[1 - (1 - 2\zeta^2)]^2 + 4\zeta^2(1 - 2\zeta^2)]}} = \frac{1}{2\zeta\sqrt{1 - \zeta^2}}$$

3.12 A 120-kg machine is mounted at the midspan of a 1.5-m-long simply supported beam of elastic modulus $E = 200 \times 10^9 \text{ N/m}^2$ and cross-section moment of inertia $I = 1.53 \times 10^{-6} \text{ m}^4$. An experiment is run on the system during which the machine is subject to a harmonic excitation of magnitude 2000 N at a variety of excitation frequencies. The largest steady-state amplitude recorded during the experiment is 2.5 mm. Estimate the damping ratio of the system.

The stiffness of the beam is

$$k = \frac{48EI}{L^3} = \frac{48\left(200 \times 10^9 \dfrac{\text{N}}{\text{m}^2}\right)(1.53 \times 10^{-6} \text{ m}^4)}{(1.5 \text{ m})^3} = 4.35 \times 10^6 \frac{\text{N}}{\text{m}}$$

The system's natural frequency is

$$\omega_n = \sqrt{\frac{k}{m}} = \sqrt{\frac{4.35 \times 10^6 \dfrac{\text{N}}{\text{m}}}{120 \text{ kg}}} = 190.4 \frac{\text{rad}}{\text{s}}$$

The maximum value of the magnification factor is

$$M_{max} = \frac{m\omega_n^2 X_{max}}{F_0} = \frac{(120 \text{ kg})\left(190.4 \dfrac{\text{rad}}{\text{s}}\right)^2 (0.0025 \text{ m})}{2000 \text{ N}} = 5.44$$

Equation (3.14) can be rearranged as

$$\zeta^4 - \zeta^2 + \frac{1}{4M_{max}^2} = 0$$

which is a quadratic equation in ζ^2 whose roots are

$$\zeta = \left[\frac{1}{2} \left(1 \pm \sqrt{1 - \frac{1}{M_{max}^2}} \right) \right]^{1/2}$$

Substituting $M_{max} = 5.44$ and noting that use of the positive sign in the $\pm$ choice leads to a damping ratio greater than $1/\sqrt{2}$ leads to $\zeta = 0.092$.

3.13   An 82-kg machine tool is mounted on an elastic foundation. An experiment is run to determine the stiffness and damping properties of the foundation. When the tool is excited with a harmonic force of magnitude 8000 N at a variety of frequencies, the maximum steady-state amplitude obtained is 4.1 m at a frequency of 40 Hz. Use this information to estimate the stiffness and damping ratio of the foundation.

Using Eqs. (3.10) and (3.14), the maximum steady-state amplitude is related to the damping ratio by

$$M_{max} = \frac{1}{2\zeta\sqrt{1 - \zeta^2}} = \frac{m\omega_n^2 X_{max}}{F_0} \tag{3.47}$$

Then from Eq. (3.15) the natural frequency and the frequency at which the maximum steady-state amplitude occurs are related by

$$r_{m_M} = \sqrt{1 - 2\zeta^2} = \frac{\omega_{max}}{\omega_n}$$

$$\omega_n = \frac{\omega_{max}}{\sqrt{1 - 2\zeta^2}}$$

which when substituted into Eq. (3.47), leads to

$$\frac{m\omega_{max}^2 X_{max}}{(1 - 2\zeta^2)F_0} = \frac{1}{2\zeta\sqrt{1 - \zeta^2}}$$

$$\frac{(82 \text{ kg})\left[\left(40 \frac{\text{cycle}}{\text{s}} \right)\left(2\pi \frac{\text{rad}}{\text{cycle}} \right) \right]^2 (0.0041 \text{ m})}{(1 - 2\zeta^2)(8000 \text{ N})} = \frac{1}{2\zeta\sqrt{1 - \zeta^2}}$$

Substituting given and calculated values and rearranging leads to

$$28.20\zeta^2(1 - \zeta^2) = (1 - 2\zeta^2)^2$$

$$\zeta^4 - \zeta^2 + 0.03107 = 0$$

$$\zeta = 0.179, 0.984$$

However, since a maximum steady-state amplitude is attained only for $\zeta < 1/\sqrt{2}$, $\zeta = 0.179$. Eq. (3.15) is used to determine the natural frequency as

$$\omega_n = \frac{\omega}{\sqrt{1 - \zeta^2}} = \frac{\left(40 \frac{\text{cycle}}{\text{s}} \right)\left(2\pi \frac{\text{rad}}{\text{cycle}} \right)}{\sqrt{1 - (0.179)^2}} = 255.5 \frac{\text{rad}}{\text{s}}$$

from which the foundation's stiffness is calculated:

$$k = m\omega_n^2 = (82 \text{ kg})\left(255.5 \frac{\text{rad}}{\text{s}} \right)^2 = 5.35 \times 10^6 \frac{\text{N}}{\text{m}}$$

3.14   A 35-kg electric motor that operates at 60 Hz is mounted on an elastic foundation of stiffness 3×10^6 N/m. The phase difference between the excitation and the steady-state response is 21°. What is the damping ratio of the system?

The natural frequency and frequency ratio are

$$\omega_n = \sqrt{\frac{k}{m}} = \sqrt{\frac{3 \times 10^6 \, \frac{N}{m}}{35 \, \text{kg}}} = 292.8 \, \frac{\text{rad}}{\text{s}}$$

$$r = \frac{\omega}{\omega_n} = \frac{\left(60 \, \frac{\text{cycle}}{\text{s}}\right)\left(2\pi \, \frac{\text{rad}}{\text{cycle}}\right)}{292.8 \, \frac{\text{rad}}{\text{s}}} = 1.288$$

Equation (3.13) can be rearranged to solve for ζ as:

$$\zeta = \frac{1 - r^2}{2r} \tan \phi$$

However, since the frequency ratio is greater than 1, the response leads the excitation, and if the phase angle is taken to be between 0 and 180°, the appropriate value is $\phi = 180° - 21° = 159°$. Thus

$$\zeta = \frac{1 - (1.288)^2}{2(1.288)} \tan^{-1}(159°) = 0.0982$$

3.15   The machine of mass m, of Fig. 3-15, is mounted on an elastic foundation modeled as a spring of stiffness k in parallel with a viscous damper of damping coefficient c. The machine has an unbalanced component rotating at a constant speed ω. The unbalance can be represented by a particle of mass m_0, a distance e from the axis of rotation. Derive the differential equation governing the machine's displacement, and determine its steady-state amplitude.

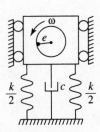

Fig. 3-15

Free body diagrams of the machine at an arbitrary instant are shown in Fig. 3-16. Summing forces in the vertical direction

$$\left(\sum F\right)_{\text{ext}} = \left(\sum F\right)_{\text{eff}}$$

and noting that gravity cancels with the static spring force leads to

$$-kx - c\dot{x} = (m - m_0)\ddot{x} + m_0 e\omega^2 \sin \theta + m_0 \ddot{x} \qquad (3.48)$$

Since ω is constant,

$$\theta = \omega t + \theta_0 \qquad (3.49)$$

Substituting Eq. (3.49) into Eq. (3.48) and rearranging leads to

$$m\ddot{x} + c\dot{x} + kx = -m_0 e\omega^2 \sin(\omega t + \theta_0) = m_0 e\omega^2 \sin(\omega t + \psi) \qquad (3.50)$$

where

$$\psi = \theta_0 + \pi$$

Thus the response of a system due to a rotating unbalance is that of a system excited by a frequency squared harmonic excitation. The constant of proportionality defined in Eq. (3.16) is

$$A = m_0 e$$

This application of Eq. (3.17) leads to

$$\frac{mX}{m_0 e} = \Lambda(r, \zeta) = \frac{r^2}{\sqrt{(1 - r^2)^2 + (2\zeta r)^2}} \qquad (3.51)$$

where

$$r = \frac{\omega}{\omega_n}, \qquad \omega_n = \sqrt{\frac{k}{m}}, \qquad \zeta = \frac{c}{2m\omega_n}$$

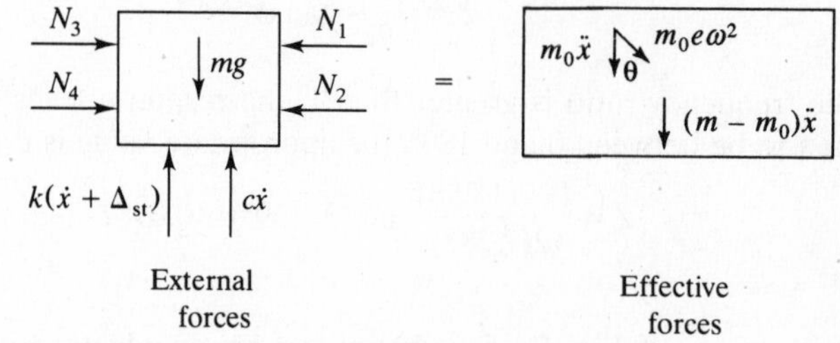

Fig. 3-16

3.16 A 65-kg industrial sewing machine has a rotating unbalance of 0.15 kg-m. The machine operates at 125 Hz and is mounted on a foundation of equivalent stiffness 2×10^6 N/m and damping ratio 0.12. What is the machine's steady-state amplitude?

The natural frequency and frequency ratio of the system are

$$\omega_n = \sqrt{\frac{k}{m}} = \sqrt{\frac{2 \times 10^6 \, \dfrac{N}{m}}{65 \text{ kg}}} = 175.4 \, \frac{\text{rad}}{\text{s}}$$

$$r = \frac{\omega}{\omega_n} = \frac{\left(125 \, \dfrac{\text{cycle}}{\text{s}}\right)\left(2\pi \, \dfrac{\text{rad}}{\text{cycle}}\right)}{175.4 \, \dfrac{\text{rad}}{\text{s}}} = 4.48$$

From the results of Problem 3.15, the excitation provided to the machine by the rotating unbalance is a frequency squared harmonic excitation with $A = m_0 e$, the magnitude of the rotating unbalance. Thus using Eq. (3.51) of Problem 3.15,

$$\frac{mX}{m_0 e} = \Lambda(4.48, 0.12) = \frac{(4.48)^2}{\sqrt{[1 - (4.48)^2]^2 + [2(0.12)(4.48)]^2}} = 1.051$$

$$X = \frac{1.051(0.15 \text{ kg-m})}{65 \text{ kg}} = 2.43 \text{ mm}$$

3.17 An 80-kg reciprocating machine is placed on a thin, massless beam. A frequency sweep is run to determine the magnitude of the machine's rotating unbalance and the beam's equivalent stiffness. As the speed of the machine is increased, the following is noted:

(a) The steady-state amplitude of the machine at a speed of 65 rad/s is 7.5 mm.

(b) The maximum steady-state amplitude occurs for a speed less than 65 rad/s.

(c)   As the speed is greatly increased, the steady-state amplitude approaches 5 mm.

Assume the system is undamped.

Problem 3.15 illustrates that a machine with a rotating unbalance experiences a frequency squared harmonic excitation with $A = m_0e$, the magnitude of the rotating unbalance. Figure 3-8 shows that as the frequency ratio grows large, $\Lambda \to 1$. Thus from condition (c)

$$\frac{(80 \text{ kg})(0.005 \text{ m})}{m_0e} = 1$$

$$m_0e = 0.4 \text{ kg-m}$$

Since the maximum steady-state amplitude occurs for a speed less than 65 rad/s, it is probable that 65 rad/s corresponds to a frequency ratio greater than 1. Thus, for an undamped system with $r > 1$,

$$\Lambda = \frac{r^2}{r^2 - 1}$$

For $\omega = 65$ rad/s,

$$\Lambda = \frac{mX}{m_0e} = \frac{(80 \text{ kg})(0.0075 \text{ m})}{0.04 \text{ kg-m}} = 1.5$$

Thus

$$1.5 = \frac{r^2}{r^2 - 1} \to r = 1.73 \to \omega_n = \frac{65 \frac{\text{rad}}{\text{s}}}{1.73} = 37.6 \frac{\text{rad}}{\text{s}}$$

$$\to k = m\omega_n^2 = (80 \text{ kg})\left(37.6 \frac{\text{rad}}{\text{s}}\right)^2 = 1.13 \times 10^5 \frac{\text{N}}{\text{m}}$$

3.18   A 500-kg tumbler has an unbalance of 1.26 kg, 50 cm from its axis of rotation. For what stiffnesses of an elastic mounting of damping ratio 0.06 will the tumbler's steady-state amplitude be less than 2 mm at all speeds between 200 and 600 r/min?

The results of Problem 3.15 show that a machine with a rotating unbalance is subject to a frequency squared excitation with $A = m_0e$. Thus in order for the steady-state amplitude to be less than 2 mm when the tumbler is installed on the mounting, the largest allowable value of Λ is

$$\Lambda_{\text{all}} = \frac{mX_{\max}}{m_0e} = \frac{(500 \text{ kg})(0.002 \text{ m})}{(1.26 \text{ kg})(0.5 \text{ m})} = 1.587$$

From Eq. (3.19), $\Lambda_{\max}(\zeta = 0.06) = 8.36 > \Lambda_{\text{all}}$. Then from Fig. 3-8, since $\Lambda_{\text{all}} > 1$ and $\zeta < 1/\sqrt{2}$, there are two values of r such that $\Lambda(r, 0.06) = 1.587$. In order for $\Lambda < 1.587$, the frequency ratio cannot be between these two values, which are obtained by solving

$$1.587 = \frac{r^2}{\sqrt{(1 - r^2)^2 + (0.12r)^2}}$$

Squaring the above equation, multiplying through by the denominator of the right-hand side, and rearranging leads to

$$1.519r^4 - 5.001r^2 + 2.519 = 0$$

which is a quadratic equation in r^2 and can be solved using the quadratic formula. The resulting allowable frequency ranges correspond to

$$r < 0.788 \quad \text{or} \quad r > 1.634$$

In order for $r < 0.788$ over the entire frequency range, $r = 0.788$ should correspond to a frequency less than 600 r/min. Thus

$$\omega_n > \frac{\omega}{r} = \frac{\left(600\ \frac{r}{min}\right)\left(2\pi\ \frac{rad}{r}\right)\left(\frac{1\ min}{60\ s}\right)}{0.788} = 79.73\ \frac{rad}{s}$$

$$k_{min} = (500\ kg)\left(79.73\ \frac{rad}{s}\right)^2 = 3.18 \times 10^6\ \frac{N}{m}$$

In order for $r > 1.634$ over the entire frequency range, $r = 1.684$ should correspond to a frequency grater than 200 r/min. Thus

$$\omega_n < \frac{\left(200\ \frac{r}{min}\right)\left(2\pi\ \frac{rad}{r}\right)\left(\frac{1\ min}{60\ s}\right)}{1.634} = 12.82\ \frac{rad}{s}$$

$$k_{max} = (500\ kg)\left(12.82\ \frac{rad}{s}\right)^2 = 8.21 \times 10^4\ \frac{N}{m}$$

Hence the acceptable mounting stiffnesses are

$$k < 8.21 \times 10^4\ \frac{N}{m} \quad and \quad k > 3.18 \times 10^6\ \frac{N}{m}$$

3.19 A 40-kg fan has a rotating unbalance of magnitude 0.1 kg-m. The fan is mounted on the beam of Fig. 3-17. The beam has been specially treated to add viscous damping. As the speed of the fan is varied, it is noted that its maximum steady-state amplitude is 20.3 mm. What is the fan's steady-state amplitude when it operates at 1000 r/min?

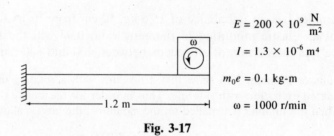

$E = 200 \times 10^9\ \frac{N}{m^2}$

$I = 1.3 \times 10^{-6}\ m^4$

$m_0 e = 0.1\ kg\text{-}m$

$\omega = 1000$ r/min

Fig. 3-17

The maximum value of Λ is

$$\Lambda_{max} = \frac{mX_{max}}{m_0 e} = \frac{(40\ kg)(0.0203\ m)}{0.1\ kg\text{-}m} = 8.12$$

The damping ratio is determined using Eq. (3.19):

$$8.12 = \frac{1}{2\zeta\sqrt{1-\zeta^2}}$$

$$\zeta = 0.0617$$

The beam's stiffness is

$$k = \frac{3EI}{L^3} = \frac{3\left(200 \times 10^9\ \frac{N}{m^2}\right)(1.3 \times 10^{-6}\ m^4)}{(1.2\ m)^3} = 4.51 \times 10^5\ \frac{N}{m}$$

and the system's natural frequency is

$$\omega_n = \sqrt{\frac{k}{m}} = \sqrt{\frac{4.51 \times 10^5 \dfrac{N}{m}}{40 \text{ kg}}} = 106.2 \frac{\text{rad}}{\text{s}}$$

The frequency ratio is

$$r = \frac{\omega}{\omega_n} = \frac{\left(1000 \dfrac{\text{r}}{\text{min}}\right)\left(2\pi \dfrac{\text{rad}}{\text{r}}\right)\left(\dfrac{1 \text{ min}}{60 \text{ s}}\right)}{106.2 \dfrac{\text{rad}}{\text{s}}} = 0.986$$

The steady-state amplitude is calculated by

$$X = \frac{m_0 e}{m} \Lambda(0.986, 0.0617)$$

$$= \frac{0.1 \text{ kg-m}}{40 \text{ kg}} \frac{(0.986)^2}{\sqrt{[1 - (0.986)^2]^2 + [2(0.0617)(0.986)]^2}}$$

$$= 19.48 \text{ mm}$$

3.20 The fan of Problem 3.19 is to operate at 1000 r/min, 1250 r/min, 1500 r/min, 1750 r/min, and 2000 r/min. What is the minimum mass that should be added to the fan such that its steady-state amplitude is less than 10 mm at all operating speeds?

Adding mass to the fan decreases the system's natural frequency, thus increasing the frequency ratio at each operating speed. With no additional mass, $r = 0.986$ for $\omega = 1000$ r/min. Adding mass will probably lead to a frequency ratio greater than 1 for $\omega = 1000$ r/min. Figure 3-8 shows that for $r > 1$, the steady-state amplitude for a frequency squared excitation decreases with increasing excitation frequency. Thus if $X < 10$ mm for $\omega = 1000$ r/min, then $X < 10$ mm for all $\omega > 1000$ r/min. The desired magnification factor for $\omega = 1000$ r/min is

$$M = \frac{m\omega_n^2 X}{F_0} = \frac{kX}{m_0 e \omega^2} = \frac{\left(4.51 \times 10^5 \dfrac{N}{m}\right)(0.01 \text{ m})}{(0.1 \text{ kg-m})\left(104.7 \dfrac{\text{rad}}{\text{s}}\right)^2} = 4.11$$

Thus

$$4.11 = \frac{1}{\sqrt{(1 - r^2)^2 + [2(0.0617)r]^2}}$$

Solving for r leads to $r = 1.096$. Thus

$$\omega_n = \frac{104.7 \dfrac{\text{rad}}{\text{s}}}{1.096} = 95.5 \frac{\text{rad}}{\text{s}}$$

$$m = \frac{k}{\omega_n^2} = \frac{4.51 \times 10^5 \dfrac{N}{m}}{\left(95.5 \dfrac{\text{rad}}{\text{s}}\right)^2} = 49.5 \text{ kg}$$

Thus the minimum mass that should be added to the machine is 9.5 kg.

3.21 The tail rotor section of the helicopter of Fig. 3-18 consists of four blades, each of mass 2.3 kg, and an engine box of mass 28.5 kg. The center of gravity of each blade is 170 mm from the rotational axis. The tail section is connected to the main body of the helicopter by an elastic structure. The natural frequency of the tail section is observed as 135 rad/s. During flight, the rotor operates at 900 r/min. What is the vibration amplitude of the tail section if one of the blades falls off during flight? Assume a damping ratio of 0.05.

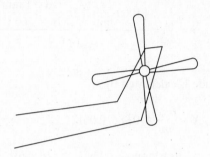

Fig. 3-18

The total mass of the rotor is

$$m = 4(2.3 \text{ kg}) + 28.5 \text{ kg} = 37.7 \text{ kg}$$

The equivalent stiffness of the tail section is

$$k_{eq} = m\omega_n^2 = (37.7 \text{ kg})\left(135 \frac{\text{rad}}{\text{s}}\right)^2 = 6.87 \times 10^5 \frac{\text{N}}{\text{m}}$$

If a blade falls off during flight, the rotor is unbalanced and leads to harmonic excitation of the tail section. The magnitude of the rotating unbalance is

$$m_0 e = (2.3 \text{ kg})(0.170 \text{ m}) = 0.391 \text{ kg-m}$$

The natural frequency of the rotor after one blade falls off is

$$\omega_n = \sqrt{\frac{k_{eq}}{m}} = \sqrt{\frac{6.87 \times 10^5 \frac{\text{N}}{\text{m}}}{37.7 \text{ kg} - 2.3 \text{ kg}}} = 139.3 \frac{\text{rad}}{\text{s}}$$

The frequency ratio is

$$r = \frac{\omega}{\omega_n} = \frac{\left(900 \frac{\text{r}}{\text{min}}\right)\left(2\pi \frac{\text{rad}}{\text{r}}\right)\left(\frac{1 \text{ min}}{60 \text{ s}}\right)}{139.3 \frac{\text{rad}}{\text{s}}} = 0.677$$

The steady-state amplitude is calculated using Eq. (3.17):

$$X = \frac{m_0 e}{m} \Lambda(0.677, 0.05)$$

$$= \frac{0.391 \text{ kg-m}}{35.4 \text{ kg}} \frac{(0.677)^2}{\sqrt{[1 - (0.677)^2]^2 + [2(0.05)(0.677)]^2}}$$

$$= 9.27 \text{ mm}$$

3.22 When a circular cylinder of length L and diameter D is placed in a steady flow of mass density ρ and velocity v, vortices are shed alternately from the upper and lower surfaces of the cylinder, leading to a net harmonic force acting on the cylinder of the form of Eq. (3.5). The frequency at which vortices are shed is related to the Strouhal number (S) by

$$S = \frac{\omega D}{2\pi v} \qquad (3.52)$$

The excitation amplitude is related to the drag coefficient C_D by

$$C_D = \frac{F_0}{\frac{1}{2}\rho v^2 DL} \qquad (3.53)$$

The drag coefficient and Strouhal number vary little with the Reynolds number Re for $1 \times 10^3 < Re < 2 \times 10^5$. These approximately constant values are

$$S = 0.2, \qquad C_D = 1.0$$

In this case, show that the amplitude of excitation is proportional to the square of the frequency, and determine the constant of proportionality.

Solving for v from Eq. (3.52) and setting $S = 0.2$

$$v = \frac{\omega D}{0.4\pi} \qquad (3.54)$$

Substituting Eq. (3.54) into Eq. (3.53) with $C_D = 1.0$ leads to

$$C_D = \frac{F_0}{\frac{1}{2}\rho\left(\dfrac{\omega D}{0.4\pi}\right)^2 DL}$$

which leads to

$$F_0 = 0.317\rho D^3 L\omega^2$$

3.23   As a publicity stunt, a 120-kg man is camped on the end of the flagpole of Fig. 3-19. What is the amplitude of vortex-induced vibration to which the man is subject in a 5 m/s wind? Assume a damping ratio of 0.02 and the mass density of air as 1.2 kg/m³.

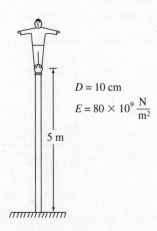

$D = 10$ cm

$E = 80 \times 10^9 \dfrac{N}{m^2}$

5 m

Fig. 3-19

The flagpole is modeled as a cantilever beam of stiffness

$$k = \frac{3EI}{L^3} = \frac{3\left(80 \times 10^9 \, \frac{N}{m^2}\right) \frac{\pi}{4}(0.05 \, m)^4}{(5 \, m)^2} = 9.42 \times 10^3 \, \frac{N}{m}$$

The natural frequency of the man is

$$\omega_n = \sqrt{\frac{k}{m}} = \sqrt{\frac{9.42 \times 10^3 \, \frac{N}{m}}{120 \, kg}} = 8.86 \, \frac{rad}{s}$$

The vortex shedding frequency is

$$\omega = \frac{0.4\pi v}{D} = \frac{0.4\pi\left(5 \, \frac{m}{s}\right)}{0.1 \, m} = 62.8 \, \frac{rad}{s}$$

Hence the frequency ratio is

$$r = \frac{\omega}{\omega_n} = \frac{62.8 \, \frac{rad}{s}}{8.86 \, \frac{rad}{s}} = 7.09$$

Using the results of Problem 3.22, it is noted that vortex shedding provides a frequency squared excitation with

$$A = 0.317\rho D^3 L = 0.317\left(1.2 \, \frac{kg}{m^3}\right)(0.1 \, m)^3(5 \, m) = 1.9 \times 10^{-3} \, kg\text{-}m$$

Then using Eq. (3.17),

$$X = \frac{A}{m}\Lambda(7.09, 0.02)$$

$$= \frac{1.9 \times 10^{-3} \, kg\text{-}m}{120 \, kg} \frac{(7.09)^2}{\sqrt{[1 - (7.09)^2]^2 + [2(0.02)(7.09)]^2}}$$

$$= 1.67 \times 10^{-5} \, m$$

3.24 A 35-kg block is connected to a support through a spring of stiffness 1.4×10^6 N/m in parallel with a dashpot of damping coefficient 1.8×10^3 N-s/m. The support is given a harmonic displacement of amplitude 10 mm at a frequency of 35 Hz. What is the steady-state amplitude of the absolute displacement of the block?

The natural frequency, damping ratio, and frequency ratio are

$$\omega_n = \sqrt{\frac{k}{m}} = \sqrt{\frac{1.4 \times 10^6 \, \frac{N}{m}}{35 \, kg}} = 200 \, \frac{rad}{s}$$

$$\zeta = \frac{c}{2m\omega_n} = \frac{1.8 \times 10^3 \, \frac{N\text{-}s}{m}}{2(35 \, kg)\left(200 \, \frac{rad}{s}\right)} = 0.129$$

$$r = \frac{\omega}{\omega_n} = \frac{\left(35 \, \frac{cycle}{s}\right)\left(2\pi \, \frac{rad}{cycle}\right)}{200 \, \frac{rad}{s}} = 1.10$$

The amplitude of absolute acceleration is obtained using Eq. (3.27) as

$$X = YT(1.10, 0.129)$$

$$= (0.01 \text{ m}) \sqrt{\frac{1 + [2(0.129)(1.10)]^2}{[1 - (1.10)^2]^2 + [2(0.129)(1.10)]^2}}$$

$$= 29.4 \text{ mm}$$

3.25 For the system of Problem 3.24 determine the steady-state amplitude of the displacement of the block relative to its support.

The displacement of the block relative to its support is obtained using Eq. (3.26):

$$Z = Y\Lambda(1.10, 0.129)$$

$$= (0.01 \text{ m}) \frac{(1.10)^2}{\sqrt{[1 - (1.10)^2]^2 + [2(1.10)(0.129)]^2}}$$

$$= 34.3 \text{ mm}$$

3.26 A 35-kg flow monitoring device is placed on a table in a laboratory. A pad of stiffness 2×10^5 N/m and damping ratio 0.08 is placed between the apparatus and the table. The table is bolted to the laboratory floor. Measurements indicate that the floor has a steady-state vibration amplitude of 0.5 mm at a frequency of 30 Hz. What is the amplitude of acceleration of the flow monitoring device?

The natural frequency and frequency ratio are

$$\omega_n = \sqrt{\frac{k}{m}} = \sqrt{\frac{2 \times 10^5 \frac{\text{N}}{\text{m}}}{35 \text{ kg}}} = 75.6 \frac{\text{rad}}{\text{s}}$$

$$r = \frac{\omega}{\omega_n} = \frac{\left(30 \frac{\text{cycle}}{\text{s}}\right)\left(2\pi \frac{\text{rad}}{\text{cycle}}\right)}{75.6 \frac{\text{rad}}{\text{s}}} = 2.49$$

The amplitude of absolute displacement of the flow measuring device is calculated using Eq. (3.27):

$$X = YT(2.49, 0.08)$$

$$= (0.0005 \text{ m}) \sqrt{\frac{1 + [2(0.08)(2.49)]^2}{[1 - (2.49)^2]^2 + [2(0.08)(2.49)]^2}}$$

$$= 1.03 \times 10^{-4} \text{ m}$$

The acceleration amplitude is

$$A = \omega^2 X = \left[\left(30 \frac{\text{cycle}}{\text{s}}\right)\left(2\pi \frac{\text{rad}}{\text{cycle}}\right)\right]^2 (1.03 \times 10^{-4} \text{ m}) = 3.66 \frac{\text{m}}{\text{s}^2}$$

3.27 What is the maximum deflection of the elastic mounting between the flow measuring device and the table of Problem 3.26?

The elastic mounting is placed between the flow measuring device and the table. Hence its deflection is the deflection of the flow measuring device relative to the table. The amplitude of relative displacement is calculated using Eq. (3.26):

$$Z = Y\Lambda(2.49, 0.08)$$

$$= (0.0005 \text{ m}) \frac{(2.49)^2}{\sqrt{[1 - (2.49)^2]^2 + [2(0.08)(2.49)]^2}}$$

$$= 5.94 \times 10^{-4} \text{ m}$$

3.28 A simplified model of a vehicle suspension system is shown in Fig. 3-20. The body of a 500-kg vehicle is connected to the wheels through a suspension system that is modeled as a spring of stiffness 4×10^5 N/m in parallel with a viscous damper of damping coefficient 3000 N-s/m. The wheels are assumed to be rigid and follow the road contour. The contour of the road traversed by the vehicle is shown in Fig. 3-21. If the vehicle travels at a constant speed of 52 m/s, what is the acceleration amplitude of the vehicle?

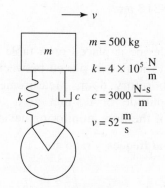

Fig. 3-20

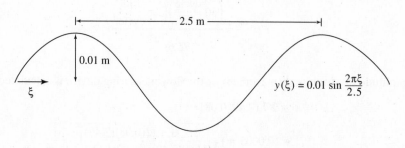

Fig. 3-21

The natural frequency and damping ratio of the system are

$$\omega_n = \sqrt{\frac{k}{m}} = \sqrt{\frac{4 \times 10^5 \frac{\text{N}}{\text{m}}}{500 \text{ kg}}} = 28.3 \frac{\text{rad}}{\text{s}}$$

$$\zeta = \frac{c}{2m\omega_n} = \frac{3000 \frac{\text{N-s}}{\text{m}}}{2(500 \text{ kg})\left(28.3 \frac{\text{rad}}{\text{s}}\right)} = 0.106$$

The mathematical description of the road contour is

$$y(\xi) = 0.01 \sin(0.8\pi\xi) \text{ m}$$

If the vehicle travels with a constant horizontal velocity, $\xi = vt$. Thus the time-dependent vertical displacement of the wheel is

$$y(t) = 0.01 \sin[0.8\pi vt]$$

Since the wheel follows the road contour, it acts as a harmonic base displacement for the body of the vehicle. The frequency of the displacement is

$$\omega = 0.8\pi v = 0.8\pi\left(52 \,\frac{\text{m}}{\text{s}}\right) = 130.7 \,\frac{\text{rad}}{\text{s}}$$

Hence the frequency ratio is

$$r = \frac{\omega}{\omega_n} = \frac{130.7 \,\dfrac{\text{rad}}{\text{s}}}{28.3 \,\dfrac{\text{rad}}{\text{s}}} = 4.62$$

The amplitude of absolute displacement of the vehicle is calculated using Eq. (3.27):

$$X = YT(4.62, 0.106)$$

$$= (0.01 \text{ m})\sqrt{\frac{1 + [2(0.106)(4.62)]^2}{[1 - (4.62)^2]^2 + [2(0.106)(4.62)]^2}}$$

$$= 6.87 \times 10^{-4} \text{ m}$$

The vehicle's acceleration amplitude is

$$A = \omega^2 X = \left(130.7 \,\frac{\text{rad}}{\text{s}}\right)^2 (6.87 \times 10^{-4} \text{ m}) = 11.7 \,\frac{\text{m}}{\text{s}^2}$$

3.29 Let A be the amplitude of the absolute acceleration of the vehicle of Problem 3.28. Show that

$$\frac{A}{\omega_n^2 Y} = R(r, \zeta) = r^2 \sqrt{\frac{1 + (2\zeta r)^2}{(1 - r^2)^2 + (2\zeta r)^2}}$$

where Y is the amplitude of the road contour.

The amplitude of acceleration is $\omega^2 X$ where X is the amplitude of absolute displacement of the vehicle. From Eq. (3.27),

$$\frac{\omega^2 X}{\omega^2 Y} = T(r, \zeta)$$

$$\omega_n^2 \frac{A}{\omega_n^2 \omega^2 Y} = T(r, \zeta)$$

$$\frac{A}{\omega_n^2 Y} = \frac{\omega^2}{\omega_n^2} T(r, \zeta) = r^2 \sqrt{\frac{1 + (2\zeta r)^2}{(1 - r^2)^2 + (2\zeta r)^2}}$$

3.30 Plot $R(r, \zeta)$ from Problem 3.29 as a function of r for the value of ζ obtained in Problem 3.28. At what vehicle speeds do the relative maximum and minimum of R occur?

The plot of $R(r, 0.106)$ is shown in Fig. 3-22. The values of r for which the maximum and minimum of $R(r, \zeta)$ for a given ζ occur are obtained by setting $dR^2/d\mu = 0$ where $\mu = r^2$. To this end

$$R^2 = \frac{\mu^2 + 4\zeta^2\mu^3}{\mu^2 + (4\zeta^2 - 2)\mu + 1}$$

and using the quotient rule for differentiation,

$$\frac{dR^2}{d\mu} = \frac{(2\mu + 12\zeta^2\mu^2)[\mu^2 + (4\zeta^2 - 2)\mu + 1] - (\mu^2 + 4\zeta^2\mu^3)[(2\mu + (4\zeta^2 - 2)]}{[\mu^2 + (4\zeta^2 - 2)\mu + 1]^2}$$

Setting the numerator to zero leads to

$$4\zeta^2\mu^3 + (32\zeta^4 - 16\zeta^2)\mu^2 + (16\zeta^2 - 2)\mu + 2 = 0$$

Substituting $\zeta = 0.106$ and rearranging leads to

$$\mu^3 - 3.909\mu^2 - 40.5\mu + 44.5 = 0$$

whose positive roots are

$$\mu = 1.025, 8.190 \rightarrow r = 1.012, 2.862$$

The vehicle speeds for which the maximum and minimum steady-state amplitudes occur are given by

$$v = \frac{r\omega_n}{0.8\pi} = \frac{\left(28.3\,\frac{\text{rad}}{\text{s}}\right)r}{0.8\pi} = 11.26r$$

$$v_{max} = 11.26(1.012) = 11.40\,\frac{\text{m}}{\text{s}}$$

$$v_{min} = 11.26(2.862) = 32.2\,\frac{\text{m}}{\text{s}}$$

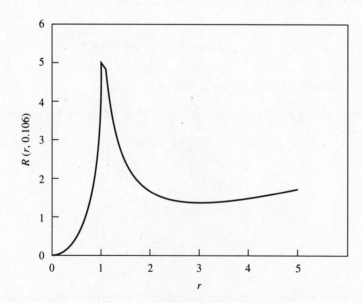

Fig. 3-22

3.31 Determine the form of $W(r, \zeta)$ such that $X/Y = W(r, \zeta)$ for the system of Fig. 3-23. What is W_{max}?

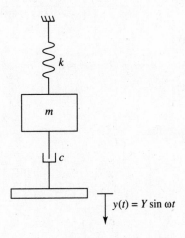

Fig. 3-23

Free body diagrams of the block are shown at an arbitrary instant in Fig. 3-24. Summation of forces

$$\left(\sum F\right)_{ext} = \left(\sum F\right)_{eff}$$

leads to

$$-kx - c(\dot{x} - \dot{y}) = m\ddot{x}$$

$$m\ddot{x} + c\dot{x} + kx = c\dot{y} = c\omega Y \cos(\omega t) \tag{3.55}$$

$$\ddot{x} + 2\zeta\omega_n\dot{x} + \omega_n^2 x = \frac{c}{m}\omega t \cos \omega t = 2\zeta\omega_n Y \cos \omega t$$

where

$$\omega_n = \sqrt{\frac{k}{m}} \qquad \zeta = \frac{c}{2m\omega_n}$$

Equation (3.55) is of the form of Eq. (3.4) with the excitation of Eq. (3.5) where

$$F_0 = c\omega Y \qquad \psi = \frac{\pi}{2}$$

Thus the steady-state amplitude is obtained using Eq. (3.10) as

$$\frac{m\omega_n^2 X}{c\omega Y} = M(r, \zeta)$$

$$\frac{m\omega_n^2 X}{2\zeta m\omega_n \omega Y} = M(r, \zeta)$$

$$\frac{\omega_n X}{2\zeta\omega Y} = M(r, \zeta)$$

$$\frac{X}{Y} = W(r, \zeta) = 2\zeta r M(r, \zeta) = \frac{2\zeta r}{\sqrt{(1 - r^2)^2 + (2\zeta r)^2}}$$

The value of r for which the maximum of W is obtained by setting $dW^2/dr = 0$. The quotient rule for differentiation is applied, giving

$$\frac{dW^2}{dr} = \frac{8\zeta^2 r[(1 - r^2)^2 + (2\zeta r)^2] - 4\zeta^2 r^2[2(1 - r^2)(-2r) + 2(2\zeta r)(2\zeta)]}{[(1 - r^2)^2 + (2\zeta r)^2]^2}$$

Setting $dW^2/dr = 0$ leads to

$$2 - 2r^4 = 0 \rightarrow r = 1$$

and

$$W_{\text{max}} = 1$$

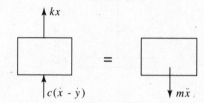

Fig. 3-24

3.32 Determine the steady-state amplitude of angular oscillation for the system of Fig. 3-25.

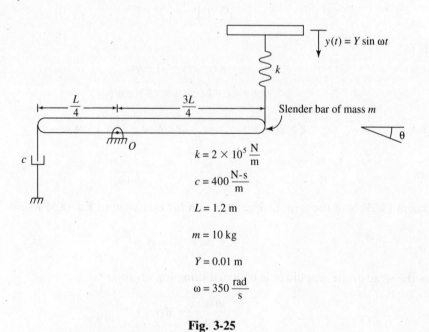

$$k = 2 \times 10^5 \, \frac{\text{N}}{\text{m}}$$

$$c = 400 \, \frac{\text{N-s}}{\text{m}}$$

$$L = 1.2 \, \text{m}$$

$$m = 10 \, \text{kg}$$

$$Y = 0.01 \, \text{m}$$

$$\omega = 350 \, \frac{\text{rad}}{\text{s}}$$

Fig. 3-25

Free body diagrams of the system at an arbitrary instant are shown in Fig. 3-26. Summing moments about 0,

$$-k\left(\frac{3}{4}L\theta - y\right)\left(\frac{3}{4}L\right) - \frac{1}{4}cL\dot{\theta}\left(\frac{L}{4}\right) = \frac{1}{12}mL^2\ddot{\theta} + \frac{1}{4}mL\ddot{\theta}\left(\frac{1}{4}L\right)$$

$$\frac{7}{48}mL^2\ddot{\theta} + \frac{1}{16}cL^2\dot{\theta} + \frac{9}{16}kL^2\theta = \frac{3}{4}kLy(t) = \frac{3}{4}kYL \sin \omega t$$

(3.56)

The natural frequency and damping ratio are

$$\omega_n = \sqrt{\frac{\frac{9}{16}kL^2}{\frac{7}{48}mL^2}} = \sqrt{\frac{27k}{7m}} = \sqrt{\frac{27\left(2\times10^5\ \frac{N}{m}\right)}{7(10\ \text{kg})}} = 277.8\ \frac{\text{rad}}{\text{s}}$$

$$2\zeta\omega_n = \frac{\frac{1}{16}cL^2}{\frac{7}{48}mL^2}$$

$$\zeta = \frac{3c}{14m\omega_n} = \frac{3\left(400\ \frac{\text{N-s}}{\text{m}}\right)}{14(10\ \text{kg})\left(277.7\ \frac{\text{rad}}{\text{s}}\right)} = 0.0309$$

Equation (3.56) is of the form of Eq. (3.4) with the excitation of Eq. (3.5) where

$$F_0 = \frac{3}{4}kLY = \frac{3}{4}\left(2\times10^5\ \frac{N}{m}\right)(1.2\ \text{m})(0.01\ \text{m}) = 1800\ \text{N-m}$$

$$m_{\text{eq}} = \tfrac{7}{48}mL^2 = \tfrac{7}{48}(10\ \text{kg})(1.2\ \text{m})^2 = 2.1\ \text{kg-m}^2$$

The frequency ratio for the system is

$$r = \frac{\omega}{\omega_n} = \frac{350\ \frac{\text{rad}}{\text{s}}}{277.8\ \frac{\text{rad}}{\text{s}}} = 1.26$$

The system's magnification factor is

$$M(1.26, 0.0309) = \frac{1}{\sqrt{[1-(1.26)^2]^2 + [2(0.0309)(1.26)]^2}} = 1.69$$

The steady-state amplitude is obtained using Eq. (3.10) as

$$\frac{m_{\text{eq}}\omega_n^2\Theta}{F_0} = M(1.26, 0.0309)$$

$$\Theta = \frac{F_0 M(1.26, 0.0309)}{m_{\text{eq}}\omega_n^2} = \frac{(1800\ \text{N-m})(1.69)}{(2.1\ \text{kg-m}^2)\left(277.8\ \frac{\text{rad}}{\text{s}}\right)^2}$$

$$= 0.0188\ \text{rad} = 1.08°$$

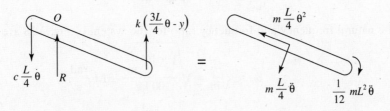

External forces   =   Effective forces

Fig. 3-26

3.33 Determine the steady-state amplitude for the machine in the system of Fig. 3-27.

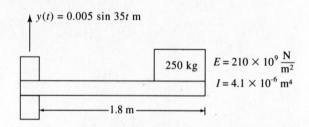

Fig. 3-27

The system is modeled as a 250-kg block attached through a spring of stiffness

$$k = \frac{3EI}{L^3} = \frac{3\left(210 \times 10^9 \ \frac{N}{m^2}\right)(4.1 \times 10^{-6} \ m^4)}{(1.8 \ m)^3} = 4.43 \times 10^5 \ \frac{N}{m}$$

to a support undergoing harmonic motion. The system is undamped with a natural frequency of

$$\omega_n = \sqrt{\frac{k}{m}} = \sqrt{\frac{4.43 \times 10^5 \ \frac{N}{m}}{250 \ kg}} = 42.1 \ \frac{rad}{s}$$

and frequency ratio

$$r = \frac{\omega}{\omega_n} = \frac{35 \ \frac{rad}{s}}{42.1 \ \frac{rad}{s}} = 0.831$$

The steady-state amplitude is

$$X = YT(0.831, 0) = \frac{0.005 \ m}{1 - (0.831)^2} = 0.0162 \ m$$

3.34 Approximate the steady-state amplitude of the block in the system of Fig. 3-28.

Fig. 3-28

The natural frequency and frequency ratio for the system of Fig. 3-28 are

$$\omega_n = \sqrt{\frac{k}{m}} = \sqrt{\frac{1 \times 10^5 \ \frac{N}{m}}{100 \ kg}} = 31.6 \ \frac{rad}{s}$$

$$r = \frac{\omega}{\omega_n} = \frac{40 \ \frac{rad}{s}}{31.6 \ \frac{rad}{s}} = 1.27$$

The system's force ratio is

$$\iota = \frac{\mu m g}{F_0} = \frac{0.08(100 \text{ kg})\left(9.81 \frac{\text{m}}{\text{s}^2}\right)}{300 \text{ N}} = 0.262$$

The magnification factor is determined using Eq. (3.43):

$$M = \sqrt{\frac{1 - \left(\frac{4\iota}{\pi}\right)^2}{(1 - r^2)^2}} = \sqrt{\frac{1 - \left[\frac{4(0.262)}{\pi}\right]^2}{[1 - (1.27)^2]^2}} = 1.538$$

The steady-state amplitude is

$$X = \frac{F_0 M}{m \omega_n^{\,2}} = \frac{(300 \text{ N})(1.538)}{(100 \text{ kg})\left(31.6 \frac{\text{rad}}{\text{s}}\right)^2} = 4.61 \text{ mm}$$

3.35 When a free vibration test is run on the system of Fig. 3-29, the ratio of amplitudes on successive cycles is 2.5 to 1. Determine the response of the machine due to a rotating unbalance of magnitude 0.25 kg-m when the machine operates at 2000 r/min and the damping is assumed to be viscous.

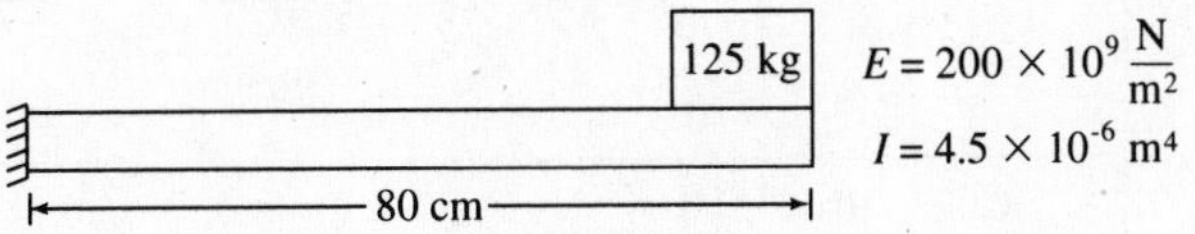

Fig. 3-29

The system's equivalent stiffness, natural frequency, and frequency ratio are

$$k = \frac{3EI}{L^3} = \frac{3\left(200 \times 10^9 \frac{\text{N}}{\text{m}^2}\right)(4.5 \times 10^{-6} \text{ m}^4)}{(0.8 \text{ m})^3} = 5.27 \times 10^6 \frac{\text{N}}{\text{m}}$$

$$\omega_n = \sqrt{\frac{k}{m}} = \sqrt{\frac{5.27 \times 10^6 \frac{\text{N}}{\text{m}}}{125 \text{ kg}}} = 205.3 \frac{\text{rad}}{\text{s}}$$

$$r = \frac{\omega}{\omega_n} = \frac{\left(2000 \frac{\text{r}}{\text{min}}\right)\left(2\pi \frac{\text{rad}}{\text{r}}\right)\left(\frac{1 \text{ min}}{60 \text{ s}}\right)}{205.4 \frac{\text{rad}}{\text{s}}} = 1.02$$

The logarithmic decrement for underdamped vibrations is

$$\delta = \ln(2.5) = 0.916$$

from which the viscous damping ratio is calculated as

$$\zeta = \frac{\delta}{\sqrt{4\pi^2 + \delta^2}} = \frac{0.916}{\sqrt{4\pi^2 + (0.916)^2}} = 0.144$$

Noting from Problem 3.15 that the rotating unbalance provides a frequency squared excitation with $A = m_0 e$, and using Eq. (3.17):

$$X = \frac{m_0 e}{m} \Lambda(1.02, 0.144)$$

$$= \frac{0.25 \text{ kg-m}}{125 \text{ kg}} \frac{(1.02)^2}{\sqrt{[1 - (1.02)^2]^2 + [2(0.144)(1.02)]^2}} = 7.02 \text{ mm}$$

3.36 Repeat Problem 3.35 if the damping is assumed to be hysteretic.

Equation (3.44) is used to determine the hysteretic damping coefficient from the logarithmic decrement

$$h = \frac{\delta}{\pi} = \frac{\ln{(2.5)}}{\pi} = 0.292$$

The steady-state amplitude is obtained using Eqs. (3.18) and (3.46) and calculated by

$$X = \frac{m_0 e}{m} \frac{r^2}{\sqrt{(1-r^2)^2 + h^2}}$$

$$= \frac{0.25 \text{ kg-m}}{125 \text{ kg}} \frac{(1.02)^2}{\sqrt{[1-(1.02)^2]^2 + (0.292)^2}} = 7.06 \text{ mm}$$

3.37 Determine the Fourier series representation for the periodic excitation of Fig. 3-30.

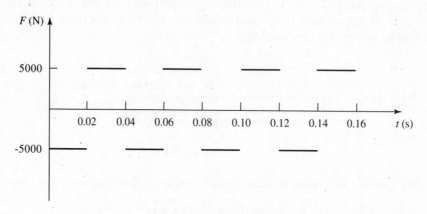

Fig. 3-30

The excitation of Fig. 3-30 is an odd excitation of a period 0.04 s. Thus $a_i = 0$, $i = 0, 1, 2, \ldots$. The Fourier sine coefficients are calculated by

$$b_i = \frac{2}{T} \int_0^T F(t) \sin{\frac{2\pi i}{T}} t \, dt$$

$$= \frac{2}{0.04} \left[\int_0^{0.02} (-5000) \sin{50\pi i t} \, dt + \int_{0.02}^{0.04} (5000) \sin{50\pi i t} \, dt \right]$$

$$= (50)(5000)\left(\frac{-1}{50\pi i}\right)[-\cos{\pi i} + \cos{0} + \cos{2\pi i} - \cos{\pi i}]$$

$$= -\frac{10,000}{\pi i}[1 - (-1)^i]$$

Thus the Fourier series representation for $F(t)$ is

$$F(t) = -\frac{10,000}{\pi} \sum_{i=1}^{\infty} \frac{1}{i}[(1) - (-1)^i] \sin{50\pi i t}$$

$$= -\frac{20,000}{\pi} \sum_{i=1,3,5,\ldots}^{\infty} \frac{1}{i} \sin{50\pi i t}$$

3.38 Determine the Fourier series representation for the excitation of Fig. 3-31.

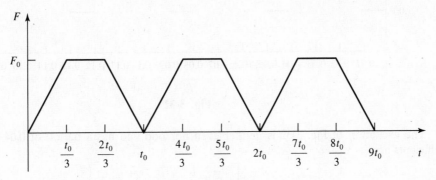

Fig. 3-31

The excitation of Fig. 3-31 is an even excitation of period t_0. Hence $b_i = 0$, $i = 1, 2, \ldots$. The Fourier cosine coefficients are

$$a_0 = \frac{2}{t_0} \int_0^{t_0} F(t)\, dt$$

$$= \frac{2}{t_0} \left[\int_0^{t_0/3} \left(\frac{3F_0}{t_0}\right) t\, dt + \int_{t_0/3}^{(2/3)t_0} F_0\, dt + \int_{(2/3)t_0}^{t_0} 3F_0\left(1 - \frac{t}{t_0}\right) dt \right]$$

$$= \frac{4}{3} F_0$$

$$a_i = \frac{2}{t_0} \int_0^{t_0} F(t) \cos \frac{2\pi i}{t_0} t\, dt$$

$$= \frac{2}{t_0} \left[\int_0^{(1/3)t_0} \left(\frac{3F_0}{t_0}\right) t \cos \frac{2\pi i}{t_0} t\, dt + \int_{(1/3)t_0}^{(2/3)t_0} F_0 \cos \frac{2\pi i}{t_0} t\, dt + \int_{(2/3)t_0}^{t_0} 3F_0\left(1 - \frac{t}{t_0}\right) \cos \frac{2\pi i}{t_0} t\, dt \right]$$

$$= \frac{3F_0}{i^2 \pi^2} \left(\frac{1}{2} \cos \frac{2\pi i}{3} + \frac{1}{2} \cos \frac{4\pi i}{3} - 1 \right)$$

$$= \begin{cases} -\dfrac{9F_0}{2i^2 \pi^2} & i = 1, 2, 4, 5, 7, 8, \ldots \\[2mm] 0 & i = 3, 6, 9, 12, \ldots \end{cases}$$

Thus the Fourier Series representation for $F(t)$ is

$$F(t) = \frac{2}{3} F_0 - \frac{9F_0}{2\pi^2} \sum_{i=1,2,4,5,7,8}^{\infty} \frac{1}{i^2} \cos \frac{2\pi i}{t_0} t$$

3.39 Determine the Fourier series representation for the excitation of Fig. 3-32.

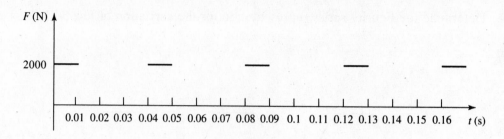

Fig. 3-32

The excitation of Fig. 3-32 is neither even nor odd and has a period of 0.04 s. The Fourier coefficients are

$$a_0 = \frac{2}{0.04}\left[\int_0^{0.01} (2000)\,dt + \int_{0.01}^{0.04} (0)\,dt\right]$$

$$= 1000 \text{ N}$$

$$a_i = \frac{2}{0.04}\left[\int_0^{0.01} (2000)\cos\frac{2\pi i}{0.04}t\,dt + \int_{0.01}^{0.04} (0)\cos\frac{2\pi i}{0.04}t\,dt\right]$$

$$= (50)(2000)\left(\frac{1}{50\pi i}\right)\left(\sin\frac{\pi}{2}i - \sin 0\right)$$

$$= \frac{2000}{\pi i}\sin\frac{\pi i}{2}$$

$$b_i = \frac{2}{0.04}\left[\int_0^{0.01} (2000)\sin\frac{2\pi i}{0.04}t\,dt + \int_{0.01}^{0.04} (0)\sin\frac{2\pi i}{t_0}t\,dt\right]$$

$$= (50)(2000)\left(-\frac{1}{50\pi i}\right)\left(\cos\frac{\pi}{2}i - \cos 0\right)$$

$$= \frac{2000}{\pi i}\left(1 - \cos\frac{\pi}{2}i\right)$$

Then

$$c_i = \sqrt{a_i^2 + b_i^2} = \frac{2000}{\pi i}\sqrt{\sin^2\frac{\pi}{2}i + \left(1 - \cos\frac{\pi}{2}i\right)^2}$$

$$= \frac{4000}{\pi i}\left|\sin\frac{\pi}{4}i\right|$$

and

$$\kappa_i = \tan^{-1}\left(\frac{\sin\frac{\pi}{2}i}{1 - \cos\frac{\pi}{2}i}\right)$$

Thus the Fourier series representation for $F(t)$ is

$$F(t) = 500 + \frac{4000}{\pi}\sum_{i=1}^{\infty}\frac{1}{i}\left|\sin\left(\pi\frac{i}{4}\right)\right|\sin(50\pi i + \kappa_i)t$$

3.40 A 200-kg press is subject to the time-dependent excitation of Problem 3.39 and Fig.

3-32. The machine sits on an elastic foundation of stiffness 1.8×10^7 N/m and damping
ratio 0.06. Determine the steady-state response of the machine, and approximate its
maximum displacement from equilibrium.

The natural frequency of the system is

$$\omega_n = \sqrt{\frac{k}{m}} = \sqrt{\frac{1.8 \times 10^7 \, \frac{\text{N}}{\text{m}}}{200 \text{ kg}}} = 300 \, \frac{\text{rad}}{\text{s}}$$

The system response is obtained using Eq. (3.39) as

$$x(t) = \frac{1}{1.8 \times 10^7} \left[500 + \frac{4000}{\pi} \sum_{i=1}^{\infty} \frac{1}{i} \left| \sin\left(\frac{\pi}{4} i\right) \right| M(r_i, \zeta) \sin(50\pi i + \kappa_i - \phi_i)t \right]$$

where

$$r_i = \frac{\omega_i}{\omega_n} = \frac{50\pi i}{300 \, \frac{\text{rad}}{\text{s}}}$$

and

$$M(r_i, \zeta) = \frac{1}{\sqrt{(1 - r_i^2)^2 + (0.12 r_i)^2}}$$

Table 3-1 illustrates the evaluation of the response. Then

$$x_{\text{max}} < \frac{1}{1.8 \times 10^7} \left[500 + \sum_{i=1}^{\infty} c_i M(r_i, \zeta) \right] = 3.34 \times 10^{-4} \text{ m}$$

Table 3-1

i	ω_i	r_i	c_i	M_i	$c_i M_i$	κ_i	ϕ_i
1	157.1	0.523	900.2	1.37	1233	0.785	0.086
2	314.1	1.047	636.6	6.31	4017	0	−0.915
3	471.2	1.57	300	0.68	202.7	−0.785	−0.128
4	628.3	2.094	0	0.29	0	1.571	−0.074
5	785.4	2.672	179.8	0.17	30.7	0.785	−0.054
6	942.5	3.141	212.0	0.11	23.9	0	−0.042
7	1099.6	3.665	128.6	0.080	10.3	−0.785	0.035
8	1256.6	4.188	0	0.06	0	1.571	−0.030

Supplementary Problems

3.41 A 100-kg machine is attached to a spring of stiffness 2×10^5 N/m and is subject to a harmonic
excitation of magnitude 700 N and period 0.1 s. What is the machine's amplitude of forced
vibration?

Ans. 3.59 mm

3.42 A 185-kg machine is attached to the midspan of a simply supported beam of length 1.5 m, elastic modulus 210×10^9 N/m², and cross-sectional moment of inertia 3×10^{-6} m⁴. What is the steady-state amplitude of the machine when it is subject to a harmonic excitation of magnitude 4×10^4 N and frequency 125 rad/s?

Ans.   6.59 mm

3.43   A 45-kg machine is to be placed at the end of a 2.5-m steel ($E = 210 \times 10^9$ N/m²) cantilever beam. The machine is to be subject to a harmonic excitation of magnitude 1000 N at 40 rad/s. For what values of the beam's cross-sectional moment of inertia will the machine's steady-state amplitude be limited to 15 mm?

Ans. $I < 1.32 \times 10^{-7}$ m⁴ or $I > 3.44 \times 10^{-6}$ m⁴

3.44   At what speeds will the steady-state amplitude of torsional oscillations of the disk of the system of Fig. 3-33 be less than 2°?

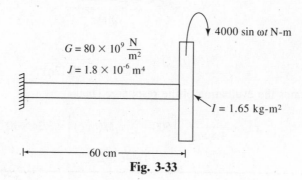

$G = 80 \times 10^9 \frac{N}{m^2}$

$J = 1.8 \times 10^{-6}$ m⁴

4000 sin ωt N-m

$I = 1.65$ kg-m²

60 cm

Fig. 3-33

Ans. $\omega < 275.1$ rad/s and $\omega > 463.4$ rad/s

3.45   When a 50-kg machine, placed on an undamped isolator, is subject to a harmonic excitation at 125 Hz, its steady-state amplitude is observed as 1.8 mm. When the machine is attached to two of these isolators in series and subjected to the same excitation, its steady-state amplitude is 1.2 mm. What is the stiffness of one of these isolators?

Ans. 1.54×10^7 N/m

3.46   What is the diameter of the shaft of Fig. 3-34 if, when subject to the harmonic excitation shown, beating occurs with a period of oscillation of 0.082 s?

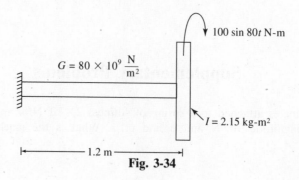

$G = 80 \times 10^9 \frac{N}{m^2}$

100 sin 80t N-m

$I = 2.15$ kg-m²

1.2 m

Fig. 3-34

Ans.   18.2 mm

3.47 Repeat Problem 3.41 as if the spring were in parallel with a viscous damper of damping coefficient 1200 N-s/m.

Ans. 3.35 mm

3.48 Repeat Problem 3.42 as if the beam had a viscous damping ratio of 0.05 and the excitation frequency was 200 rad/s.

Ans. 22.7 mm

3.49 For what excitation frequencies will the steady-state amplitude of the machine of Fig. 3-35 be less than 1.5 mm?

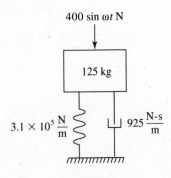

Fig. 3-35

Ans. $\omega < 18.7$ rad/s and $\omega > 67.5$ rad/s

3.50 If $\omega = 100$ rad/s and $\ell = 20$ cm, what is the steady-state amplitude of angular oscillation of the bar of Fig. 3-36?

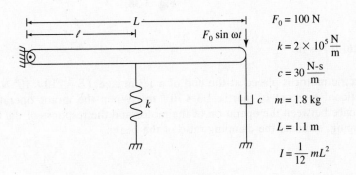

Fig. 3-36

Ans. 1.7°

3.51 If $\omega = 150$ rad/s, for what values of ℓ will the steady-state amplitude of the bar of Fig. 3.36 be 1°?

Ans. $\ell = 0.255$ m, 0.314 m

3.52 When the system of Fig. 3-37 is subjected to a harmonic excitation of magnitude 100 N but varying excitation frequencies, the maximum steady-state displacement of the machine is observed as 1.5 mm. What is the value of c?

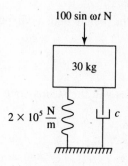

100 sin ωt N

30 kg

$2 \times 10^5 \dfrac{\text{N}}{\text{m}}$ c

Fig. 3-37

Ans. 827.9 N-s/m

3.53 For what values of c_t will the steady-state amplitude of the system of Fig. 3-38 be less than 1.5°?

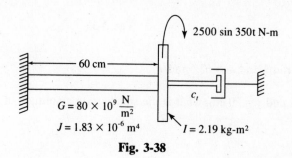

2500 sin 350t N-m

—— 60 cm ——

$G = 80 \times 10^9 \dfrac{\text{N}}{\text{m}^2}$

c_t

$J = 1.83 \times 10^{-6}$ m⁴ $I = 2.19$ kg-m²

Fig. 3-38

Ans. $c_t > 261.0$ N-s-m

3.54 A 65-kg electric motor is placed at the end of a 1.3-m steel ($E = 210 \times 10^9$ N/m²) cantilever beam of cross-sectional moment of inertia 1.3×10^{-6} m⁴. When the motor operates at 200 r/min, the phase difference between the operation of the motor and the response of the beam is 5°. Assuming viscous damping, estimate the damping ratio of the beam.

Ans. 0.146

3.55 Derive Eq. (3.19) from Eq. (3.18).

3.56 A 300-kg machine is attached to an elastic foundation of stiffness 3.1×10^6 N/m and damping ratio 0.06. When excited by a frequency squared excitation at very large speeds, the machine's steady-state amplitude is 10 mm. What is the maximum steady-state amplitude the machine would experience at lower speeds?

Ans. 83.5 mm

3.57 What is the steady-state amplitude of a 100-kg machine with a 0.25 kg-m rotating unbalance operating at 2000 r/min when the machine is placed on an isolator of stiffness 4.5×10^6 N/m and damping ratio 0.03?

Ans. 37.9 mm

3.58 As the operating speed of a 75-kg reciprocating machine with a rotating unbalance is increased, its steady-state amplitude approaches 1.78 mm. What is the magnitude of the rotating unbalance?

Ans. 0.134 kg-m

3.59 A 400-kg tumbler with a 0.45-kg-m rotating unbalance operates at speeds between 400 and 600 r/min. If the tumbler is placed on an elastic foundation of stiffness 1×10^6 N/m and damping ratio 0.1, what is the maximum steady-state amplitude of the tumbler over its operating range?

Ans. 5.65 mm

3.60 Repeat Problem 3.59 as if the tumbler's operating range were from 1000 to 1350 r/min.

Ans. 1.45 mm

3.61 For what speeds will the steady-state amplitude of the tumbler of Problem 3.59 be less than 1.9 mm?

Ans. $\omega < 40.3$ rad/s and $\omega > 77.2$ rad/s

3.62 Determine the required stiffness of an undamped elastic mounting for an 80-kg compressor with a 0.2-kg-m rotating unbalance such that its steady-state amplitude is less than 3.1 mm at all speeds between 300 and 600 r/min.

Ans. $k > 5.71 \times 10^5$ N/m and $k < 1.53 \times 10^4$ N/m

3.63 Repeat Problem 3.62 as if the mounting had a damping ratio of 0.07.

Ans. $k > 5.64 \times 10^5$ N/m and $k < 1.55 \times 10^4$ N/m

3.64 A 500-kg block is connected through a spring of stiffness 1.3×10^5 N/m in parallel with a viscous damper of damping coefficient 1800 N-s/m to a massless base. The base is given a prescribed harmonic displacement of amplitude 2 mm and frequency 15.0 rad/s. What is the steady-state amplitude of the block's displacement relative to the base?

Ans. 6.98 mm

3.65 Determine the steady-state amplitude of absolute acceleration of the block of Problem 3.64.

Ans. 1.85 m/s^2

3.66 A 300-kg vehicle traverses a road whose contour is approximately sinusoidal of amplitude 2.5 mm and period 2.6 m. Use the simplified suspension system model of Problem 3.28 with $k = 2.5 \times 10^5$ N/m and $\zeta = 0.3$ to predict the acceleration amplitude of the vehicle as it travels at 30 m/s.

Ans. 4.31 m/s^2

3.67   A 10-kg computer system, used for data acquisition and data reduction in a laboratory, is placed on a table which is bolted to the floor. Due to operation of rotating equipment, the floor has a vibration amplitude of 0.2 mm at a frequency of 30 Hz. If the table is modeled as a spring of stiffness 1.3×10^6 N/m with a damping ratio of 0.04, what is the steady-state acceleration amplitude of the computer?

   Ans.   9.77 m/s^2

3.68   If the table of Problem 3.67 is assumed to be rigid, what is the maximum stiffness of an undamped isolator placed between the computer and the table such that the steady-state amplitude of the computer is less than 6 m/s^2?

   Ans. 1.63×10^5 N/m

3.69   Determine the function $V(r, \zeta)$ such that $X/Y = V(r, \zeta)$ for the system of Fig. 3-39.

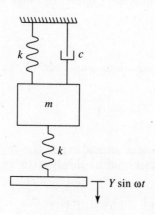

Fig. 3-39

   Ans.

$$V(r, \zeta) = \frac{1}{2} M(r, \zeta) \qquad \zeta = \frac{c}{2\sqrt{zmk}} \qquad r = \omega \sqrt{\frac{m}{2k}}$$

3.70   If the frequency of the base motion of the system of Fig. 3-39 and Problem 3.69 is varied, what is the maximum steady-state amplitude of the block?

   Ans.

$$\frac{Y}{4\zeta\sqrt{1 - \zeta^2}}$$

3.71   A 90-kg controller is placed at the end of a 1.5-m steel ($E = 210 \times 10^9$ N/m^2) cantilever beam ($I = 1.53 \times 10^{-6}$ m^4). The base of the beam is given a harmonic motion of amplitude 1.5 mm. For what frequencies will the controller's acceleration be limited to 12 m/s^2?

   Ans. $\omega > 72.5$ rad/s or $\omega < 47.7$ rad/s

3.72 Repeat Problem 3.42 as if the beam's amplitude of free vibrations decays to 1/3 of its value in 10 cycles, the damping is assumed to be hysteretic, and the excitation frequency is 200 rad/s.

Ans.   25.1 mm

3.73 A 120-kg machine is placed at the midspan of a 85-cm aluminum ($E = 100 \times 10^9$) N/m² simply supported beam ($I = 4.56 \times 10^{-6}$ m⁴). When the machine, which has a rotating unbalance of 0.68 kg-m, operates at 458 rad/s, its steady-state amplitude is measured as 13.2 mm. If the damping is assumed to be hysteretic, determine the beam's hysteretic damping coefficient.

Ans.   0.060

3.74 The steady-state amplitude of the system of Fig. 3-40 is 1.21 mm. What is the coefficient of friction between the block and the surface?

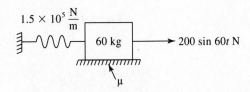

Fig. 3-40

Ans.   0.245

3.75 A 200-kg press is mounted on an elastic pad of stiffness 3.62×10^6 N/m and damping ratio 0.1. The press is used in a plant whose floor vibrations are measured as

$$y(t) = 0.0014 \sin 100t + 0.0006 \sin (200t - 0.12) \text{ m}$$

Determine the steady-state displacement of the press relative to the floor.

Ans.

$$0.00297 \sin (100t - 0.320) + 0.00048 \sin (200t - 2.78) \text{ m}$$

3.76 Determine the Fourier series representation for the periodic excitation of Fig. 3-41.

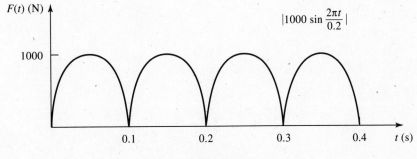

Fig. 3-41

Ans.

$$\frac{4000}{\pi} + \frac{4000}{\pi} \sum_{\ell=1}^{\infty} \frac{1}{1 - 4\ell^2} \cos 20\pi \ell t$$

3.77   Determine the Fourier series representation for the periodic excitation of Fig. 3-42.

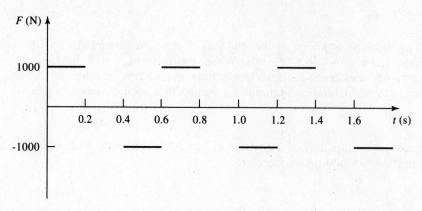

Fig. 3-42

Ans.

$$\frac{3000}{\pi} \sum_{\ell=1}^{\infty} \frac{1}{\ell} \sin \frac{10}{3} \pi \ell t$$

3.78   A 50-kg block is attached to a spring of stiffness 3.16×10^6 N/m in parallel with a viscous damper such that the system's damping ratio is 0.12. The block is excited by the periodic excitation of Fig. 3-42. Approximate the maximum displacement of the block in the steady-state.

Ans.   5.17 mm

Chapter 4

General Forced Response
of 1-Degree-of-Freedom Systems

4.1 GENERAL DIFFERENTIAL EQUATION

The general form of the differential equation governing the motion of a forced 1-degree-of-freedom system with viscous damping is

$$\ddot{x} + 2\zeta\omega_n\dot{x} + \omega_n^2 x = \frac{1}{m_{eq}}F(t) \tag{4.1}$$

4.2 CONVOLUTION INTEGRAL

The *convolution integral* provides the general solution of Eq. (4.1) subject to $x(0) = 0$ and $\dot{x}(0) = 0$. For an arbitrary $F(t)$, the convolution integral response is

$$x(t) = \int_0^t F(\tau)h(t - \tau)\,d\tau \tag{4.2}$$

where $h(t)$ is the response of the system due to a unit impulse applied at $t = 0$. For a system whose free vibrations are underdamped,

$$h(t) = \frac{1}{m_{eq}\omega_n}e^{-\zeta\omega_n t}\sin\omega_d t \tag{4.3}$$

where
$$\omega_d = \omega_n\sqrt{1 - \zeta^2} \tag{4.4}$$

is the damped natural frequency. Thus the response of an underdamped system is

$$x(t) = \frac{1}{m_{eq}\omega_d}\int_0^t F(\tau)e^{-\zeta\omega_n(t-\tau)}\sin\omega_d(t - \tau)\,d\tau \tag{4.5}$$

4.3 LAPLACE TRANSFORM SOLUTIONS

The Laplace transform of a function $x(t)$ is defined as

$$\mathcal{L}\{x(t)\} = \bar{x}(s) = \int_0^\infty e^{-st}x(t)\,dt \tag{4.6}$$

Tables of Laplace transforms and properties of Laplace transforms follows.

Table 4.1

Number	$f(t)$	$\bar{f}(s)$
1	1	$\dfrac{1}{s}$
2	t^n	$\dfrac{n!}{s^{n+1}}$
3	$e^{\alpha t}$	$\dfrac{1}{s-\alpha}$
4	$\sin \omega t$	$\dfrac{\omega}{s^2+\omega^2}$
5	$\cos \omega t$	$\dfrac{s}{s^2+\omega^2}$
6	$\delta(t-a)$	e^{-as}
7	$u(t-a)$	$\dfrac{e^{-as}}{s}$

Table 4.2

Property name	Formula
Definition of transform	$\bar{f}(s) = \displaystyle\int_0^\infty e^{-st}(t)\,dt$
Linearity	$\mathscr{L}\{\alpha f(t) + \beta g(t)\} = \alpha\bar{f}(s) + \beta\bar{g}(s)$
Transform of derivatives	$\mathscr{L}\left\{\dfrac{d^n f}{dt^n}\right\} = s^n\bar{f}(s) - s^{n-1}f(0) - \cdots - f^{(n-1)}(0)$
First shifting theorem	$\mathscr{L}\{e^{-at}f(t)\} = \bar{f}(s+a)$
Second shifting theorem	$\mathscr{L}\{f(t-a)u(t-a)\} = e^{-as}\bar{f}(s)$
Inverse transform	$f(t) = \dfrac{1}{2\pi i}\displaystyle\int_{\gamma-i\infty}^{\gamma+i\infty} \bar{f}(s)e^{st}\,ds$

The properties of the Laplace transform are used to transform Eq. (4.1) into an algebraic equation whose solution is

$$\bar{x}(s) = \frac{\dfrac{\bar{F}(s)}{m_{eq}} + (s + 2\zeta\omega_n)x(0) + \dot{x}(0)}{s^2 + 2\zeta\omega_n s + \omega_n^2} \tag{4.7}$$

Inverting Eq. (4.7) for $\zeta < 1$ leads to

$$x(t) = e^{-\zeta\omega_n t}\left[x(0)\cos\omega_d t + \frac{\dot{x}(0) + \zeta\omega_n x(0)}{\omega_d}\sin\omega_d t\right]$$

$$+ \frac{1}{m_{eq}}\mathscr{L}^{-1}\left\{\frac{\bar{F}(s)}{s^2 + 2\zeta\omega_n s + \omega_n^2}\right\} \tag{4.8}$$

4.4 UNIT IMPULSE FUNCTION AND UNIT STEP FUNCTION

The *unit impulse function* $\delta(t - t_0)$ is the mathematical representation of the force applied to a system resulting in a unit impulse applied to the system at $t = t_0$. Its mathematical definition is

$$\delta(t - t_0) = \begin{cases} 0 & t \neq t_0 \\ \infty & t = t_0 \end{cases} \tag{4.9}$$

but

$$\int_0^\infty \delta(t - t_0)\, dt = 1 \tag{4.10}$$

The *unit step function* $u(t - t_0)$ is related to the unit impulse function by

$$u(t - t_0) = \int_0^t \delta(\tau - t_0)\, d\tau \tag{4.11}$$

leading to

$$u(t - t_0) = \begin{cases} 0 & t \leq t_0 \\ 1 & t > t_0 \end{cases} \tag{4.12}$$

The unit step function may be used to develop a unified mathematical expression for an excitation force whose mathematical form changes at discrete times.

An important integral formula is

$$\int_0^t F(\tau)\delta(\tau - t_0)\, d\tau = F(t_0)u(t - t_0) \tag{4.13}$$

4.5 NUMERICAL METHODS

While the convolution integral provides a solution to Eq. (4.1) for an arbitrary $F(t)$, it is not always possible to evaluate the convolution integral in closed form. This is the case, for example, if $F(t)$ is known empirically, rather than by a mathematical expression. In these cases, the solution of Eq. (4.1) can be approximated using numerical methods.

One form of numerical approximation of the solution of Eq. (4.1) is numerical integration of the convolution integral. The function $F(t)$ can be interpolated by an interpolation function $\hat{F}(t)$ such that when $F(t)$ is replaced by $\hat{F}(t)$ in Eq. (4.1), the integral has a closed form evaluation. Often the interpolating function is defined piecewise. That is, its form changes at discrete values of time.

A second form of numerical approximation to the solution of Eq. (4.1) is direct numerical simulation of Eq. (4.1) using a self-starting method such as the Adams method or a Runge-Kutta method.

4.6 RESPONSE SPECTRUM

Let t_0 be a characteristic time in the definition of an excitation, and let F_0 be the maximum value of the excitation. The *response spectrum* is a plot of the nondimensional parameter $(m\omega_n^2 x_{max})/F_0$ versus the nondimensional parameter $(\omega_n t_0)/(2\pi)$. The response spectrum can be developed for any damping ratio.

Solved Problems

4.1 Use the convolution integral to determine the response of an undamped 1-degree-of-freedom system of natural frequency ω_n and mass m when subject to a constant force of magnitude F_0. The system is at rest in equilibrium at $t = 0$.

Substituting $F(t)$ into Eq. (4.5) with $\zeta = 0$ leads to

$$x(t) = \frac{1}{m\omega_n} \int_0^t F_0 \sin \omega_n(t - \tau)\, d\tau$$

$$= \frac{F_0}{m\omega_n^2} \cos \omega_n(t - \tau) \, \Big|_{\tau=0}^{\tau=t}$$

$$= \frac{F_0}{m\omega_n^2} (1 - \cos \omega_n t)$$

4.2 Use the convolution integral to determine the response of an underdamped 1-degree-of-freedom system of natural frequency ω_n, damping ratio ζ, and mass m when subject to a constant force of magnitude F_0. The system is at rest in equilibrium at $t = 0$.

Substituting for $F(t)$ in Eq. (4.5) leads to

$$x(t) = \frac{1}{m\omega_d} \int_0^t F_0 e^{-\zeta\omega_n(t-\tau)} \sin \omega_d(t - \tau)\, d\tau$$

Let $v = t - \tau$. Then

$$x(t) = \frac{F_0}{m\omega_d} \int_{v=t}^{v=0} e^{-\zeta\omega_n v} \sin \omega_d v \,(-dv)$$

$$= -\frac{F_0}{m\omega_d\omega_n^2} e^{-\zeta\omega_n v}(\zeta\omega_n \sin \omega_d v + \omega_d \cos \omega_d v) \, \Big|_{v=0}^{v=t}$$

$$= \frac{F_0}{m\omega_n^2} \left[1 - e^{-\zeta\omega_n t}\left(\frac{\zeta}{\sqrt{1 - \zeta^2}} \sin \omega_d t + \cos \omega_d t \right) \right]$$

4.3 Use the convolution integral to determine the response of an undamped 1-degree-of-freedom system of natural frequency ω_n and mass m when subject to a time-dependent excitation of the form $F(t) = F_0 e^{-\alpha t}$. The system is at rest in equilibrium at $t = 0$.

Mathcad

Substituting for $F(t)$ in Eq. (4.5) leads to

$$x(t) = \frac{1}{m\omega_n} \int_0^t F_0 e^{-\alpha\tau} \sin \omega_n(t - \tau)\, d\tau$$

Let $v = t - \tau$. Then

$$x(t) = \frac{F_0}{m\omega_n} \int_{v=t}^{v=0} e^{-\alpha(t-v)} \sin \omega_n v \,(-dv)$$

$$= \frac{F_0 e^{-\alpha t}}{m\omega_n} \int_{v=0}^{v=t} e^{\alpha v} \sin \omega_n v \, dv$$

$$= \frac{F_0}{m\omega_n(\alpha^2 + \omega_n^2)} e^{\alpha v}(\alpha \sin \omega_n v - \omega_n \cos \omega_n v) \, \Big|_{v=0}^{v=t}$$

$$= \frac{F_0}{m\omega_n(\alpha^2 + \omega_n^2)} (\alpha \sin \omega_n t - \omega_n \cos \omega_n t + \omega_n e^{-\alpha t})$$

4.4   Use the convolution integral to determine the time-dependent response of an undamped 1-degree-of-freedom system of natural frequency ω_n and mass m when subject to a harmonic excitation of the form $F(t) = F_0 \sin \omega t$ with $\omega \neq \omega_n$.

Substituting for $F(t)$ in Eq. (4.5) with $\zeta = 0$ leads to

$$x(t) = \frac{1}{m\omega_n} \int_0^t F_0 \sin \omega\tau \sin \omega_n(t - \tau)\, d\tau$$

Use of a trigonometric identity for the product of sine functions of different arguments leads to

$$x(t) = \frac{1}{2m\omega_n} \int_0^t F_0\{\cos\left[(\omega + \omega_n)\tau - \omega_n t\right] - \cos\left[(\omega - \omega_n)\tau + \omega_n t\right]\}\, d\tau$$

$$= \frac{F_0}{2m\omega_n}\left\{\frac{1}{\omega + \omega_n}\sin\left[(\omega + \omega_n)\tau - \omega_n t\right] - \frac{1}{\omega - \omega_n}\sin\left[(\omega - \omega_n)\tau + \omega_n t\right]\right\}\Bigg|_{\tau=0}^{\tau=t}$$

$$= \frac{F_0}{2m\omega_n}\left\{\frac{1}{\omega + \omega_n}\left[\sin \omega t + \sin \omega_n t\right] - \frac{1}{\omega - \omega_n}\left[\sin \omega t - \sin \omega_n t\right]\right\}$$

$$= \frac{F_0}{m\omega_n(\omega^2 - \omega_n^2)}(\omega \sin \omega_n t - \omega_n \sin \omega t)$$

4.5   The differential equation governing the motion of the system of Fig. 4-1 is

$$\tfrac{7}{48}mL^2\ddot{\theta} + \tfrac{9}{16}kL^2\theta = \tfrac{3}{4}LF(t)$$

Determine the time-dependent response of the system if $F(t) = F_0 e^{-\alpha t}$.

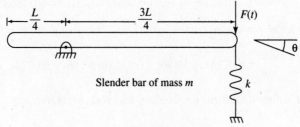

Fig. 4-1

The natural frequency of the system is

$$\omega_n = \sqrt{\frac{\frac{9}{16}kL^2}{\frac{7}{48}mL^2}} = \sqrt{\frac{27k}{7m}}$$

The governing differential equation can be put into the form of Eq. (4.1) with

$$m_{eq} = \tfrac{7}{48}mL^2 \qquad F(t) = \tfrac{3}{4}LF_0 e^{-\alpha t}$$

Substituting into the convolution integral solution, Eq. (4.5), leads to

$$\theta(t) = \frac{1}{\frac{7}{48} mL^2 \omega_n} \int_0^t \frac{3}{4} LF_0 e^{-\alpha\tau} \sin \omega_n(t-\tau)\, d\tau$$

$$= \frac{36}{7} \frac{F_0}{mL\omega_n} \int_0^t e^{-\alpha\tau} \sin \omega_n(t-\tau)\, d\tau$$

Performing the integration as in Problem 4.3 leads to

$$\theta(t) = \frac{36}{7} \frac{F_0}{mL\omega_n(\alpha^2 + \omega_n{}^2)} (\alpha \sin \omega_n t - \omega_n \cos \omega_n t + \omega_n e^{-\alpha t})$$

4.6 For the system of Problem 4.5, if $m = 20$ kg, $L = 1.4$ m, $k = 1.4 \times 10^4$ N/m, $F_0 = 100$ N, and $\alpha = 12$ (s)$^{-1}$, determine the maximum angular displacement of the bar from its equilibrium position.

The natural frequency of the system is

$$\omega_n = \sqrt{\frac{27k}{7m}} = \sqrt{\frac{27\left(1.4 \times 10^4 \frac{\text{N}}{\text{m}}\right)}{7(20 \text{ kg})}} = 51.96 \frac{\text{rad}}{\text{s}}$$

Substituting given values into the solution obtained in Problem 4.6 leads to

$$\theta(t) = \frac{36(100 \text{ N})}{7(20 \text{ kg})(1.4 \text{ m})\left(51.96 \frac{\text{rad}}{\text{s}}\right)\left[(12.0 \text{ s}^{-1})^2 + \left(51.96 \frac{\text{rad}}{\text{s}}\right)^2\right]}$$

$$\times (12.0 \sin 51.96t - 51.96 \cos 51.96t + 51.96 e^{-12t})$$

$$= 0.00149 \sin 51.96t - 0.00645 \cos 51.96t + 0.00645 e^{-12t}$$

The time at which the maximum occurs is obtained by setting $d\theta/dt = 0$. To this end

$$\frac{d\theta}{dt} = 0.0774 \cos 51.96t + 0.335 \sin 51.96t - 0.0774 e^{-12t} = 0$$

A trial-and-error solution leads to $t = 0.0538$ s and $\theta_{max} = 0.00996$ rad.

4.7 Use the convolution integral to determine the response of an undamped 1-degree-of-freedom system to the excitation of Fig. 4-2.

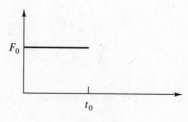

Fig. 4-2

For $t < t_0$,

$$x(t) = \frac{1}{m\omega_n} \int_0^t F_0 \sin \omega_n(t - \tau)\, d\tau$$

$$= \frac{F_0}{m\omega_n^2}(1 - \cos \omega_n t)$$

For $t > t_0$,

$$x(t) = \frac{1}{m\omega_n} \int_0^{t_0} F_0 \sin \omega_n(t - \tau)\, d\tau$$

$$= \frac{F_0}{m\omega_n^2}[\cos \omega_n(t - t_0) - \cos \omega_n t]$$

4.8   Use the convolution integral to determine the response of an undamped 1-degree-of-freedom system due to the triangular pulse of Fig. 4-3.

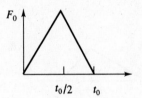

F_0

$t_0/2$ t_0

Fig. 4-3

For $t < t_0/2$,

$$x(t) = \frac{1}{m\omega_n} \int_0^t \left(\frac{2F_0}{t_0}\right)\tau \sin \omega_n(t - \tau)\, d\tau$$

$$= \frac{2F_0}{m\omega_n^2 t_0}\left(t - \frac{1}{\omega_n}\sin \omega_n t\right)$$

For $t_0/2 \le t < t_0$,

$$x(t) = \frac{1}{m\omega_n}\left[\int_0^{t_0/2}\left(2\frac{F_0}{t_0}\right)\tau \sin \omega_n(t - \tau)\, d\tau + \int_{t_0/2}^t 2F_0\left(1 - \frac{\tau}{t_0}\right)\sin \omega_n(t - \tau)\, d\tau\right]$$

$$= \frac{2F_0}{m\omega_n^2 t_0}\left\{1 - \frac{t}{t_0} + \frac{1}{\omega_n t_0}\left[2\sin \omega_n\left(t - \frac{1}{2}t_0\right) - \sin \omega_n t\right]\right\}$$

For $t \ge t_0$,

$$x(t) = \frac{1}{m\omega_n}\left[\int_0^{t_0/2}\left(\frac{2F_0}{t_0}\right)\tau \sin \omega_n(t - \tau)\, d\tau + \int_{t_0/2}^{t_0} 2F_0\left(1 - \frac{\tau}{t_0}\right)\sin \omega_n(t - \tau)\, d\tau\right]$$

$$= \frac{2F_0}{m\omega_n^3 t_0}\left[2\sin \omega_n\left(t - \frac{1}{2}t_0\right) - \sin \omega_n t - \sin \omega_n(t - t_0)\right]$$

4.9   Develop a unified mathematical expression for the triangular pulse of Fig. 4-3 using unit step functions.

     The graphical breakdown of the triangular pulse is shown in Fig. 4-4. Using the graphical breakdown and the definition of unit step functions,

$$x(t) = 2F_0 \frac{t}{t_0} u(t) - 2F_0 \frac{t}{t_0} u\left(t - \frac{1}{2} t_0\right) + 2F_0\left(1 - \frac{t}{t_0}\right) u\left(t - \frac{1}{2} t_0\right) - 2F_0\left(1 - \frac{t}{t_0}\right) u(t - t_0)$$

$$= 2F_0\left[\frac{t}{t_0} u(t) + \left(1 - 2\frac{t}{t_0}\right) u\left(t - \frac{1}{2} t_0\right) - \left(1 - \frac{t}{t_0}\right) u(t - t_0)\right]$$

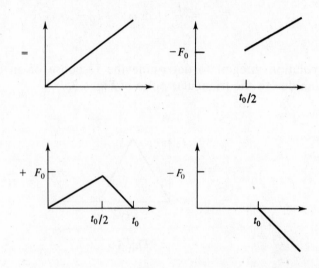

Fig. 4-4

4.10   Show that

$$\int_0^t u(\tau - t_0) g(t, \tau)\, d\tau = u(t - t_0) \int_{t_0}^t g(t, \tau)\, d\tau$$

     Note that for $\tau < t_0$, $u(\tau - t_0) = 0$. Thus for $t < t_0$, the integrand is identically zero. Then for $t > t_0$,

$$\int_0^t u(\tau - t_0) g(t, \tau)\, d\tau = \int_0^{t_0} u(\tau - t_0) g(t, \tau)\, d\tau + \int_{t_0}^t u(\tau - t_0) g(t, \tau)\, d\tau$$

$$= \int_{t_0}^t g(t, \tau)\, d\tau$$

Thus

$$\int_0^t u(\tau - t_0) g(t, \tau)\, d\tau = \begin{cases} 0 & t < t_0 \\ \int_{t_0}^t g(t, \tau)\, d\tau & t > t_0 \end{cases} = u(t - t_0) \int_{t_0}^t g(t, \tau)\, d\tau$$

4.11   Use the results of Problems 4.9 and 4.10 to develop a unified mathematical expression for the response of an undamped 1-degree-of-freedom system due to the triangular pulse of Fig. 4-3.

Substitution of the unified mathematical representation of the triangular pulse developed in Problem 4.9 into the convolution integral, Eq. (4.5) with $\zeta = 0$, leads to

$$x(t) = \frac{1}{m\omega_n} \int_0^t 2F_0\left[\frac{\tau}{t_0}u(\tau) + \left(1 - 2\frac{\tau}{t_0}\right)u\left(\tau - \frac{1}{2}t_0\right) - \left(1 - \frac{\tau}{t_0}\right)u(\tau - t_0)\right] \sin \omega_n(t - \tau)\, d\tau$$

Using the integral formula of Problem 4.10,

$$x(t) = \frac{2F_0}{m\omega_n}\left[u(t)\int_0^t \frac{\tau}{t_0}\sin \omega_n(t - \tau)\, d\tau + u\left(t - \frac{1}{2}t_0\right)\int_{(1/2)t_0}^t \left(1 - 2\frac{\tau}{t_0}\right)\sin \omega_n(t - \tau)\, d\tau\right.$$

$$\left. - u(t - t_0)\int_{t_0}^t \left(1 - \frac{\tau}{t_0}\right)\sin \omega_n(t - \tau)\, d\tau\right]$$

Evaluation of the integrals yields

$$x(t) = \frac{2F_0}{m\omega_n^2}\left\{u(t)\left(\frac{t}{t_0} - \frac{1}{\omega_n t_0}\sin \omega_n t\right) + u\left(t - \frac{1}{2}t_0\right)\left[1 - 2\frac{t}{t_0} + \frac{2}{\omega_n t_0}\sin \omega_n\left(t - \frac{1}{2}t_0\right)\right]\right.$$

$$\left. - u(t - t_0)\left[1 - \frac{t}{t_0} + \frac{1}{\omega_n t_0}\sin \omega_n(t - t_0)\right]\right\}$$

4.12 Use the convolution integral and unit step functions to develop a unified mathematical expression for the response of an undamped 1-degree-of-freedom system to the excitation of Fig. 4-5.

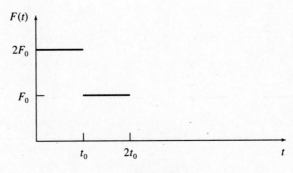

Fig. 4-5

The unified mathematical representation of the excitation of Fig. 4-5 is

$$F(t) = 2F_0[u(t) - u(t - t_0)] + F_0[u(t - t_0) - u(t - 2t_0)]$$

$$= 2F_0 u(t) - F_0 u(t - t_0) - F_0 u(t - 2t_0)$$

Substitution into the convolution integral, Eq. (4.5), with $\zeta = 0$ leads to

$$x(t) = \frac{F_0}{m\omega_n} \left\{ \int_0^t [2u(\tau) - u(\tau - t_0) - u(\tau - 2t_0)] \sin \omega_n(t - \tau) \, d\tau \right\}$$

$$= \frac{F_0}{m\omega_n} \left[2u(t) \int_0^t \sin \omega_n(t - \tau) \, d\tau - u(t - t_0) \int_{t_0}^t \sin \omega_n(t - \tau) \, d\tau - u(t - 2t_0) \int_{2t_0}^t \sin \omega_n(t - \tau) \, d\tau \right]$$

$$x(t) = \frac{F_0}{m\omega_n^2} \{ u(t)(1 - \cos \omega_n t) - u(t - t_0)[1 - \cos \omega_n(t - t_0)] - u(t - 2t_0)[1 - \cos \omega_n(t - 2t_0)] \}$$

4.13 Determine a unified mathematical expression for the response of an undamped 1-degree-of-freedom system of $\omega_n = 100$ rad/s and a mass of 10 kg subject to a rectangular pulse of magnitude 2000 N and duration 0.1 s followed by an impulse of magnitude 200 N-s applied at $t = 0.25$ s.

The mathematical representation of the excitation is

$$F(t) = 2000[u(t) - u(t - 0.1)] + 20\delta(t - 0.25)$$

Substitution into Eq. (4.5) with $\zeta = 0$ leads to

$$x(t) = \frac{1}{(10 \text{ kg})\left(100 \frac{\text{rad}}{\text{s}}\right)} \left\{ 2000 \int_0^t [u(\tau) - u(\tau - 0.1)] \sin 100(t - \tau) \, d\tau \right.$$

$$\left. + 20 \int_0^t \delta(\tau - 0.25) \sin 100(t - \tau) \, d\tau \right\}$$

$$= 2u(t) \int_0^t \sin 100(t - \tau) \, d\tau - 2u(t - 0.1) \int_{0.1}^t \sin 100(t - \tau) \, d\tau + 0.02 \sin 100(t - 0.25)u(t - 0.25)$$

$$= 0.02u(t)(1 - \cos 100t) - 0.02u(t - 0.1)[1 - \cos (100t - 10)] + 0.02u(t - 0.25) \sin (100t - 25)$$

4.14 Use unit step functions to develop an infinite series representation for the periodic function of Fig. 4-6.

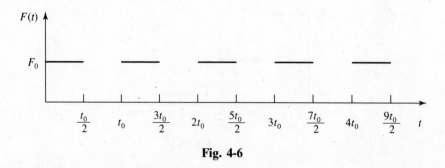

Fig. 4-6

The graphical breakdown of the excitation of Fig. 4-6 is shown in Fig. 4-7. The representation of $F(t)$ in terms of unit step functions is

$$F(t) = F_0[u(t) - u(t - \tfrac{1}{2}t_0)] + F_0[u(t - t_0) - u(t - \tfrac{3}{2}t_0)] + F_0[u(t - 2t_0) - u(t - \tfrac{5}{2}t_0)] + \cdots +$$

which can be written as

$$F(t) = F_0 \sum_{i=0}^{\infty} \{ u(t - it_0) - u[t - \tfrac{1}{2}(2i - 1)t_0] \}$$

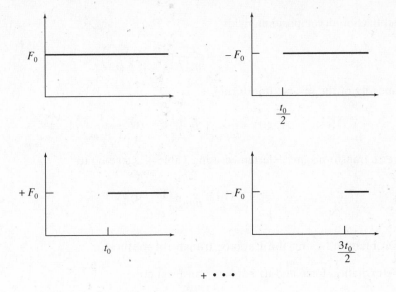

Fig. 4-7

4.15 Use the convolution integral to determine a mathematical representation for the response of an undamped 1-degree-of-freedom system due to the periodic excitation of Fig. 4-6.

Substitution of the mathematical form of the excitation developed in Problem 4.14 into the convolution integral, Eq. (4.5), with $\zeta = 0$ leads to

$$x(t) = \frac{1}{m\omega_n} \int_0^t F_0 \sum_{i=1}^{\infty} \left\{ u(\tau - it_0) - u\left[\tau - \frac{1}{2}(2i-1)t_0\right] \right\} \sin \omega_n(t-\tau)\, d\tau$$

The order of integration and summation can be interchanged assuming the infinite series converges for all t. This leads to

$$x(t) = \frac{F_0}{m\omega_n} \sum_{i=0}^{\infty} \left\{ \int_0^t u(\tau - it_0) \sin \omega_n(t-\tau)\, d\tau - \int_0^t u\left[\tau - \frac{1}{2}(2i-1)t_0\right] \sin \omega_n(t-\tau)\, d\tau \right\}$$

The integrals are evaluated using the integral formula developed in Problem 4.10:

$$x(t) = \frac{F_0}{m\omega_n^2} \sum_{i=0}^{\infty} \left\{ u(t-it_0) \cos \omega_n(t-\tau) \Big|_{\tau=it_0}^{\tau=t} - u\left[t - \frac{1}{2}(2i-1)t_0\right] \cos \omega_n(t-\tau) \Big|_{\tau=\frac{1}{2}(2i-1)t_0}^{\tau=t} \right\}$$

$$= \frac{F_0}{m\omega_n^2} \sum_{i=0}^{\infty} \left\{ u(t-it_0)[1 - \cos \omega_n(t-it_0)] - u\left[t - \frac{1}{2}(2i-1)t_0\right]\left\{1 - \cos \omega_n\left[t - \frac{1}{2}(2i-1)t_0\right]\right\} \right\}$$

4.16 Solve Problem 4.1 using the Laplace transform method.

The excitation force and its Laplace transform are

$$F(t) = F_0$$

$$\bar{F}(s) = \frac{F_0}{s}$$

Substituting into Eq. (4.7) with $x(0) = 0$ and $\dot{x}(0) = 0$ and $\zeta = 0$ leads to

$$\bar{x}(s) = \frac{F_0}{m} \frac{1}{s(s^2 + \omega_n^2)}$$

A partial fraction decomposition yields

$$\bar{x}(s) = \frac{F_0}{m\omega_n^2}\left(\frac{1}{s} - \frac{s}{s^2 + \omega_n^2}\right)$$

Using linearity of the inverse transform,

$$x(t) = \frac{1}{m\omega_n^2}\left(\mathscr{L}^{-1}\left\{\frac{1}{s}\right\} - \mathscr{L}^{-1}\left\{\frac{1}{s^2 + \omega_n^2}\right\}\right)$$

The inverse transforms are determined using Table 4.1, leading to

$$x(t) = \frac{1}{m\omega_n^2}(1 - \cos\omega_n t)$$

4.17 Solve Problem 4.2 using the Laplace transform method.

The excitation force and its Laplace transform are

$$F(t) = F_0$$

$$\bar{F}(s) = \frac{F_0}{s}$$

Substitution into Eq. (4.7) with $x(0) = 0$ and $\dot{x}(0) = 0$ leads to

$$\bar{x}(s) = \frac{F_0}{m}\frac{1}{s(s^2 + 2\zeta\omega_n s + \omega_n^2)}$$

A partial fraction decomposition leads to

$$\bar{x}(s) = \frac{1}{m\omega_n^2}\left(\frac{1}{s} - \frac{s + 2\zeta\omega_n}{s^2 + 2\zeta\omega_n s + \omega_n^2}\right)$$

Completing the square of the quadratic denominator and using linearity of the inverse transform leads to

$$x(t) = \frac{1}{m\omega_n^2}\left(\mathscr{L}^{-1}\left\{\frac{1}{s}\right\} - \mathscr{L}^{-1}\left\{\frac{s + \zeta\omega_n}{(s + \zeta\omega_n)^2 + \omega_d^2}\right\} - \zeta\frac{\omega_n}{\omega_d}\mathscr{L}^{-1}\left\{\frac{\omega_d}{(s + \zeta\omega_n)^2 + \omega_d^2}\right\}\right)$$

The first shifting theorem of Table 4-2 is used to obtain

$$x(t) = \frac{1}{m\omega_n^2}\left[1 - e^{-\zeta\omega_n t}\left(\mathscr{L}^{-1}\left\{\frac{s}{s^2 + \omega_d^2}\right\} + \zeta\frac{\omega_n}{\omega_d}\mathscr{L}^{-1}\left\{\frac{\omega_d}{s^2 + \omega_d^2}\right\}\right)\right]$$

The transform pairs of Table 4-1 are used to obtain

$$x(t) = \frac{1}{m\omega_n^2}\left[1 - e^{-\zeta\omega_n t}\left(\cos\omega_d t + \zeta\frac{\omega_n}{\omega_d}\sin\omega_d t\right)\right]$$

4.18 Solve Problem 4.5 using the Laplace transform method.

The differential equation of Problem 4.5 can be rewritten as

$$\ddot{\theta} + \omega_n^2\theta = \frac{36F_0}{7mL}e^{-\alpha t}$$

where

$$\omega_n = \sqrt{\frac{27k}{7m}}$$

Assuming the system is at rest in equilibrium at $t = 0$ and taking the Laplace transform of the differential equation leads to

$$\bar{\theta}(s) = \frac{36F_0}{7mL} \frac{1}{(s + \alpha)(s + \omega_n^2)}$$

Partial fraction decomposition yields

$$\bar{\theta}(s) = \frac{36F_0}{7mL(\alpha^2 + \omega_n^2)} \left(\frac{1}{s + \alpha} + \frac{\alpha - s}{s^2 + \omega_n^2} \right)$$

Use of linearity of the inverse transform leads to

$$\theta(t) = \frac{36F_0}{7mL(\alpha^2 + \omega_n^2)} \left(\mathscr{L}^{-1}\left\{\frac{1}{s + \alpha}\right\} + \frac{\alpha}{\omega_n} \mathscr{L}^{-1}\left\{\frac{\omega_n}{s^2 + \omega_n^2}\right\} - \mathscr{L}^{-1}\left\{\frac{s}{s^2 + \omega_n^2}\right\} \right)$$

Use of Table 4-1 leads to

$$\theta(t) = \frac{12F_0}{7mL(\alpha^2 + \omega_n^2)} \left(e^{-\alpha t} + \frac{\alpha}{\omega_n} \sin \omega_n t - \cos \omega_n t \right)$$

4.19 Use the Laplace transform method to determine the response of an undamped 1-degree-of-freedom system of natural frequency ω_n and mass m, initially at rest in equilibrium and subject to the triangular pulse of Fig. 4-3.

From the results of Problem 4.5, the mathematical expression for the triangular pulse is

$$F(t) = 2F_0 \left[\frac{t}{t_0} u(t) + \left(1 - 2\frac{t}{t_0}\right)u\left(t - \frac{1}{2}t_0\right) - \left(1 - \frac{t}{t_0}\right)u(t - t_0) \right]$$

and its Laplace transform is obtained using the second shifting theorem and Table 4-1 as

$$\bar{F}(s) = 2\frac{F_0}{t_0} \left(\frac{1}{s^2} - \frac{2}{s^2} e^{-s(t_0/2)} + \frac{1}{s^2} e^{-st_0} \right)$$

Substitution into Eq. (4.7) leads to

$$\bar{x}(s) = \frac{2F_0}{mt_0} \frac{1 - 2e^{-s(t_0/2)} + e^{-st_0}}{s^2(s^2 + \omega_n^2)}$$

A partial fraction decomposition yields

$$\bar{x}(s) = \frac{2F_0}{m\omega_n^2 t_0} (1 - 2e^{-s(t_0/2)} + e^{-st_0}) \left(\frac{1}{s^2} - \frac{1}{s^2 + \omega_n^2} \right)$$

The system response is obtained by application of the second shifting theorem and the transform pairs of Table 4-1. Thus

$$x(t) = \frac{2F_0}{m\omega_n^2 t_0} \left\{ t - \frac{1}{\omega_n} \sin \omega_n t - 2u\left(t - \frac{1}{2}t_0\right)\left[\left(t - \frac{1}{2}t_0\right) - \frac{1}{\omega_n} \sin \omega_n\left(t - \frac{1}{2}t_0\right)\right] \right.$$
$$\left. + u(t - t_0)\left[(t - t_0) - \frac{1}{\omega_n} \sin \omega_n(t - t_0)\right] \right\}$$

4.20 Use the Laplace transform method to determine the response of an undamped 1-degree-of-freedom system of natural frequency ω_n and mass m, initially at rest in equilibrium and subject to the periodic excitation of Fig. 4-6.

From Problem 4.16, the mathematical representation of the periodic function of Fig. 4-6 is

$$F(t) = F_0 \sum_{i=0}^{\infty} \left[u(t - it_0) - u\left(t - \frac{1}{2}(2i-1)t_0\right) \right]$$

The second shifting theorem is used to obtain

$$\bar{F}(s) = \frac{F_0}{s} \sum_{u=0}^{\infty} \left(e^{-ist_0} - e^{-\frac{1}{2}(2i-1)st_0} \right)$$

Substitution into Eq. (4.7) with $x(0) = 0$, $\dot{x}(0) = 0$, and $\zeta = 0$ leads to

$$\bar{x}(s) = \frac{F_0}{m} \frac{1}{s(s^2 + \omega_n^2)} \sum_{i=0}^{\infty} \left[e^{-ist_0} - e^{-\frac{1}{2}(2i-1)st_0} \right]$$

$$= \frac{F_0}{m\omega_n^2} \left(\frac{1}{s} - \frac{s}{s^2 + \omega_n^2} \right) \sum_{i=0}^{\infty} \left[e^{-ist_0} - e^{-\frac{1}{2}(2i-1)st_0} \right]$$

Inversion of the transform is performed using the second shifting theorem:

$$x(t) = \frac{F_0}{m\omega_n^2} \sum_{i=0}^{\infty} \left\{ [1 - \cos \omega_n(t - it_0)]u(t - it_0) \right\} - \left\{ 1 - \cos \omega_n \left[1 - \frac{1}{2}(2i-1)t_0 \right] \right\} u \left[t - \frac{1}{2}(2i-1)t_0 \right]$$

4.21 Use the Laplace transform method to determine the response of an underdamped 1-degree-of-freedom system of damping ratio ζ, natural frequency ω_n, and mass m, initially at rest in equilibrium and subject to a series of applied impulses, each of magnitude I, beginning at $t = 0$, and each a time t_0 apart.

The mathematical form of the excitation and its Laplace transform are

$$F(t) = I \sum_{i=0}^{\infty} \delta(t - it_0)$$

$$\bar{F}(s) = I \sum_{i=0}^{\infty} e^{-ist_0}$$

Substitution into Eq. (4.7) with $x(0) = 0$ and $\dot{x}(0) = 0$ leads to

$$\bar{x}(s) = \frac{I}{m} \frac{1}{s^2 + 2\zeta\omega_n s + \omega_n^2} \sum_{i=0}^{\infty} e^{-ist_0}$$

$$= \frac{I}{m} \frac{1}{(s + \zeta\omega_n)^2 + \omega_d^2} \sum_{i=0}^{\infty} e^{-ist_0}$$

Inversion of the above transform is achieved using both shifting theorems:

$$x(t) = \frac{I}{m\omega_d} \sum_{i=0}^{\infty} e^{-\zeta\omega_n(t - it_0)} \sin \omega_d(t - it_0)u(t - it_0)$$

4.22 Let $v = \dot{x}$. Rewrite Eq. (4.1) as a system of two first-order ordinary differential equations with t as the independent variable and x and v as dependent variables.

From the definition of v,

$$\dot{x} = v \qquad\qquad (4.14)$$

and from Eq. (4.1),

$$\dot{v} = \frac{1}{m_{eq}} F(t) - 2\zeta\omega_n v - \omega_n^2 x \qquad\qquad (4.15)$$

Equations (4.14) and (4.15) form a set of first-order linear simultaneous equations to solve for x and v.

4.23 Direct numerical simulation of Eq. (4.1) often involves rewriting Eq. (4.1) as two first-order differential equations, as in Problem 4.22. The time interval over which a solution is desired is discretized, and recurrence relations are developed for approximations to the dependent variables at the discrete times. Let t_1, t_2, ..., be the discrete times at equal intervals Δt. The *Euler method* is an implicit method using a first-order Taylor series expansion to approximate the time derivatives of the independent variables. Develop the recurrence relations for a 1-degree-of-freedom system using the Euler method.

Let $f(t)$ be a continuously differentiable function of a single variable. Its Taylor series expansion is

$$f(t + \Delta t) = f(t) + (\Delta t)\dot{f}(t) + \tfrac{1}{2}(\Delta t)^2 \ddot{f}(t) + \tfrac{1}{6}(\Delta t)^3 \dddot{f}(t) + \cdots +$$

Truncation after the linear term and rearranging leads to

$$\dot{f}(t) = \frac{f(t + \Delta t) - f(t)}{\Delta t} + O(\Delta t) \qquad (4.16)$$

Define $x_i = x(i \, \Delta t)$ and $v_i = v(i \, \Delta t)$. Application of Eq. (4.16) to Eqs. (4.14) and (4.15) of Problem 4.22 at $t = t_i = i \, \Delta t$ leads to

$$\frac{x_{i+1} - x_i}{\Delta t} + O(\Delta t) = v_i$$

$$\frac{v_{i+1} - v_i}{\Delta t} + O(\Delta t) = \frac{1}{m_{eq}} F(t_i) - 2\zeta\omega_n v_i - \omega_n^2 x_i$$

or

$$x_{i+1} = x_i + (\Delta t)v_i + O(\Delta t)$$

$$v_{i+1} = v_i + \Delta t\left(\frac{1}{m_{eq}} F(t_i) - 2\zeta\omega_n v_i - \omega_n^2 x_i\right) + O(\Delta t)$$

4.24 Illustrate the application of the explicit Euler method to a 1-degree-of-freedom system of mass 10 kg, natural frequency 100 rad/s, and damping ratio 0.1 subject to a constant force of magnitude 100 N. The system is at rest in equilibrium at $t = 0$. Use a time increment of 0.001 s, and compare with the exact solution of Problem 4.2.

The calculations are illustrated in Table 4-3.

4.25 A numerical approximation to an integral

$$I(t) = \int_0^t f(\tau) \, d\tau$$

is of the form

$$I(t_j) = \sum_{i=0}^{j} \alpha_i f(t_i) \, \Delta t \qquad (4.17)$$

where $t_1, t_2, \ldots, t_j$ are called *knots*, the intermediate values at which the integrand is evaluated. The values of α_i are specific to the numerical method used. Develop a form of Eq. (4.17) that can be used to approximate the convolution integral, Eq. (4.2).

The extension of Eq. (4.17) to the convolution integral is

$$x(t_j) = \sum_{i=0}^{j} \alpha_i F(t_i) h(t_j - t_i) \, \Delta t$$

Table 4.3

t_i	$x(t_i)$	$v(t_i)$	$F(t_i)/m$	$2\zeta\omega_n v_i$	$\omega_n^2 x_i$	$x(t_i + \Delta)$	$v(t_i + \Delta)$	$x(t)$ (exact)
0	0	0	10	0	0	0	0.01	0
0.001	0	0.01	10	0.2	0	0.00001	0.0198	4.96E-06
0.002	0.00001	0.0198	10	0.396	0.1	2.98E-05	0.029304	1.97E-05
0.003	2.98E-05	0.029304	10	0.58608	0.298	5.91E-05	0.03842	4.38E-05
0.004	5.91E-05	0.03842	10	0.768398	0.59104	9.75E-05	0.04706	7.69E-05
0.005	9.75E-05	0.04706	10	0.94121	0.975239	0.000145	0.055144	0.000118
0.006	0.000145	0.055144	10	1.102881	1.445844	0.0002	0.062595	0.000168
0.007	0.0002	0.062595	10	1.251906	1.997284	0.000262	0.069346	0.000225
0.008	0.000262	0.069346	10	1.386922	2.623237	0.000332	0.075336	0.000288
0.009	0.000332	0.075336	10	1.506719	3.316699	0.000407	0.080513	0.000357
0.01	0.000407	0.080513	10	1.610251	4.070058	0.000488	0.084832	0.000431
0.011	0.000488	0.084832	10	1.696645	4.875184	0.000572	0.08826	0.000509
0.012	0.000572	0.08826	10	1.765208	5.723506	0.000661	0.090772	0.000591
0.013	0.000661	0.090772	10	1.815434	6.60611	0.000751	0.09235	0.000675
0.014	0.000751	0.09235	10	1.847003	7.513827	0.000844	0.092989	0.00076
0.015	0.000844	0.092989	10	1.859786	8.437328	0.000937	0.092692	0.000846
0.016	0.000937	0.092692	10	1.853844	9.367221	0.001029	0.091471	0.000932
0.017	0.001029	0.091471	10	1.829423	10.29414	0.001121	0.089348	0.001017
0.018	0.001121	0.089348	10	1.786951	11.20885	0.00121	0.086352	0.0011
0.019	0.00121	0.086352	10	1.727035	12.10233	0.001297	0.082522	0.001181
0.02	0.001297	0.082522	10	1.650448	12.96585	0.001379	0.077906	0.001258

Assuming $t_k = k\,\Delta t$,

$$x(t_j) = \sum_{i=0}^{j} \alpha_i F(i\,\Delta t) h[(j-i)\,\Delta t]\,\Delta t$$

4.26 For the *trapezoidal rule*, the values of α_i in Eq. (4.17) of Problem 4.25 are $\alpha_i = \alpha_j = 0.5$, $\alpha_2 = \alpha_3 = \cdots = \alpha_{j-1} = 1$. Illustrate the use of the trapezoidal rule to approximate the time-dependent response of a system of mass 10 kg, natural frequency 50 rad/s, and damping ratio 0.05 subject to the time-dependent excitation of Fig. 4-8. Use $\Delta t = 0.01$ s, and approximate $x(0.01)$, $x(0.02)$, and $x(0.03)$.

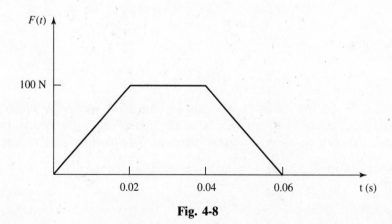

Fig. 4-8

Note that

$$h(k \, \Delta t) = \frac{1}{m\omega_d} e^{-\zeta\omega_n k \, \Delta t} \sin(\omega_d k \, \Delta t)$$

$$= 2.003 \times 10^{-3} e^{-0.025k} \sin(0.4995k)$$

$$h(0 \, \Delta t) = 0, \qquad h(\Delta t) = 9.36 \times 10^{-4}$$

$$h(2 \, \Delta t) = 1.56 \times 10^{-3}, \qquad h(3 \, \Delta t) = 1.84 \times 10^{-3}$$

Then, using the trapezoidal rule

$$x(0.01) = [0.5F(0)h(\Delta t) + 0.5F(0.01)h(0)] \, \Delta t$$

$$= [0.5(0)(9.36 \times 10^{-4}) + 0.5(50)(0)](0.04) = 0$$

$$x(0.02) = [0.5F(0)h(2 \, \Delta t) + F(0.01)h(\Delta t) + 0.5F(0.02)h(0)] \, \Delta t$$

$$= [0.5(0)(1.56 \times 10^{-3}) + (50)(9.36 \times 10^{-4}) + 0.5(100(0)](0.01) = 4.68 \times 10^{-4} \text{ m}$$

$$x(0.03) = [0.5F(0)h(3 \, \Delta t) + F(0.01)h(2 \, \Delta t) + F(0.02)h(\Delta t) + 0.5F(0.03)h(0)] \, \Delta t$$

$$= 1.72 \times 10^{-3} \text{ m}$$

4.27 Define $t_j = j \, \Delta t$. Show that Eq. (4.5) can be rewritten as

$$x(t_k) = \frac{1}{m\omega_d} e^{-\zeta\omega_n t_k} \left(\sin \omega_d t_k \sum_{j=1}^{k} G_{1j} - \cos \omega_d t_k \sum_{j=1}^{k} G_{2j} \right)$$

where

$$G_{1j} = \int_{t_{j-1}}^{t_j} F(\tau) e^{\zeta\omega_n \tau} \cos \omega_d \tau \, d\tau$$

$$G_{2j} = \int_{t_{j-1}}^{t_j} F(\tau) e^{\zeta\omega_n \tau} \sin \omega_d \tau \, d\tau$$

Note that

$$\sin \omega_d(t - \tau) = \sin \omega_d t \cos \omega_d \tau - \cos \omega_d t \sin \omega_d \tau$$

Substituting the previous trigonometric identity into Eq. (4.5) and rearranging leads to

$$x(t) = \frac{e^{-\zeta\omega_n t}}{m\omega_d} \left[\sin \omega_d t \int_0^t F(\tau) e^{\zeta\omega_n \tau} \cos \omega_d \tau \, d\tau - \cos \omega_d t \int_0^t F(\tau) e^{\zeta\omega_n \tau} \sin \omega_d \tau \, d\tau \right]$$

Noting that

$$\int_0^t g(\tau) \, d\tau = \int_0^{t_1} g(\tau) \, d\tau + \int_{t_1}^{t_2} g(\tau) \, d\tau + \cdots + \int_{t_{k-1}}^{t_k} g(\tau) \, d\tau$$

$$= \sum_{j=1}^{k} \int_{t_{j-1}}^{t_j} g(\tau) \, d\tau$$

leads to

$$x(t_k) = \frac{e^{-\zeta\omega_n t_k}}{m\omega_d} \left[\sin \omega_d t_k \sum_{j=1}^{k} \int_{t_{j-1}}^{t_j} F(\tau) e^{\zeta\omega_n \tau} \cos \omega_d \tau \, d\tau - \cos \omega_d t_k \sum_{j=1}^{k} \int_{t_{j-1}}^{t_j} F(\tau) e^{\zeta\omega_n \tau} \sin \omega_d \tau \, d\tau \right]$$

4.28 One method of approximating the convolution integral is to interpolate $F(t)$ piecewise and exactly integrate the interpolation times the trigonometric functions using Eq. (4.17) of Problem 4.27. Suppose $F(t)$ is interpolated by piecewise constants chosen such that the constant interpolate on the interval from t_{j-1} to t_j is equal to $F(t)$ evaluated at the midpoint of the interval. Determine the appropriate forms of G_{1j} and G_{2j} for a piecewise constant interpolation of $F(t)$.

Let $F_j = F[(t_j + t_{j-1})/2]$. Then

$$G_{ij} = \int_{t_{j-1}}^{t_j} F_j e^{\zeta \omega_n \tau} \cos \omega_n \tau \, d\tau$$

$$= \frac{F_j(1 - \zeta^2)}{\omega_d} \left[e^{\zeta \omega_n t_j} \left(\sin \omega_d t_j + \zeta \frac{\omega_n}{\omega_d} \cos \omega_d t_j \right) - e^{\zeta \omega_n t_{j-1}} \left(\sin \omega_d t_{j-1} + \zeta \frac{\omega_n}{\omega_d} \cos \omega_d t_{j-1} \right) \right]$$

and $$G_{2j} = \int_{t_{j-1}}^{t_j} F_j e^{\zeta \omega_n \tau} \sin \omega_d \tau \, d\tau$$

$$= \frac{F_j(1 - \zeta^2)}{\omega_d} \left[e^{\zeta \omega_n t_j} \left(-\cos \omega_d t_j + \zeta \frac{\omega_n}{\omega_d} \sin \omega_d t_j \right) - e^{\zeta \omega_n t_{j-1}} \left(-\cos \omega_d t_{j-1} + \zeta \frac{\omega_n}{\omega_d} \sin \omega_d t_{j-1} \right) \right]$$

4.29 Discuss the advantages and disadvantages of using each of the following numerical methods to provide a numerical approximation to the solution of Eq. (4.1): (*a*) implicit Euler method, (*b*) fourth-order Runge-Kutta method, (*c*) numerical integration of the convolution integral using the trapezoidal rule (Problem 4.26), and (*d*) numerical integration of the convolution integral using piecewise constant interpolate for $F(t)$ (Problem 4.28).

(*a*) The implicit Euler method is only first-order accurate. Its application leads to a pair of recurrence relations from which the approximations are successively determined. Application of the recurrence relations does not require evaluation of the excitation at times other than those for which approximations are obtained.

(*b*) The fourth-order Runge-Kutta method is fourth-order accurate. It is an explicit method in that its recurrence relations are used to determine the approximations successively. However, their application requires evaluation of the excitation at times other than those at which approximations are obtained.

(*c*) The trapezoidal rule is a numerical integration of the convolution integral. It provides a linear interpolation to the integrand between two knots. Its application does not lead to a recursion relation for the approximation at subsequent times. The formula must be repeated for each time the approximation is required, but an efficient algorithm is easy to develop.

(*d*) The excitation is interpolated by piecewise constants, multiplied by appropriate trigonometric functions, and the approximate integrand exactly integrated. An algorithm using this method to approximate the convolution integral is easy to develop in that the approximation for the response at an arbitrary time can be calculated as the response at the previous time plus the approximation to the integral between the two times.

4.30 To protect a computer during a move, the computer is placed in a cushioned crate. The motion of the computer in the crate can be modeled as a 1-degree-of-freedom mass-spring-viscous damper system. During the loading phase of the move, the crate is subject to the velocity of Fig. 4-9. Determine the displacement of the computer relative to the crate.

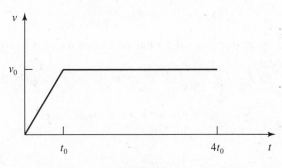

Fig. 4-9

Let y be the acceleration to which the crate is subjected. The differential equation governing the displacement of the computer relative to its crate is Eq. (3.24),

$$\ddot{z} + 2\zeta\omega_n\dot{z} + \omega_n^2 z = -\ddot{y}$$

whose convolution integral solution is

$$z(t) = -\frac{1}{\omega_d}\int_0^t \ddot{y}(\tau)e^{-\zeta\omega_n(t-\tau)}\sin\omega_d(t-\tau)\,d\tau$$

The time-dependent velocity and acceleration are

$$v = v_0\frac{t}{t_0}[u(t) - u(t - t_0)] + v_0[u(t - t_0) - u(t - 4t_0)]$$

$$\ddot{y} = \frac{dv}{dt} = \frac{v_0}{t_0}[u(t) - u(t - t_0)] + v_0\frac{t}{t_0}[\delta(t) - \delta(t - t_0)] + v_0[\delta(t - t_0) - \delta(t - 4t_0)]$$

The relative displacement is

$$z(t) = -\frac{v_0}{\omega_d}\int_0^t \left\{\frac{1}{t_0}[u(\tau) - u(\tau - t_0)] + \frac{\tau}{t_0}[\delta(\tau) - \delta(\tau - t_0)\right.$$

$$+ [\delta(\tau - t_0) - \delta(\tau - 4t_0)]\}e^{-\zeta\omega_n(t-\tau)}\sin\omega_d(t-\tau)\,d\tau$$

$$= \frac{v_0}{\omega_d t_0}\left\{\frac{u(t)}{\omega_n^2}[e^{-\zeta\omega_n t}(\zeta\omega_n\sin\omega_d t + \omega_d\cos\omega_d t) - \omega_d]\right.$$

$$- \frac{u(t - t_0)}{\omega_n^2}[e^{-\zeta\omega_n(t-t_0)}[\zeta\omega_n\sin\omega_d(t - t_0) + \omega_d\cos\omega_d(t - t_0)] - \omega_d]$$

$$\left. - t_0 u(t - 4t_0)e^{-\zeta\omega_n(t-4t_0)}\sin\omega_d(t - 4t_0)\right\}$$

4.31 Develop the response spectrum of an undamped system subject to the rectangular pulse of Fig. 4-2.

Problem 4.7 as

The response of a 1-degree-of-freedom system due to the rectangular pulse is determined in

$$x(t) = \frac{F_0}{m\omega_n^2}\begin{cases} 1 - \cos\omega_n t & t < t_0 \\ \cos\omega_n(t - t_0) - \cos\omega_n t & t > t_0 \end{cases}$$

Whether the maximum occurs for $t < t_0$ or for $t > t_0$ depends upon the system parameters. The maximum response for $t < t_0$ occurs either at $t = t_0$ or when $\cos \omega_n t = -1$ ($\omega_n t = \pi$). Thus if $t_0 < \pi/\omega_n$, the maximum is not achieved for $t < t_0$. For $t_0 > \pi/\omega_n$,

$$x_{max} = \frac{2F_0}{m\omega_n^2}$$

For $t_0 < \pi/\omega_n$, x_{max} occurs for $t > t_0$ at a value of t when $dx/dt = 0$. To this end, for $t > t_0$,

$$\frac{dx}{dt} = \frac{F_0}{m\omega_n^2}[-\omega_n \sin \omega_n(t - t_0) + \omega_n \sin \omega_n t]$$

$$\frac{dx}{dt} = 0 \rightarrow \sin \omega_n t = \sin \omega_n(t - t_0)$$

whose solution is

$$t = \frac{1}{2}\left[t_0 + \frac{(2n - 1)\pi}{2\omega_n}\right] \qquad n = 1, 2, \ldots$$

This leads to

$$\frac{m\omega_n^2 x_{max}}{F_0} = \begin{cases} 2 \sin \dfrac{\omega_n t_0}{2} & t_0 < \dfrac{\pi}{\omega_n} \\ 2 & t_0 > \dfrac{\pi}{\omega_n} \end{cases}$$

which is plotted in Fig. 4-10.

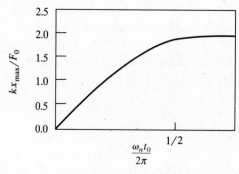

Fig. 4-10

4.32 Devise an algorithm to numerically develop the response spectrum for a 1-degree-of-freedom system with viscous damping subject to an arbitrary excitation. Assume that an integration method such as Euler's method or a fourth-order Runge-Kutta is used to solve Eq. (4.1).

1. Equation (4.1) can be nondimensionalized by introducing

$$x^* = \frac{m\omega_n^2 x}{F_0}, \qquad t^* = \frac{\omega_n t}{2\pi}$$

where F_0 is the maximum value of $F(t)$. In terms of these nondimensional variables, Eq. (4.1) becomes

$$\ddot{x}^* + 4\pi\zeta\dot{x}^* + 4\pi^2 x^* = \frac{4\pi^2}{F_0} F(t^*) \tag{4.18}$$

Note that the natural period in nondimensional time is 1.

2. Define $\alpha = \omega t_0/(2\pi)$, which can be viewed as a nondimensional value of the pulse duration. Let α range from $\Delta\alpha$ to 2.5 in increments of $\Delta\alpha = 0.1$. Equation (4.18) should be solved using, say Runge-Kutta for each value of α. For a particular value of α, a time increment and a final time for the Runge-Kutta simulation must be chosen. It is important to set the nondimensional time increment small enough such that enough integration steps are used over the duration of the pulse and over 1 natural period. Also the final time must be chosen large enough such that the integration is carried out sufficiently beyond the duration of the pulse and over several natural periods. Note that for $\alpha < 1$, the duration of the pulse is less than the natural period. Possible choices for the time increment and final time are

$$\Delta t^* = \frac{\alpha}{20}, \qquad t_f^* = 2.5$$

For $\alpha > 1$, the natural period is less than the duration of the pulse. Possible choices for the time increment and final time are

$$\Delta t^* = \tfrac{1}{20}, \qquad t_f^* = 2.5\alpha$$

3. Numerical simulation of Eq. (4.18) is developed for each value of α, as described in step 2. The maximum value of the nondimensional response $x_{\max}^*$ is recorded. The response spectrum is a plot of $x_{\max}^*$ versus α.

4.33 The force exerted on a structure due to a shock or impact is often modeled by the excitation of Fig. 4-11. The response spectrum for this type of excitation for several damping ratios is shown in Fig. 4-12. What is the maximum displacement of an undamped 1000-kg structure of stiffness 5×10^6 N/m subject to such a blast with $F_0 = 1500$ N and $t_0 = 0.05$ s?

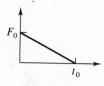

Fig. 4-11

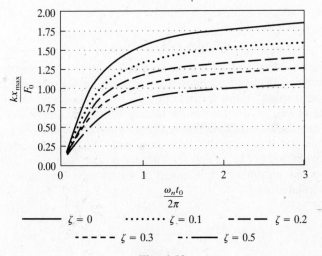

Fig. 4-12

The natural frequency of the structure is

$$\omega_n = \sqrt{\frac{k}{m}} = \sqrt{\frac{5 \times 10^6 \, \dfrac{\text{N}}{\text{m}}}{1000 \text{ kg}}} = 70.7 \, \frac{\text{rad}}{\text{s}}$$

The value of the nondimensional parameter on the horizontal scale of the response spectrum is

$$\frac{\omega_n t_0}{2\pi} = \frac{\left(70.7 \, \dfrac{\text{rad}}{\text{s}}\right)(0.05 \text{ s})}{2\pi} = 0.563$$

From Fig. 4-12, for $\zeta = 0$,

$$\frac{kx_{\max}}{F_0} = 1.2$$

from which the maximum displacement is calculated as

$$x_{\max} = \left(\frac{kx_{\max}}{F_0}\right)\frac{F_0}{k} = 1.2 \frac{5000 \text{ N}}{5 \times 10^6 \, \dfrac{\text{N}}{\text{m}}} = 1.2 \text{ mm}$$

4.34   For what diameters of the circular bar of the system of Fig. 4-13 will the maximum displacement of the block be less than 18 mm when subjected to a blast modeled by Fig. 4-11 with $F_0 = 10,000$ N and $t_0 = 0.1$ s?

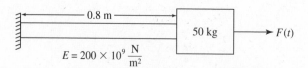

Fig. 4-13

Let k be the stiffness of the system. Since it is desired to limit $x < 18$ mm,

$$\frac{kx_{\max}}{F_0} < \frac{k(0.018 \text{ m})}{10,000 \text{ N}} = 1.8 \times 10^{-6}k \tag{4.19}$$

Also

$$\frac{\omega_n t_0}{2\pi} = \sqrt{\frac{k}{50 \text{ kg}}}\frac{0.1 \text{ s}}{2\pi} = 2.25 \times 10^{-3}\sqrt{k} \tag{4.20}$$

Since k is unknown, a trial-and-error procedure using the response spectrum of Fig. 4-12 is used to determine allowable values of k. Suppose $k = 5 \times 10^5$ N/m, thus using Eq. (4.20), $\omega_n t_0/(2\pi) = 1.59$. From the response spectrum this leads to $kx_{\max}/F_0 = 1.7$. However, from Eq. (4.20), the maximum allowable value of $kx_{\max}/F_0$ for $k = 5 \times 10^5$ N/m is 0.9. Inspection of Eqs. (4.19) and (4.20) shows that $\omega_n t_0/(2\pi)$ increases slower than the maximum allowable value of $kx_{\max}/F_0$. Thus an increase in k is tried. Note that if $k = 1 \times 10^6$ N/m, then using Eq. (4.20) leads to $\omega_n t_0/(2\pi) = 2.25$. The response spectrum yields $kx_{\max}/F_0 = 1.8$, which is the same as using Eq. (4.19). Thus if $k > 1 \times 10^6$ N/m, the maximum displacement of the block is less than 18 mm. The minimum area of the bar is calculated as

$$A = \frac{kL}{E} = \frac{\left(1 \times 10^6 \, \dfrac{\text{N}}{\text{m}}\right)(0.8 \text{ m})}{200 \times 10^9 \, \dfrac{\text{N}}{\text{m}^2}} = 4 \times 10^{-6} \text{ m}^2$$

Hence the minimum diameter is

$$D = \sqrt{\frac{4A}{\pi}} = \sqrt{\frac{4(4 \times 10^{-6}\ \text{m}^2)}{\pi}} = 2.26 \times 10^{-3}\ \text{m}$$

Supplementary Problems

4.35 Use the convolution integral to develop the response of an undamped 1-degree-of-freedom system of mass m and natural frequency ω_n subject to an excitation of the form $F(t) = F_0 \sin \omega_n t$. The system is at rest in equilibrium at $t = 0$.

Ans.

$$\frac{F_0}{2m\omega_n^2} \sin \omega_n t - \frac{F_0}{2m\omega_n} t \cos \omega_n t$$

4.36 Use the convolution integral to develop the response of an undamped 1-degree-of-freedom system subject to an excitation of the form $F(t) = F_0(1 - e^{-\alpha t})$. The system is at rest in equilibrium at $t = 0$.

Ans.

$$-\frac{F_0}{m\omega_n^2(\alpha^2 + \omega_n^2)}[\omega_n^2 e^{-\alpha t} - \alpha^2 - \omega_n^2 + \alpha^2 \cos \omega_n t + \alpha \omega_n \sin \omega_n t]$$

4.37 Use the convolution integral to determine the response of a 1-degree-of-freedom system of mass m, damping ratio ζ, and natural frequency ω_n subject to an excitation of the form $F(t) = F_0 \sin \omega t$ for $\omega \neq \omega_n$. The system is at rest in equilibrium at $t = 0$.

Ans.

$$\frac{F_0}{m[(\omega_n^2 - \omega^2)^2 + (2\zeta\omega\omega_n)^2]}\left[2\zeta\omega\omega_n(\cos \omega_n t - \cos \omega t) + (\omega_n^2 - \omega^2)\left(\sin \omega t - \frac{\omega}{\omega_n} \sin \omega_n t\right)\right]$$

4.38 Use the convolution integral to determine the time-dependent response of the system of Fig. 4-14.

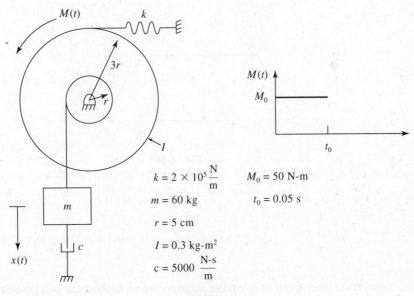

$$k = 2 \times 10^5\ \frac{\text{N}}{\text{m}} \qquad M_0 = 50\ \text{N-m}$$

$$m = 60\ \text{kg} \qquad t_0 = 0.05\ \text{s}$$

$$r = 5\ \text{cm}$$

$$I = 0.3\ \text{kg-m}^2$$

$$c = 5000\ \frac{\text{N-s}}{\text{m}}$$

Fig. 4-14

Ans.

$$0.0111[1 - e^{-13.9t}(0.140 \sin 99.03t + \cos 99.03t)] \qquad t < 0.05s$$

$$0.0111e^{-13.9t}[0.140 \sin 99.03t + \cos 99.03t - 0.280 \sin (99.03t - 4.95) - 2.00 \cos (99.03t - 4.95)]$$

$$t > 0.05s$$

4.39   An undamped 1-degree-of-freedom system is subject to the excitation of Fig. 4-15. Use the convolution integral to determine the system response for $t > t_0$.

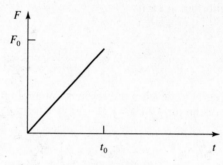

Fig. 4-15

Ans.

$$\frac{F_0}{m\omega_n^2 t_0}\left[t_0 \cos \omega_n(t - t_0) + \frac{1}{\omega_n}\sin (t - t_0) - \frac{1}{\omega_n}\sin \omega_n t\right]$$

4.40   Develop a unified mathematical expression for the excitation of Fig. 4-16 using unit step functions.

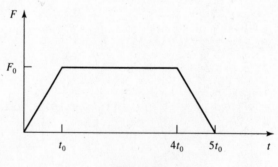

Fig. 4-16

Ans.

$$F_0\frac{t}{t_0}u(t) + F_0\left(1 - \frac{t}{t_0}\right)u(t - t_0) + F_0\left(4 - \frac{t}{t_0}\right)u(t - 4t_0) + F_0\left(5 - \frac{t}{t_0}\right)u(t - 5t_0)$$

4.41   Use the convolution integral to develop the response of an undamped 1-degree-of-freedom system subject to the excitation of Fig. 4-16. The system is at rest in equilibrium at $t = 0$.

Ans.

$$\frac{F_0}{m\omega_n{}^2 t_0}\left[\left(t-\frac{1}{\omega_n}\sin\omega_n t\right)u(t)-\left(t-t_0-\frac{1}{\omega_n}\sin\omega_n(t-t_0)\right)u(t-t_0)\right.$$

$$\left.-\left(t-4t_0-\frac{1}{\omega_n}\sin\omega_n(t-4t_0)\right)u(t-4t_0)-\left(t-5t_0-t_0\cos\omega_n(t-5t_0)-\frac{1}{\omega_n}\sin(t-5t_0)\right)u(t-5t_0)\right]$$

4.42   Use the convolution integral to develop the response of an undamped 1-degree-of-freedom system due to the excitation of Fig. 4-17. The system is at rest in equilibrium at $t=0$.

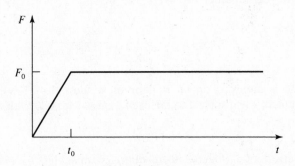

Fig. 4-17

Ans.

$$\frac{F_0}{m\omega_n{}^2 t_0}\left[\left(t-\frac{1}{\omega_n}\sin\omega_n t\right)u(t)-\left(t-t_0-\frac{1}{\omega_n}\sin\omega_n(t-t_0)\right)u(t-t_0)\right]$$

4.43   Solve Problem 4.35 using the Laplace transform method.

4.44   Solve Problem 4.37 using the Laplace transform method.

4.45   Solve Problem 4.41 using the Laplace transform method.

4.46   Solve Problem 4.42 using the Laplace transform method.

4.47   A 50-kg block is attached to a spring of stiffness 2×10^6 N/m. The block is subject to an impulse of magnitude 25 N-s at $t=0$ and an impulse of magnitude 15 N-s at $t=0.1$ s. What is the maximum displacement of the block?

Ans.   6.28 mm

4.48   Use the Laplace transform method to determine the response of a 1-degree-of-freedom system with damping ratio ζ and natural frequency ω_n when subject to the periodic excitation of Fig. 4-18.

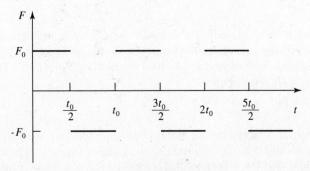

Fig. 4-18

Ans.

$$\frac{F_0}{m\omega_n{}^2}\left[1 - e^{-\zeta\omega_n t}\left(\cos \omega_n t + \frac{\zeta}{\sqrt{1-\zeta^2}}\sin \omega_n t\right)\right]u(t)$$

$$+2\sum_{i=1}^{\infty}(-1)^i\left\{1 - e^{-\zeta\omega_n(t-it_0)}\left[\cos \omega_n(t - it_0) + \frac{\zeta}{\sqrt{1-\zeta^2}}\sin \omega_n(t - it_0)\right]\right\}u(t - it_0)$$

4.49 Determine an equation defining the response spectrum for an undamped 1-degree-of-freedom system subject to the excitation of Fig. 4-17.

Ans.

$$\frac{kx_{max}}{F_0} = 1 + \frac{1}{\omega_n t_0}\sqrt{2(1 - \cos \omega_n t_0)}$$

4.50 A 100-kg machine is mounted on an isolator of stiffness 5×10^5 N/m. What is the machine's maximum displacement when subject to the excitation of Fig. 4-17 with $F_0 = 1000$ N and $t_0 = 0.11$ s?

Ans. 2.50 mm

4.51 The response spectrum of a 1-degree-of-freedom system subject to a sinusoidal pulse is shown in Fig. 4-19. A 25-kg block is attached to a spring of stiffness 5×10^6 N/m and subject to a sinusoidal pulse of magnitude 1250 N and duration 0.02 s. What is the maximum displacement of the block?

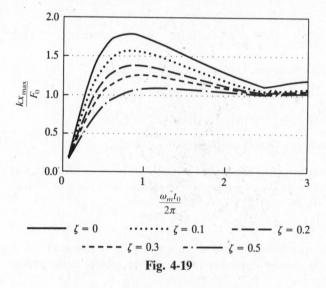

Fig. 4-19

Ans. 0.4 mm

4.52 A 200-kg machine is to be placed on a vibration isolator of damping ratio 0.1. For what range of isolator stiffness will the maximum displacement of the machine be less than 2 mm when it is subject to a sinusoidal pulse of magnitude 1000 N and duration 0.04 s? The response spectrum for a sinusoidal pulse is shown in Fig. 4-19.

Ans. $k > 5\times10^5$ N/m

4.53 A 500-kg machine is attached to an isolator of stiffness 3×10^5 N/m and damping ratio 0.05. During startup it is subject to an excitation of the form of Fig. 4-17 with $t_0 = 0.1$ s and $F_0 = 5000$ N. Use the trapezoidal rule for numerical integration of the convolution integral to approximate the machine's displacement. Use $\Delta = 0.04$ s.

4.54   Consider the method for numerical evaluation of the convolution integral introduced in Problems 4.27 and 4.28. Develop an expression for G_{1j} if $F(t)$ is interpolated by a series of impulses. That is, on the interval from t_j to t_{j+1}, $F(t)$ is interpolated by an impulse of magnitude $F(t_j^*)\,\Delta t$ applied at t_j^* where t_j^* is the midpoint of the interval.

Ans.

$$F(t_j^*)\,\Delta t e^{\zeta\omega_n t_j^*}\cos\omega_d t_j^*$$

4.55   Use numerical integration of the convolution integral to approximate the response of a machine of mass 250 kg attached to an isolator of stiffness 2×10^6 N/m and damping ratio 0.05 when subject to the excitation

$$F(t) = 1000e^{t^2}\ \text{N}$$

Use the method of Problem 4.28.

4.56   Repeat Problem 4.55 using the interpolations of Problem 4.55.

Chapter 5

Free Vibrations of Multi-Degree-of-Freedom Systems

5.1 LAGRANGE'S EQUATIONS

Let $x_1, x_2, x_3, \ldots, x_n$ be a set of generalized coordinates for an n-degree-of-freedom system. The motion of the system is governed by a set of n ordinary differential equations with the generalized coordinates as the dependent variables and time as the independent variable. One method of deriving the differential equations, referred to as the *free body diagram method*, involves applying conservation laws to free body diagrams of the system drawn at an arbitrary instant.

An energy method provides an alternative to derive the differential equations governing the vibrations of a multi-degree-of-freedom system. Let $V(x_1, x_2, \ldots, x_n)$ be the potential energy of the system at an arbitrary instant. Let $T(x_1, x_2, \ldots, x_n, \dot{x}_1, \dot{x}_2, \ldots, \dot{x}_n)$ be the kinetic energy of the system at the same arbitrary instant. The *lagrangian* $L(x_1, x_2, \ldots, x_n, \dot{x}_1, \dot{x}_2, \ldots, \dot{x}_n)$ is defined as

$$L = T - V \tag{5.1}$$

The lagrangian is viewed as a function of $2n$ independent variables, with the time derivatives of the generalized coordinates assumed to be independent of the generalized coordinates.

Let $\delta x_1, \delta x_2, \ldots, \delta x_n$ be variations of the generalized coordinates. The virtual work δW done by the nonconservative forces in the system due to the variations of the generalized coordinates can be written as

$$\delta W = \sum_{i=1}^{n} Q_i \, \delta x_i \tag{5.2}$$

Lagrange's equations are

$$\frac{d}{dt}\left(\frac{\partial L}{\partial \dot{x}_1}\right) - \frac{\partial L}{\partial x_i} = Q_i, \qquad i = 1, 2, \ldots, n \tag{5.3}$$

Application of Lagrange's equations leads to a set of n independent differential equations.

5.2 MATRIX FORMULATION OF DIFFERENTIAL EQUATIONS FOR LINEAR SYSTEMS

For a linear system, the potential and kinetic energies have quadratic forms:

$$V = \tfrac{1}{2} \sum_{i=1}^{n} \sum_{j=1}^{n} k_{ij} x_i x_j \tag{5.4}$$

$$T = \tfrac{1}{2} \sum_{i=1}^{n} \sum_{j=1}^{n} m_{ij} \dot{x}_i \dot{x}_j \tag{5.5}$$

If viscous damping and externally applied forces, independent of the generalized coordinates, are the only nonconservative forces, the virtual work can be expressed as

$$\delta W = \sum_{i=1}^{n} \sum_{j=1}^{n} c_{ij} \dot{x}_j \, \delta x_j + \sum_{i=1}^{n} F_i \, \delta x_i \tag{5.6}$$

136

Application of Lagrange's equations to the lagrangian developed using Eqs. (5.4) and (5.5) and the virtual work of Eq. (5.6) leads to

$$\mathbf{M}\ddot{\mathbf{x}} + \mathbf{C}\dot{\mathbf{x}} + \mathbf{K}\mathbf{x} = \mathbf{F} \tag{5.7}$$

where $\mathbf{M}$ is the $n \times n$ *mass matrix* whose elements are m_{ij}, $\mathbf{K}$ is the $n \times n$ *stiffness matrix* whose elements are k_{ij}, $\mathbf{C}$ is the $n \times n$ *viscous damping matrix* whose elements are c_{ij}, $\mathbf{x}$ is the $n \times 1$ *displacement vector* whose elements are x_i, and $\mathbf{F}$ is the $n \times 1$ *force vector* whose elements are F_i. The matrices are *symmetric*. For example, $m_{ij} = m_{ji}$.

5.3   STIFFNESS INFLUENCE COEFFICIENTS

Stiffness influence coefficients are used to sequentially calculate the columns of the stiffness matrix for a linear system. Imagine the system in static equilibrium with $x_j = 1$ and $x_i = 0$ for $i \neq j$. The jth column of the stiffness matrix, the stiffness influence coefficients $k_{1j}, k_{2j}, \ldots, k_{nj}$, are the forces that must be applied to the particles whose displacements are described by the generalized coordinates to maintain the system in equilibrium in the prescribed position. The forces are assumed positive in the positive direction of the generalized coordinates. If x_i is an angular coordinate, then k_{ij} is an applied moment. *Maxwell's reciprocity relation* implies that $k_{ij} = k_{ji}$.

5.4   FLEXIBILITY MATRIX

The *flexibility matrix* $\mathbf{A}$ is the inverse of the stiffness matrix. *Flexibility influence coefficients* can be used to sequentially calculate the columns of the flexibility matrix. The jth column of the flexibility matrix is the column of values of the generalized coordinates induced by static application of a unit load to the particle whose displacement is described by x_j. If x_j is an angular coordinate, then a unit moment is applied. The reciprocity relation implies that the flexibility matrix is symmetric, $a_{ij} = a_{ji}$.

The flexibility matrix is easier to calculate than the stiffness matrix for most structural systems that are modeled using a finite number of degrees of freedom. The differential equations governing the motion of a linear n-degree-of-freedom system can be written using the flexibility matrix as

$$\mathbf{A}\mathbf{M}\ddot{\mathbf{x}} + \mathbf{A}\mathbf{C}\dot{\mathbf{x}} + \mathbf{x} = \mathbf{A}\mathbf{F} \tag{5.8}$$

5.5   NORMAL MODE SOLUTION

Introduction of the *normal mode solution*

$$\mathbf{x} = \mathbf{X}e^{i\omega t} \tag{5.9}$$

into Eq. (5.7) with $\mathbf{C} = \mathbf{0}$ and $\mathbf{F} = \mathbf{0}$ leads to the following matrix *eigenvalue-eigenvector problem* for the natural frequencies ω and their corresponding *mode shape vectors* $\mathbf{X}$:

$$\mathbf{M}^{-1}\mathbf{K}\mathbf{X} = \omega^2\mathbf{X} \tag{5.10}$$

The natural frequencies, the square roots of the eigenvalues of $\mathbf{M}^{-1}\mathbf{K}$, are obtained by setting

$$\det|\mathbf{M}^{-1}\mathbf{K} - \omega^2\mathbf{I}| = 0 \tag{5.11}$$

or alternately

$$\det|\mathbf{K} - \omega^2\mathbf{M}| = 0 \tag{5.12}$$

If the flexibility matrix is known rather than the stiffness matrix, the natural frequencies are the reciprocals of the square roots of the eigenvalues of $\mathbf{AM}$ and are calculated from

$$\det |\omega^2 \mathbf{AM} - \mathbf{I}| = 0 \tag{5.13}$$

Use of Eq. (5.11) or (5.13) leads to an nth-order algebraic equation in ω^2 with real coefficients. Since $\mathbf{M}$ and $\mathbf{K}$ are symmetric, the n roots are real, yielding the system's n natural frequencies, $\omega_1 \leq \omega_2 \leq \cdots \leq \omega_n$. If the system is stable, then $\mathbf{K}$ is nonnegative definite and the roots are nonnegative. An *unrestrained system* has a rigid body mode corresponding to a natural frequency of zero.

5.6 MODE SHAPE ORTHOGONALITY

Let $\mathbf{X}_i$ and $\mathbf{X}_j$ be mode shape vectors for an n-degree-of-freedom system corresponding to distinct natural frequencies ω_i and ω_j, respectively. These mode shapes satisfy the following *orthogonality* conditions:

$$\mathbf{X}_i^T \mathbf{MX}_j = 0 \tag{5.14}$$
$$\mathbf{X}_i^T \mathbf{KX}_j = 0 \tag{5.15}$$

The mode shape vector $\mathbf{X}_i$, when determined as an eigenvector of $\mathbf{M}^{-1}\mathbf{K}$ or $\mathbf{AM}$, is unique only to a multiplicative constant. The nonuniqueness is alleviated by requiring the mode shape to satisfy a *normalization condition*, usually specified as

$$\mathbf{X}_i^T \mathbf{MX}_i = 1 \tag{5.16}$$

If Eq. (5.16) is used as a normalization condition, then

$$\mathbf{X}_i^T \mathbf{KX}_i = \omega_i^2 \tag{5.17}$$

5.7 MATRIX ITERATION

Numerical procedures are often used to calculate natural frequencies of systems with a large number of degrees of freedom. *Matrix iteration* is a numerical procedure that allows determination of a system's natural frequencies and mode shapes successively, beginning with the smallest natural frequency. Let $\mathbf{u}_0$ be an arbitrary $n \times 1$ vector. The sequence of vectors

$$\mathbf{u}_i = \mathbf{AM\tilde{u}}_{i-1}, \qquad \tilde{\mathbf{u}}_i = \frac{\mathbf{u}_i}{|\mathbf{u}_i|_{\max}} \tag{5.18}$$

converges to $\mathbf{X}_1$, the mode shape corresponding to the lowest natural frequency. Also, $|\mathbf{u}_i|_{\max}$, the largest absolute value of an element of $\mathbf{u}_i$, converges to $1/\omega_1^2$. Matrix iteration can be used to determine natural frequencies and mode shapes for higher modes by using a trial vector orthogonal, with respect to the mass matrix, to all previously determined mode shapes.

5.8 DAMPED SYSTEMS

The determination of the free and forced response of a multi-degree-of-freedom system is significantly more difficult than for an undamped system. A special case, which is relatively easy to handle, is *proportional damping* and occurs when constants α and β exist such that

$$\mathbf{C} = \alpha \mathbf{K} + \beta \mathbf{M} \tag{5.19}$$

For proportional damping, the normal mode solution Eq. (5.9) is applicable. If $\omega_1, \omega_2, \ldots, \omega_n$ are the natural frequencies corresponding to the undamped system, then the values of ω that satisfy Eq. (5.9) are

$$\tilde{\omega}_i = i\omega_i \zeta_i \pm \omega_i \sqrt{1 - \zeta_i^2} \tag{5.20}$$

where ζ_i, a modal damping ratio is

$$\zeta_i = \frac{1}{2}\left(\alpha\omega_i + \frac{\beta}{\omega_i}\right) \tag{5.21}$$

For more general forms of **C**, it is convenient to rewrite Eq. (5.7) as

$$\tilde{\mathbf{M}}\dot{\mathbf{y}} + \tilde{\mathbf{K}}\mathbf{y} = 0 \tag{5.22}$$

where

$$\tilde{\mathbf{M}} = \begin{bmatrix} \mathbf{0} & \mathbf{M} \\ \mathbf{M} & \mathbf{C} \end{bmatrix}, \qquad \tilde{\mathbf{K}} = \begin{bmatrix} -\mathbf{M} & \mathbf{0} \\ \mathbf{0} & \mathbf{K} \end{bmatrix} \tag{5.23}$$

are symmetric $2n \times 2n$ matrices and

$$\mathbf{y} = \begin{bmatrix} \dot{\mathbf{x}} \\ \mathbf{x} \end{bmatrix} \tag{5.24}$$

is a $2n \times 1$ column vector. A solution to Eq. (5.23) is assumed as

$$\mathbf{y} = \mathbf{\Phi}e^{-\gamma t} \tag{5.25}$$

The values of γ are the complex conjugate eigenvalues of $\tilde{\mathbf{M}}^{-1}\tilde{\mathbf{K}}$, and $\mathbf{\Phi}$ is a corresponding eigenvector. The eigenvectors satisfy the orthogonality relation

$$\mathbf{\Phi}_i^T\tilde{\mathbf{M}}\mathbf{\Phi}_j = 0 \qquad i \neq j \tag{5.26}$$

Solved Problems

5.1   Use the free body diagram method to derive the differential equations governing the motion of the system of Fig. 5-1 using x_1, x_2, and x_3 as generalized coordinates.

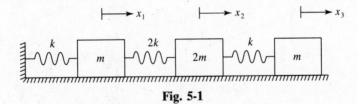

Fig. 5-1

Free body diagrams of each of the blocks of the system of Fig. 5-1 are shown in Fig. 5-2 at an arbitrary instant. Application of Newton's law to each of the free body diagrams leads to

$$-kx_1 + 2k(x_2 - x_1) = m\ddot{x}_1 \quad \rightarrow \quad m\ddot{x}_1 + 3kx_1 - 2kx_2 = 0$$

$$-2k(x_2 - x_1) + k(x_3 - x_2) = 2m\ddot{x}_2 \quad \rightarrow \quad 2m\ddot{x}_2 - 2kx_1 + 3kx_2 - kx_3 = 0$$

$$-k(x_3 - x_2) = 2m\ddot{x}_3 \quad \rightarrow \quad 2m\ddot{x}_3 - kx_2 + kx_3 = 0$$

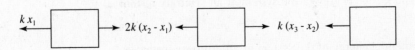

Fig. 5-2

5.2   Use the free body diagram method to derive the differential equations governing the motion of the system of Fig. 5-3 using x and θ as generalized coordinates.

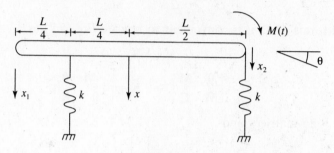

Fig. 5-3

Free body diagrams of the bar at an arbitrary instant are shown in Fig. 5-4 assuming small θ. Summing forces on the bar,

$$\left(\sum F\right)_{ext} = \left(\sum F\right)_{eff}$$

$$-k(x - \tfrac{1}{4}L\theta) - k(x + \tfrac{1}{2}L\theta) = m\ddot{x}$$

$$m\ddot{x} + 2kx + \tfrac{1}{4}kL\theta = 0$$

Summing moments about the mass center of the bar,

$$\left(\sum M_G\right)_{ext} = \left(\sum M_G\right)_{eff}$$

$$M(t) + k(x - \tfrac{1}{4}L\theta)\tfrac{1}{4}L - k(x + \tfrac{1}{2}L\theta)\tfrac{1}{2}L = I\ddot{\theta}$$

$$I\ddot{\theta} + \tfrac{1}{4}kLx + \tfrac{5}{16}kL^2\theta = M(t)$$

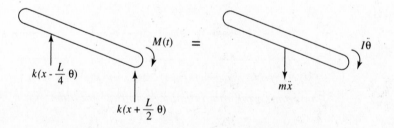

External Forces   Effective Forces

Fig. 5-4

5.3   Use Lagrange's equations to derive the differential equations governing the motion of the system of Fig. 5-1 using x_1, x_2, and x_3 as generalized coordinates. Write the differential equations in matrix form.

The kinetic energy of the system at an arbitrary instant is

$$T = \tfrac{1}{2}m\dot{x}_1^2 + \tfrac{1}{2}2m\dot{x}_2^2 + \tfrac{1}{2}2m\dot{x}_3^2$$

The potential energy of the system at an arbitrary instant is

$$V = \tfrac{1}{2}kx_1^2 + \tfrac{1}{2}2k(x_2 - x_1)^2 + \tfrac{1}{2}k(x_3 - x_2)^2$$

The lagrangian is

$$L = T - V = \tfrac{1}{2}m\dot{x}_1^2 + \tfrac{1}{2}2m\dot{x}_2^2 + \tfrac{1}{2}2m\dot{x}_3^2 - \tfrac{1}{2}kx_1^2 - \tfrac{1}{2}2k(x_2 - x_1)^2 - \tfrac{1}{2}k(x_3 - x_2)^2$$

Application of Lagrange's equations lead to

$$\frac{d}{dt}\left(\frac{\partial L}{\partial \dot{x}_1}\right) - \frac{\partial L}{\partial x_1} = 0$$

$$\frac{d}{dt}(m\dot{x}_1) + [kx_1 + 2k(x_2 - x_1)(-1)] = 0$$

$$\frac{d}{dt}\left(\frac{\partial L}{\partial \dot{x}_2}\right) - \frac{\partial L}{\partial x_2} = 0$$

$$\frac{d}{dt}(2m\dot{x}_2) + [2k(x_2 - x_1)(1) + k(x_3 - x_2)(-1)] = 0$$

$$\frac{d}{dt}\left(\frac{\partial L}{\partial \dot{x}_3}\right) - \frac{\partial L}{\partial x_3} = 0$$

$$\frac{d}{dt}(2m\dot{x}_3) + [k(x_3 - x_2)(1)] = 0$$

Rearranging and writing in matrix form leads to

$$\begin{bmatrix} m & 0 & 0 \\ 0 & 2m & 0 \\ 0 & 0 & 2m \end{bmatrix} \begin{bmatrix} \ddot{x}_1 \\ \ddot{x}_2 \\ \ddot{x}_3 \end{bmatrix} + \begin{bmatrix} 3k & -2k & 0 \\ -2k & 3k & -k \\ 0 & -k & k \end{bmatrix} \begin{bmatrix} x_1 \\ x_2 \\ x_3 \end{bmatrix} = \begin{bmatrix} 0 \\ 0 \\ 0 \end{bmatrix}$$

5.4   Use Lagrange's equations to derive the differential equations governing the motion of the system of Fig. 5-3 using x and θ as generalized coordinates. Write the differential equations in matrix form.

The kinetic energy of the system at an arbitrary instant is

$$T = \tfrac{1}{2}m\dot{x}^2 + \tfrac{1}{2}I\dot{\theta}^2$$

The potential energy of the system at an arbitrary instant is

$$V = \tfrac{1}{2}k(x - \tfrac{1}{4}L\theta)^2 + \tfrac{1}{2}k(x + \tfrac{1}{2}L\theta)^2$$

The lagrangian is

$$L = \tfrac{1}{2}m\dot{x}^2 + \tfrac{1}{2}I\dot{\theta}^2 - \tfrac{1}{2}k(x - \tfrac{1}{4}L\theta)^2 - \tfrac{1}{2}k(x + \tfrac{1}{2}L\theta)^2$$

If the variations δx and $\delta\theta$ are introduced, the virtual work done by the external moment is

$$\delta W = M(t)\,\delta\theta$$

Application of Lagrange's equations leads to

$$\frac{d}{dt}\left(\frac{\partial L}{\partial \dot{x}}\right) - \frac{\partial L}{\partial x} = Q_1$$

$$\frac{d}{dt}(m\dot{x}) + \left[k\left(x - \frac{1}{4}L\theta\right)(1) + k\left(x + \frac{1}{2}L\theta\right)(1)\right] = 0$$

$$\frac{d}{dt}\left(\frac{\partial L}{\partial \dot{\theta}}\right) - \frac{\partial L}{\partial \theta} = Q_2$$

$$\frac{d}{dt}(I\dot{\theta}) + \left[k\left(x - \frac{1}{4}L\theta\right)\left(-\frac{1}{4}L\right) + k\left(x + \frac{1}{2}L\theta\right)\left(\frac{1}{2}L\right)\right] = M(t)$$

Rearranging and rewriting in matrix form leads to

$$\begin{bmatrix} m & 0 \\ 0 & I \end{bmatrix}\begin{bmatrix} \ddot{x} \\ \ddot{\theta} \end{bmatrix} + \begin{bmatrix} 2k & \frac{1}{4}kL \\ \frac{1}{4}kL & \frac{5}{16}kL^2 \end{bmatrix}\begin{bmatrix} x \\ \theta \end{bmatrix} = \begin{bmatrix} 0 \\ M(t) \end{bmatrix}$$

5.5   Use Lagrange's equations to derive the differential equations governing the motion of the system of Fig. 5-3 using x_1 and x_2 as generalized coordinates. Write the differential equations in matrix form.

The kinetic energy of the system at an arbitrary instant is

$$T = \frac{1}{2}m\left(\frac{\dot{x}_1 + \dot{x}_2}{2}\right)^2 + \frac{1}{2}I\left(\frac{\dot{x}_2 - \dot{x}_1}{L}\right)^2$$

The potential energy of the system is

$$V = \frac{1}{2}k(\frac{3}{4}x_1 + \frac{1}{4}x_2)^2 + \frac{1}{2}kx_2^2$$

The lagrangian is

$$L = \frac{1}{2}m\left(\frac{\dot{x}_1 + \dot{x}_2}{2}\right)^2 + \frac{1}{2}I\left(\frac{\dot{x}_2 - \dot{x}_1}{L}\right)^2$$

$$- \frac{1}{2}k\left(\frac{3}{4}x_1 + \frac{1}{4}x_2\right)^2 - \frac{1}{2}kx_2^2$$

If variations δx_1 and δx_2 are introduced, the work done by the external moment is

$$\delta W = M(t)\delta\left(\frac{x_2 - x_1}{L}\right) = -\frac{1}{L}M(t)\,\delta x_1 + \frac{1}{L}M(t)\,\delta x_2$$

Application of Lagrange's equations leads to

$$\frac{d}{dt}\left(\frac{\partial L}{\partial \dot{x}_1}\right) - \frac{\partial L}{\partial x_1} = Q_1$$

$$\frac{d}{dt}\left[m\left(\frac{\dot{x}_1 + \dot{x}_2}{2}\right)\left(\frac{1}{2}\right) + I\left(\frac{\dot{x}_2 - \dot{x}_1}{L}\right)\left(-\frac{1}{L}\right)\right] + k\left(\frac{3}{4}x_1 + \frac{1}{4}x_2\right)\left(\frac{3}{4}\right) = -\frac{1}{L}M(t)$$

$$\frac{d}{dt}\left(\frac{\partial L}{\partial \dot{x}_2}\right) - \frac{\partial L}{\partial x_2} = Q_2$$

$$\frac{d}{dt}\left[m\left(\frac{\dot{x}_1 + \dot{x}_2}{2}\right)\left(\frac{1}{2}\right) + I\left(\frac{\dot{x}_2 - \dot{x}_1}{L}\right)\left(\frac{1}{L}\right)\right] + \left[k\left(\frac{3}{4}x_1 + \frac{1}{4}x_2\right)\left(\frac{1}{4}\right) + kx_2\right] = \frac{1}{L}M(t)$$

Rearranging and writing in matrix form leads to

$$\begin{bmatrix} \dfrac{m}{4} + \dfrac{I}{L^2} & \dfrac{m}{4} - \dfrac{I}{L^2} \\ \dfrac{m}{4} - \dfrac{I}{L^2} & \dfrac{m}{4} + \dfrac{I}{L^2} \end{bmatrix}\begin{bmatrix} \ddot{x}_1 \\ \ddot{x}_2 \end{bmatrix} + \begin{bmatrix} \dfrac{9}{16}k & \dfrac{3}{16}k \\ \dfrac{3}{16}k & \dfrac{17}{16}k \end{bmatrix}\begin{bmatrix} x_1 \\ x_2 \end{bmatrix} = \begin{bmatrix} -\dfrac{1}{L}M(t) \\ \dfrac{1}{L}M(t) \end{bmatrix}$$

5.6 Use Lagrange's equations to derive the differential equations governing the motion of the system of Fig. 5-5 using x_1, x_2, and θ as generalized coordinates. Write the differential equations in matrix form.

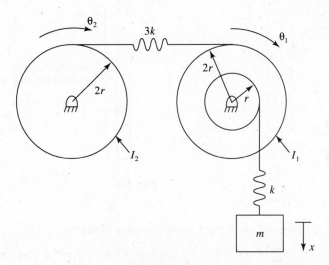

Fig. 5-5

The kinetic energy of the system at an arbitrary instant is

$$T = \tfrac{1}{2}I_1\dot{\theta}_1^2 + \tfrac{1}{2}I_2\dot{\theta}_2^2 + \tfrac{1}{2}m\dot{x}^2$$

The system's potential energy at an arbitrary instant is

$$V = \tfrac{1}{2}k(x - r\theta_1)^2 + \tfrac{1}{2}3k(2r\theta_1 - 2r\theta_2)^2$$

The lagrangian is

$$L = \tfrac{1}{2}I_1\dot{\theta}_1^2 + \tfrac{1}{2}I_2\dot{\theta}_2^2 + \tfrac{1}{2}m\dot{x}^2 - \tfrac{1}{2}k(x - r\theta_1)^2 - \tfrac{1}{2}3k(2r\theta_1 - 2r\theta_2)^2$$

Application of Lagrange's equations leads to

$$\frac{d}{dt}\left(\frac{\partial L}{\partial \dot{\theta}_1}\right) - \frac{\partial L}{\partial \theta_1} = 0$$

$$\frac{d}{dt}(I\dot{\theta}_1) + [k(x - r\theta_1)(-r) + 3k(2r\theta_1 - 2r\theta_2)(2r)] = 0$$

$$\frac{d}{dt}\left(\frac{\partial L}{\partial \dot{\theta}_2}\right) - \frac{\partial L}{\partial \theta_2} = 0$$

$$\frac{d}{dt}(I_2\dot{\theta}_2) + 3k(2r\theta_1 - 2r\theta_2)(-2r) = 0$$

$$\frac{d}{dt}\left(\frac{\partial L}{\partial \dot{x}}\right) - \frac{\partial L}{\partial x} = 0$$

$$\frac{d}{dt}(m\dot{x}) + k(x - r\theta_1)(1) = 0$$

Rearranging and writing in matrix form leads to

$$\begin{bmatrix} I_1 & 0 & 0 \\ 0 & I_2 & 0 \\ 0 & 0 & m \end{bmatrix}\begin{bmatrix} \ddot{\theta}_1 \\ \ddot{\theta}_2 \\ \ddot{x} \end{bmatrix} + \begin{bmatrix} 13kr^2 & -12kr^2 & -kr \\ -12kr^2 & 12kr^2 & 0 \\ -kr & 0 & k \end{bmatrix}\begin{bmatrix} \theta_1 \\ \theta_2 \\ x \end{bmatrix} = \begin{bmatrix} 0 \\ 0 \\ 0 \end{bmatrix}$$

5.7   Use Lagrange's equations to derive the differential equations governing the motion of the system of Fig. 5-6 using θ_1 and θ_2 as generalized coordinates.

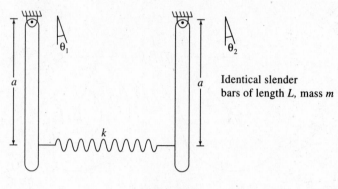

Identical slender
bars of length L, mass m

Fig. 5-6

The kinetic energy of the system at an arbitrary instant is

$$T = \tfrac{1}{2}m(\tfrac{1}{2}L\dot{\theta}_1)^2 + \tfrac{1}{2}\tfrac{1}{12}mL^2\dot{\theta}_1^2 + \tfrac{1}{2}m(\tfrac{1}{2}L\dot{\theta}_2)^2 + \tfrac{1}{2}\tfrac{1}{12}mL^2\dot{\theta}_2^2$$

The potential energy of the system at an arbitrary instant is

$$V = -mg\frac{L}{2}\cos\theta_1 - mg\frac{L}{2}\cos\theta_2 + \frac{1}{2}k(a\sin\theta_2 - a\sin\theta_1)^2$$

The lagrangian is

$$L = \frac{1}{2}\frac{1}{3}mL^2\dot{\theta}_1{}^2 + \frac{1}{2}\frac{1}{3}mL^2\dot{\theta}_2{}^2 + mg\frac{L}{2}\cos\theta_1 + mg\frac{L}{2}\cos\theta_2$$

$$-\frac{1}{2}k(a\sin\theta_2 - a\sin\theta_1)^2$$

Application of Lagrange's equations leads to

$$\frac{d}{dt}\left(\frac{\partial L}{\partial\dot{\theta}_1}\right) - \frac{\partial L}{\partial\theta_1} = 0$$

$$\frac{d}{dt}\left(\frac{1}{3}mL^2\dot{\theta}_1\right) + \left[mg\frac{L}{2}\sin\theta_1 + k(a\sin\theta_2 - a\sin\theta_1)(-a\cos\theta_1)\right] = 0$$

$$\frac{d}{dt}\left(\frac{\partial L}{\partial\dot{\theta}_2}\right) - \frac{\partial L}{\partial\theta_2} = 0$$

$$\frac{d}{dt}\left(\frac{1}{3}mL^2\dot{\theta}_2\right) + \left[mg\frac{L}{2}\sin\theta_2 + k(a\sin\theta_2 - a\sin\theta_1)(a\cos\theta_2)\right] = 0$$

Linearizing and rearranging leads to

$$\frac{1}{3}mL^2\ddot{\theta}_1 + \left(mg\frac{L}{2} + ka^2\right)\theta_1 - ka^2\theta_2 = 0$$

$$\frac{1}{3}mL^2\ddot{\theta}_2 - ka^2\theta_1 + \left(mg\frac{L}{2} + ka^2\right)\theta_2 = 0$$

5.8   The identical disks of mass m and radius r of Fig. 5-7 roll without slip. Use Lagrange's equations to derive the governing differential equations using x_1 and x_2 as generalized coordinates.

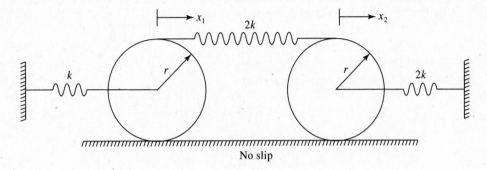

Fig. 5-7

The kinetic energy of the system at an arbitrary instant is

$$T = \frac{1}{2}m\dot{x}_1^2 + \frac{1}{2}\frac{1}{2}mr^2\left(\frac{\dot{x}_1}{r}\right)^2 + \frac{1}{2}m\dot{x}_2^2 + \frac{1}{2}\frac{1}{2}mr^2\left(\frac{\dot{x}_2}{r}\right)^2$$

The potential energy of the system at an arbitrary instant is

$$V = \tfrac{1}{2}kx_1^2 + \tfrac{1}{2}2kx_2^2 + \tfrac{1}{2}2k(2x_2 - 2x_1)^2$$

The Lagrangian is

$$L = \tfrac{1}{2}\tfrac{3}{2}m\dot{x}_1^2 + \tfrac{1}{2}\tfrac{3}{2}m\dot{x}_2^2 - \tfrac{1}{2}kx_1^2 - \tfrac{1}{2}2kx_2^2 - \tfrac{1}{2}2k(2x_2 - 2x_1)^2$$

Application of Lagrange's equations leads to

$$\frac{d}{dt}\left(\frac{3}{2}m\dot{x}_1\right) + [kx_1 + 2k(2x_2 - 2x_1)(-2)] = 0$$

$$\frac{d}{dt}\left(\frac{3}{2}m\dot{x}_2\right) + [2kx_2 + 2k(2x_2 - 2x_2)(2)] = 0$$

Rearranging and writing in matrix form yields

$$\begin{bmatrix} \frac{3}{2}m & 0 \\ 0 & \frac{3}{2}m \end{bmatrix}\begin{bmatrix} \ddot{x}_1 \\ \ddot{x}_2 \end{bmatrix} + \begin{bmatrix} 9k & -8k \\ -8k & 10k \end{bmatrix}\begin{bmatrix} x_1 \\ x_2 \end{bmatrix} = \begin{bmatrix} 0 \\ 0 \end{bmatrix}$$

5.9   Use Lagrange's equations to derive the differential equations governing the motion of the system of Fig. 5-8 using x_1, x_2, and x_3 as generalized coordinates.

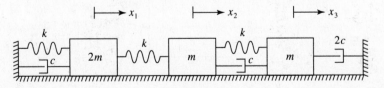

Fig. 5-8

The kinetic energy of the system at an arbitrary instant is

$$T = \tfrac{1}{2}2m\dot{x}_1^2 + \tfrac{1}{2}m\dot{x}_2^2 + \tfrac{1}{2}m\dot{x}_3^2$$

The potential energy of the system at an arbitrary instant is

$$V = \tfrac{1}{2}kx_1^2 + \tfrac{1}{2}k(x_2 - x_1)^2 + \tfrac{1}{2}k(x_3 - x_2)^2$$

If the variations δx_1, δx_2, and δx_3 are introduced when the system is in an arbitrary state, the work done by the forces in the viscous dampers is

$$\delta W = -c\dot{x}_1\,\delta x_1 - c(\dot{x}_3 - \dot{x}_2)\,\delta\,(x_3 - x_2) - 2c\dot{x}_3\,\delta x_3$$
$$= -c\dot{x}_1\,\delta x_1 - (-c\dot{x}_3 + c\dot{x}_2)\,\delta x_2 - (-c\dot{x}_2 + 3c\dot{x}_3)\,\delta x_3$$

Application of Lagrange's equations yields

$$\frac{d}{dt}\left(\frac{\partial L}{\partial \dot{x}_1}\right) - \frac{\partial L}{\partial x_1} = Q_1$$

$$\frac{d}{dt}(2m\dot{x}_1) + [kx_1 + k(x_2 - x_1)(-1)] = -c\dot{x}_1$$

$$\frac{d}{dt}\left(\frac{\partial L}{\partial \dot{x}_2}\right) - \frac{\partial L}{\partial x_2} = Q_2$$

$$\frac{d}{dt}(m\dot{x}_2) + [k(x_2 - x_1)(1) + k(x_3 - x_2)(-1)] = -c\dot{x}_2 + c\dot{x}_3$$

$$\frac{d}{dt}\left(\frac{\partial L}{\partial \dot{x}_3}\right) - \frac{\partial L}{\partial x_3} = Q_3$$

$$\frac{d}{dt}(m\dot{x}_3) + k(x_3 - x_2)(1) = c\dot{x}_2 - 3c\dot{x}_3$$

Rearranging and writing in matrix form leads to

$$\begin{bmatrix} 2m & 0 & 0 \\ 0 & m & 0 \\ 0 & 0 & m \end{bmatrix}\begin{bmatrix} \ddot{x}_1 \\ \ddot{x}_2 \\ \ddot{x}_3 \end{bmatrix} + \begin{bmatrix} c & 0 & 0 \\ 0 & c & -c \\ 0 & -c & 3c \end{bmatrix}\begin{bmatrix} \dot{x}_1 \\ \dot{x}_2 \\ \dot{x}_3 \end{bmatrix} + \begin{bmatrix} 2k & -k & 0 \\ -k & 2k & -k \\ 0 & -k & k \end{bmatrix}\begin{bmatrix} x_1 \\ x_2 \\ x_3 \end{bmatrix} = \begin{bmatrix} 0 \\ 0 \\ 0 \end{bmatrix}$$

5.10   Use Lagrange's equations to derive the differential equations governing the motion of the system of Fig. 5-9 using x and θ as generalized coordinates.

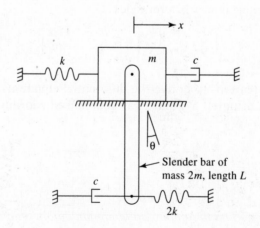

Fig. 5-9

The kinetic energy of the system at an arbitrary instant is

$$T = \tfrac{1}{2}m\dot{x}^2 + \tfrac{1}{2}2m(\dot{x} + \tfrac{1}{2}L\dot{\theta})^2 + \tfrac{1}{2}\tfrac{1}{12}2mL^2\dot{\theta}^2$$

The potential energy of the system at an arbitrary instant is

$$V = \frac{1}{2}kx^2 + \frac{1}{2}2k(x + L\theta)^2 - 2mg\frac{L}{2}\cos\theta$$

If the variations δx, and $\delta\theta$ are introduced at an arbitrary instant, the work done by the viscous damping forces is

$$\delta W = -c\dot{x}\,\delta x - c(\dot{x} + L\dot{\theta})\,\delta(x + L\theta)$$

$$= -c(2\dot{x} + L\dot{\theta})\,\delta x - cL(\dot{x} + L\dot{\theta})\,\delta\theta$$

Application of Lagrange's equations leads to

$$\frac{d}{dt}\left(\frac{\partial L}{\partial \dot{x}}\right) - \frac{\partial L}{\partial x} = 0$$

$$\frac{d}{dt}\left[m\dot{x} + 2m\left(\dot{x} + \frac{L}{2}\dot{\theta}\right)(1)\right] + [kx + 2k(x + L\theta)(1)] = -2c\dot{x} - cL\theta$$

$$\frac{d}{dt}\left(\frac{\partial L}{\partial \dot{\theta}}\right) - \frac{\partial L}{\partial \theta} = Q_2$$

$$\frac{d}{dt}\left[2m\left(\dot{x} + \frac{L}{2}\dot{\theta}\right)\left(\frac{L}{2}\right) + \frac{1}{12}2mL^2\dot{\theta}\right] + 2k(x + L\theta)(L) + mgL\sin\theta = -cL\dot{x} - cL^2\dot{\theta}$$

Rearranging and linearizing leads to

$$\begin{bmatrix} 3m & mL \\ mL & \frac{2}{3}mL^2 \end{bmatrix}\begin{bmatrix} \ddot{x} \\ \ddot{\theta} \end{bmatrix} + \begin{bmatrix} 2c & cL \\ cL & cL^2 \end{bmatrix}\begin{bmatrix} \dot{x} \\ \dot{\theta} \end{bmatrix} + \begin{bmatrix} 3k & 2kL \\ 2kL & 2kL^2 + mgL \end{bmatrix}\begin{bmatrix} x \\ \theta \end{bmatrix} = \begin{bmatrix} 0 \\ 0 \end{bmatrix}$$

5.11 Use stiffness influence coefficients to determine the stiffness matrix for the system of Fig. 5-1.

The first column of the stiffness matrix is obtained by setting $x_1 = 1$, $x_2 = 0$, and $x_3 = 0$ and solving for the applied forces as shown in Fig. 5-10a. Summing the forces to zero on each free body diagram leads to $k_{11} = 3k$, $k_{21} = -2k$, and $k_{31} = 0$. The second column is obtained by setting $x_1 = 0$, $x_2 = 1$, and $x_3 = 0$ and solving for the applied forces as shown in Fig. 5-10b. Summing fores to zero on each free body diagram of Fig. 5-10b leads to $k_{12} = -2k$, $k_{22} = 3k$, and $k_{32} = -k$. The third column is obtained by setting $x_1 = 0$, $x_2 = 0$, and $x_3 = 1$ and solving for the applied forces as shown in Fig. 5-10c. Summing forces to zero on each free body diagram of Fig. 5-10c leads to $k_{13} = 0$, $k_{23} = -k$, and $k_{33} = k$. Hence the stiffness matrix is

$$\mathbf{K} = \begin{bmatrix} 3k & -2k & 0 \\ -2k & 3k & -k \\ 0 & -k & k \end{bmatrix}$$

5.12 Use stiffness influence coefficients to determine the stiffness matrix for the system of Fig. 5-3 using x and θ as generalized coordinates.

The first column of the stiffness matrix is obtained by setting $x = 1$ and $\theta = 0$ and solving for the applied force and moment as shown in Fig. 5-11a. Summing forces to zero leads to $k_{11} = 2k$. Summing moments about the mass center to zero leads to $k_{21} + kL/4 - kL/2 = 0 \rightarrow k_{21} = kL/4$.

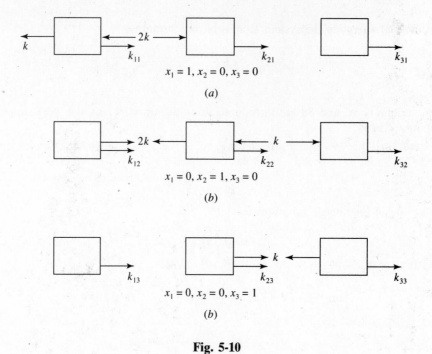

$x_1 = 1, x_2 = 0, x_3 = 0$

(a)

$x_1 = 0, x_2 = 1, x_3 = 0$

(b)

$x_1 = 0, x_2 = 0, x_3 = 1$

(b)

Fig. 5-10

The second column is obtained by setting $x = 0$ and $\theta = 1$ and solving for the applied load and moment as shown in Fig. 5-11b. Summing forces to zero leads to $k_{12} = kL/4$ while summing moments about the mass center leads to $k_{22} - kL/2(L/2) - kL/4(L/4) \rightarrow k_{22} = 5kL^2/16$. Hence the stiffness matrix is

$$\mathbf{K} = \begin{bmatrix} 2k & \frac{1}{4}kL \\ \frac{1}{4}kL & \frac{5}{16}kL^2 \end{bmatrix}$$

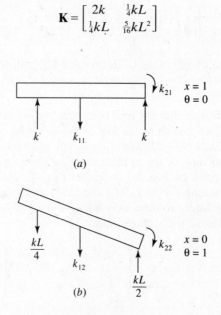

Fig. 5-11

5.13 Use stiffness influence coefficients to determine the stiffness matrix for the system of Fig. 5-3 using x_1 and x_2 as generalized coordinates.

The first column of the stiffness matrix is obtained by setting $x_1 = 1$ and $x_2 = 0$ and solving for the applied loads shown in Fig. 5-12a. Summing moments about each end of the bar to zero,

$$\sum M_A = 0 = k_{21}(L) - \tfrac{3}{4}k(\tfrac{1}{4}L) \;\rightarrow\; k_{21} = \tfrac{3}{16}k$$

$$\sum M_B = 0 = k_{11}(L) - \tfrac{3}{4}k(\tfrac{3}{4}L) \;\rightarrow\; k_{11} = \tfrac{9}{16}k$$

The second column is obtained by setting $x_1 = 0$ and $x_2 = 1$ and solving for the applied forces of Fig. 5-12b. Summing moments about each end of the bar to zero,

$$\sum M_A = 0 = k_{22}(L) - k(L) - \tfrac{1}{4}k(\tfrac{1}{4}L) \;\rightarrow\; k_{22} = \tfrac{17}{16}k$$

$$\sum M_B = 0 = k_{12}(L) - \tfrac{1}{4}k(\tfrac{3}{4}L) \;\rightarrow\; k_{12} = \tfrac{3}{16}k$$

The stiffness matrix is

$$\mathbf{K} = \begin{bmatrix} \tfrac{9}{16}k & \tfrac{3}{16}k \\ \tfrac{3}{16}k & \tfrac{17}{16}k \end{bmatrix}$$

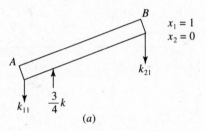

(a)

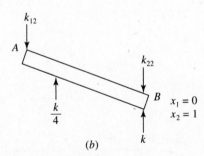

(b)

Fig. 5-12

5.14 Use stiffness influence coefficients to determine the stiffness matrix for the system of Fig. 5-5 using x, θ_1, and θ_2 as generalized coordinates.

The first column of the stiffness matrix is obtained by setting $x = 1$, $\theta_1 = 0$, and $\theta_2 = 0$ and solving for the forces and moments shown in Fig. 5-13a. Summing forces to zero on the block and moments about the pin supports to zero lead to $k_{11} = k$, $k_{21} = -kr$, and $k_{31} = 0$. The second column of the stiffness matrix is obtained by setting $x = 0$, $\theta_1 = 1$, and $\theta_2 = 0$ and solving for the forces and moments shown in Fig. 5-13b. Application of the equations of equilibrium to the free body diagram leads to $k_{12} = -kr$, $k_{22} = 13kr^2$, and $k_{32} = -12kr^2$. The third column is obtained by setting $x = 0$, $\theta_1 = 0$, and $\theta_2 = 1$ and solving for the forces and moments shown in Fig. 5-13c. Application of the equations of equilibrium to these free body diagrams leads to $k_{13} = 0$, $k_{23} = -12kr^2$, and $k_{33} = 12kr^2$. Thus the stiffness matrix is

$$\mathbf{K} = \begin{bmatrix} k & -kr & 0 \\ -kr & 13kr^2 & -12kr^2 \\ 0 & -12kr^2 & 12kr^2 \end{bmatrix}$$

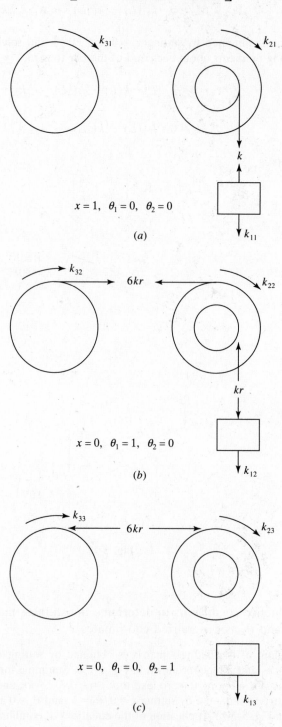

Fig. 5-13

5.15 Use stiffness influence coefficients to derive the stiffness matrix for the system of Fig. 5-9 using x and θ as generalized coordinates and assuming small θ.

The first column of the stiffness matrix is obtained by setting $x = 1$ and $\theta = 0$ and solving for the applied force and moment on the free body diagrams of Fig. 5-14a. Application of the equations of equilibrium to these free body diagrams leads to $k_{11} = 3k$ and $k_{21} = 2kL$. The second column is obtained by setting $x = 0$ and $\theta = 1$ and solving for the applied force and moment as shown on the free body diagrams of Fig. 5-14b. Application of the equations of equilibrium to these free body diagrams leads to $k_{12} = 2kL$ and $k_{22} = 2kL^2 + mgL$. Thus the stiffness matrix is

$$\mathbf{K} = \begin{bmatrix} 3k & 2kL \\ 2kL & 2kL^2 + mgL \end{bmatrix}$$

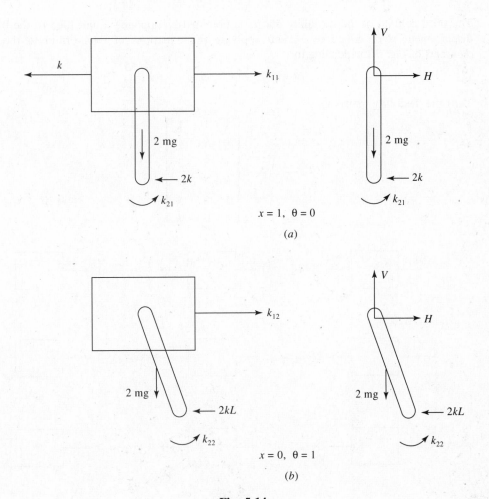

Fig. 5-14

5.16 Use flexibility influence coefficients to determine the flexibility matrix for the system of Fig. 5-1 using x_1, x_2, and x_3 as generalized coordinates.

The first column of the flexibility matrix is obtained by applying a unit load to the block whose displacement is x_1. The resulting displacements of the blocks are the flexibility influence coefficients a_{11}, a_{21}, and a_{31}. Application of the equations of static equilibrium to the free body diagrams of Fig. 5-15a leads to

$$-ka_{11} + 2k(a_{21} - a_{11}) + 1 = 0$$
$$-2k(a_{21} - a_{11}) + k(a_{31} - a_{21}) = 0$$
$$k(a_{31} - a_{21}) = 0$$

which are solved simultaneously, yielding

$$a_{11} = \frac{1}{k}, \qquad a_{21} = \frac{1}{k}, \qquad a_{31} = \frac{1}{k}$$

The second column of the flexibility matrix is obtained by applying a unit load to the middle block and applying the equations of equilibrium to the free body diagrams of Fig. 5-15b leading to

$$a_{12} = \frac{1}{k}, \qquad a_{22} = \frac{3}{2k}, \qquad a_{32} = \frac{3}{2k}$$

The third column of the flexibility matrix is obtained by applying a unit load to the block whose displacement is described by x_3 and applying the equations of equilibrium to the free body diagrams of Fig. 5-15c, leading to

$$a_{13} = \frac{1}{k}, \qquad a_{23} = \frac{3}{2k}, \qquad a_{33} = \frac{5}{2k}$$

Thus the flexibility matrix is

$$\mathbf{A} = \begin{bmatrix} \dfrac{1}{k} & \dfrac{1}{k} & \dfrac{1}{k} \\[2mm] \dfrac{1}{k} & \dfrac{3}{2k} & \dfrac{3}{2k} \\[2mm] \dfrac{1}{k} & \dfrac{3}{2k} & \dfrac{5}{2k} \end{bmatrix}$$

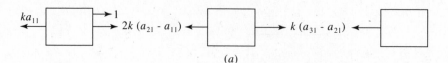

(a)

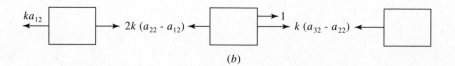

(b)

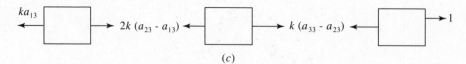

(c)

Fig. 5-15

5.17   Use flexibility influence coefficients to determine the flexibility matrix for the system of Fig. 5-3 using x and θ as generalized coordinates.

The first column of the flexibility matrix is obtained by applying a unit load to the mass center of the bar and setting $x = a_{11}$ and $\theta = a_{21}$. Application of the equations of equilibrium to the free body diagram of Fig. 5-16a leads to

$$\sum F = 0 = 1 - k\left(a_{11} - \frac{L}{4}a_{21}\right) - k\left(a_{11} + \frac{L}{2}a_{21}\right)$$

$$\sum M_G = 0 = k\left(a_{11} - \frac{L}{4}a_{21}\right)\frac{L}{4} - k\left(a_{11} + \frac{L}{2}a_{21}\right)\frac{L}{2}$$

The above equations are solved simultaneously, leading to

$$a_{11} = \frac{5}{9k}, \qquad a_{21} = -\frac{4}{9kL}$$

The second column of the flexibility matrix is obtained by applying a unit clockwise moment to the bar, applying the equations of equilibrium to the free body diagram of Fig. 5-16b, and solving simultaneously for the flexibility influence coefficients, leading to

$$a_{12} = -\frac{4}{9kL}, \qquad a_{22} = \frac{32}{9kL^2}$$

Thus the flexibility matrix is

$$\mathbf{A} = \begin{bmatrix} \dfrac{5}{9k} & -\dfrac{4}{9kL} \\ -\dfrac{4}{9kL} & \dfrac{32}{9kL^2} \end{bmatrix}$$

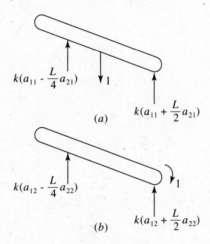

Fig. 5-16

5.18 Use flexibility influence coefficients to derive the flexibility matrix for the system of Fig. 5-17 using x_1, x_2, and θ as generalized coordinates.

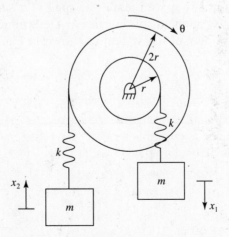

Fig. 5-17

The first column of the flexibility matrix is obtained by applying a unit load to the block whose displacement is described by x_1. Application of the equations of equilibrium to the free body diagrams of Fig. 5-18 leads to

$$\sum F = 0 = 1 - k(a_{11} - ra_{21})$$

$$\sum M_0 = 0 = k(a_{11} - ra_{21})r - k(2ra_{21} - a_{31})(2r)$$

$$\sum F = 0 = k(2ra_{21} - a_{31})$$

When a simultaneous solution of the previous equations is attempted, an inconsistency results (for example, $1 = 0$). This implies that the flexibility matrix does not exist. This is because the system is unrestrained and the stiffness matrix is singular.

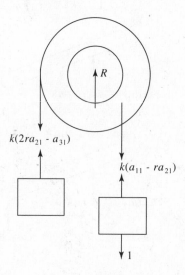

Fig. 5-18

5.19   Three machines are equally spaced along the span of a simply supported beam of elastic modulus E and mass moment of inertia I. Determine the flexibility matrix for a 3-degree-of-freedom model of the system as shown in Fig. 5-19.

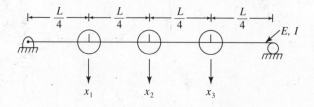

Fig. 5-19

The deflection of a particle a distance z along the neutral axis of a simply supported beam, measured from the left support, due to a concentrated unit load applied a distance a from the left support is

$$y(z) = \frac{L^3}{6EI}\left(1 - \frac{a}{L}\right)\left[\frac{a}{L}\left(2 - \frac{a}{L}\right)\frac{z}{L} - \left(\frac{z}{L}\right)^3\right]$$

for $z \leq a$. The elements of the third column of the flexibility matrix are the machine displacements induced by a unit concentrated load at $a = 3L/4$. Then

$$y(z) = \frac{L^3}{24EI}\left[\frac{15}{16}\frac{z}{L} - \left(\frac{z}{L}\right)^3\right]$$

and the flexibility influence coefficients are

$$a_{13} = y\left(\frac{L}{4}\right) = \frac{7L^3}{768EI}, \qquad a_{23} = y\left(\frac{L}{2}\right) = \frac{11L^3}{768EI}, \qquad a_{33} = y\left(\frac{3}{4}L\right) = \frac{3L^3}{256EI}$$

The second column of the flexibility matrix is determined by placing a unit concentrated load at $a = L/2$. Then

$$y(z) = \frac{L^3}{12EI}\left[\frac{3}{4}\frac{z}{L} - \left(\frac{z}{L}\right)^3\right]$$

Note that due to reciprocity and symmetry of the beam, only a_{22} must be calculated. To this end,

$$a_{22} = y\left(\frac{L}{2}\right) = \frac{L^3}{48EI}$$

Then from reciprocity, $a_{32} = a_{23}$, and from symmetry, $a_{12} = a_{32}$. Then from symmetry, $a_{11} = a_{33}$, and from reciprocity, $a_{21} = a_{12}$, and $a_{31} = a_{13}$. Thus the flexibility matrix is

$$\mathbf{A} = \frac{L^3}{EI}\begin{bmatrix} \frac{3}{256} & \frac{11}{768} & \frac{7}{768} \\ \frac{11}{768} & \frac{1}{48} & \frac{11}{768} \\ \frac{7}{768} & \frac{11}{768} & \frac{3}{256} \end{bmatrix}$$

5.20 Two degrees of freedom are to be used to model the vibrations of a fixed-fixed beam of length L, elastic modulus E, and cross-sectional moment of inertia I. Determine the flexibility matrix for this model assuming the generalized coordinates are displacements of equally spaced particles along the span of the beam.

The deflection of a particle along the neutral axis a distance z from the left support due to a concentrated unit load a distance a from the left support is

$$y(z) = \frac{L^3}{EI}\left[\frac{1}{2}\frac{a}{L}\left(1 - \frac{a}{L}\right)^2\left(\frac{z}{L}\right)^2 - \frac{1}{6}\left(1 - \frac{a}{L}\right)^2\left(1 + 2\frac{a}{L}\right)\left(\frac{z}{L}\right)^3\right]$$

for $z \leq a$. The second column of the flexibility matrix is obtained by setting $a = 2L/3$, leading to

$$y(z) = \frac{L^3}{27EI}\left[\left(\frac{z}{L}\right)^2 - \frac{7}{6}\left(\frac{z}{L}\right)^3\right]$$

Then
$$a_{12} = y\left(\frac{L}{3}\right) = \frac{11L^3}{4374EI}, \qquad a_{22} = y\left(\frac{2}{3}L\right) = \frac{16L^3}{4374EI}$$

From reciprocity, $a_{12} = a_{21}$, and from symmetry, $a_{11} = a_{22}$. Thus

$$\mathbf{A} = \frac{L^3}{4374EI}\begin{bmatrix} 16 & 11 \\ 11 & 16 \end{bmatrix}$$

5.21   A machine with a large moment of inertia is placed at the end of a cantilever beam. Because of the large moment of inertia, it is decided to include rotational effects in a model. Thus a 2-degree-of-freedom model with generalized coordinates, x, the displacement of the machine, and θ, the slope of the elastic curve at the end of the beam, are used. Determine the flexibility matrix for this model if the beam is of length L, elastic modulus E, and cross-sectional moment of inertia I.

Consider first a concentrated unit load at the end of the beam. From strength of materials, the deflection at the end of the beam is $a_{11} = L^3/(3EI)$, and the slope of the beam at its end is $a_{21} = L^2/(2EI)$. Then if a unit moment is applied to the end of the beam, the deflection at the end of the beam is $a_{12} = L^2/(2EI)$, and the slope of the elastic curve at the end is $a_{22} = L/(EI)$. Hence the flexibility matrix for this model is

$$\mathbf{A} = \frac{L}{EI}\begin{bmatrix} \dfrac{L^2}{3} & \dfrac{L}{2} \\[2ex] \dfrac{L}{2} & 1 \end{bmatrix}$$

5.22   Determine the flexibility matrix for the 4-degree-of-freedom system of Fig. 5-20.

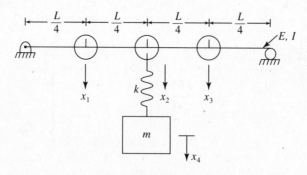

Fig. 5-20

Let x_1, x_2, and x_3 be the displacements of the particles on the beam, and let x_4 be the displacement of the particle attached to the beam through the spring. Note that the flexibility matrix for the beam without the additional mass-spring system is determined in Problem 5.19. The first column of the flexibility matrix is determined by placing a unit load acting on the first particle. Summing forces on the free body diagram of the block, shown in Fig. 5-21, shows that the force in the spring is zero. Thus the deflection of the beam is due only to a unit force applied to the beam, and the flexibility influence coefficients a_{ij}, $i = 1, 2, 3$, and $j = 1, 2, 3$, are calculated as in Problem 5.19. Also

$$a_{41} = a_{21} \qquad a_{42} = a_{22} \qquad a_{43} = a_{23}$$

Now consider a unit load applied to the hanging block. The force developed in the spring is 1. Hence the beam is analyzed as if a unit load were applied to the midspan. Also

$$k(a_{44} - a_{24}) = 1 \quad \rightarrow \quad a_{44} = \frac{1}{k} + a_{24} = \frac{1}{k} + a_{22}$$

Then using the results of Problem 5.19,

$$\mathbf{A} = \frac{L^3}{EI} \begin{bmatrix} \dfrac{3}{128} & \dfrac{11}{768} & \dfrac{7}{768} & \dfrac{11}{768} \\ \dfrac{11}{768} & \dfrac{1}{48} & \dfrac{11}{768} & \dfrac{1}{48} \\ \dfrac{7}{768} & \dfrac{11}{768} & \dfrac{3}{128} & \dfrac{11}{68} \\ \dfrac{11}{768} & \dfrac{1}{48} & \dfrac{11}{768} & \dfrac{1}{48} + \dfrac{EI}{kL^3} \end{bmatrix}$$

$$\uparrow k\,(a_{41} - a_{21}) = 0$$

Fig. 5-21

5.23 Derive the stiffness matrix for the 3-degree-of-freedom unrestrained torsional system of Fig. 5-22.

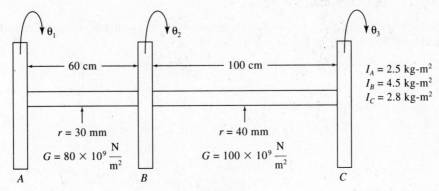

$I_A = 2.5$ kg-m^2
$I_B = 4.5$ kg-m^2
$I_C = 2.8$ kg-m^2

60 cm   100 cm

$r = 30$ mm $r = 40$ mm

$G = 80 \times 10^9 \dfrac{\text{N}}{\text{m}^2}$ $G = 100 \times 10^9 \dfrac{\text{N}}{\text{m}^2}$

A   B   C

Fig. 5-22

The torsional stiffnesses of the shafts are

$$k_{t_{AB}} = \frac{J_{AB}G_{AB}}{L_{AB}} = \frac{\dfrac{\pi}{2}(0.03\text{ m})^4 \left(80 \times 10^9\,\dfrac{\text{N}}{\text{m}^2}\right)}{0.6\text{ m}} = 1.70 \times 10^5\,\frac{\text{N-m}}{\text{rad}}$$

$$k_{t_{BC}} = \frac{J_{BC}G_{BC}}{L_{BC}} = \frac{\dfrac{\pi}{2}(0.04\text{ m})^4 \left(100 \times 10^9\,\dfrac{\text{N}}{\text{m}^2}\right)}{1.0\text{ m}} = 4.02 \times 10^5\,\frac{\text{N-m}}{\text{rad}}$$

Stiffness influence coefficients are used to show

$$\mathbf{K} = \begin{bmatrix} k_{t_{AB}} & -k_{t_{AB}} & 0 \\ -k_{t_{AB}} & k_{t_{AB}} + k_{t_{BC}} & -k_{t_{BC}} \\ 0 & -k_{t_{BC}} & k_{t_{BC}} \end{bmatrix} = 10^5 \begin{bmatrix} 1.70 & -1.70 & 0 \\ -1.70 & 5.72 & -4.02 \\ 0 & -4.02 & 4.02 \end{bmatrix}$$

5.24 The differential equations governing the motion of a 2-degree-of-freedom system are

$$\begin{bmatrix} m & 0 \\ 0 & m \end{bmatrix}\begin{bmatrix} \ddot{x}_1 \\ \ddot{x}_2 \end{bmatrix} + \begin{bmatrix} 2k & -k \\ -k & 3k \end{bmatrix}\begin{bmatrix} x_1 \\ x_2 \end{bmatrix} = \begin{bmatrix} 0 \\ 0 \end{bmatrix}$$

Determine the system's natural frequencies.

The natural frequencies are determined using Eq. (5.12):

$$\det |\mathbf{K} - \omega^2 \mathbf{M}| = 0$$

$$\det \left| \begin{bmatrix} 2k & -k \\ -k & 3k \end{bmatrix} - \omega^2 \begin{bmatrix} m & 0 \\ 0 & m \end{bmatrix} \right| = 0$$

$$\begin{vmatrix} 2k - \omega^2 m & -k \\ -k & 3k - \omega^2 m \end{vmatrix} = 0$$

$$(2k - \omega^2 m)(3k - \omega^2 m) - (-k)(-k) = 0$$

$$m^2 \omega^4 - 5km\omega^2 + 5k^2 = 0$$

The quadratic equation is used to solve for ω^2:

$$\omega^2 = \frac{5km \pm \sqrt{25k^2 m^2 - 4(m^2)(5k^2)}}{2m^2}$$

$$\omega = \left[\left(\frac{5}{2} \pm \frac{\sqrt{5}}{2} \right) \frac{k}{m} \right]^{1/2}$$

$$\omega_1 = 1.176 \sqrt{\frac{k}{m}}, \qquad \omega_2 = 1.902 \sqrt{\frac{k}{m}}$$

5.25 Determine the natural frequencies of the system of Fig. 5-3 if $m = 5$ kg, $I = 0.5$ kg-m^2, $L = 0.8$ km, and $k = 2 \times 10^9$ N/m.

Substituting the given values into the mass and stiffness matrices determined using x and θ as generalized coordinates in Problem 5.4 leads to

$$\mathbf{M} = \begin{bmatrix} 5 & 0 \\ 0 & 0.5 \end{bmatrix}, \qquad \mathbf{K} = \begin{bmatrix} 4 \times 10^5 & 4 \times 10^4 \\ 4 \times 10^4 & 4 \times 10^4 \end{bmatrix}$$

The natural frequencies are calculated using Eq. (5.12)

$$\left| \begin{bmatrix} 4 \times 10^5 & 4 \times 10^4 \\ 4 \times 10^4 & 4 \times 10^4 \end{bmatrix} - \omega^2 \begin{bmatrix} 5 & 0 \\ 0 & 0.5 \end{bmatrix} \right| = 0$$

$$\begin{vmatrix} 4 \times 10^5 - 5\omega^2 & 4 \times 10^4 \\ 4 \times 10^4 & 4 \times 10^4 - 0.5\omega^2 \end{vmatrix} = 0$$

$$(4 \times 10^5 - 5\omega^2)(4 \times 10^4 - 0.5\omega^2) - (4 \times 10^4)^2 = 0$$

$$2.5\omega^4 - 4 \times 10^5 \omega^2 + 1.44 \times 10^{10} = 0$$

$$\omega = \left[\frac{4 \times 10^5 \pm \sqrt{(4 \times 10^5)^2 - 4(2.5)(1.44 \times 10^{10})}}{2(2.5)} \right]^{1/2}$$

$$\omega_1 = 233.9 \frac{\text{rad}}{\text{s}}, \qquad \omega_2 = 324.5 \frac{\text{rad}}{\text{s}}$$

5.26 A 500-kg machine is placed 2 m from the left support of a 6-m fixed-fixed beam while a 375-kg machine is placed 4 m from the left support. Ignoring inertia effects of the beam, determine the natural frequencies of the system if $E = 200 \times 10^9$ N/m^2 and $I = 2.35 \times 10^{-6}$ m^4.

The system is modeled using 2-degrees-of-freedom with the generalized coordinates as the displacements of the machines. Using the results of Problem 5.20 with the given values substituted, the flexibility matrix for this model is

$$\mathbf{A} = \begin{bmatrix} 1.68 \times 10^{-6} & 1.16 \times 10^{-6} \\ 1.16 \times 10^{-6} & 1.68 \times 10^{-6} \end{bmatrix}$$

The appropriate mass matrix for the model is

$$\mathbf{M} = \begin{bmatrix} 500 & 0 \\ 0 & 375 \end{bmatrix}$$

The natural frequencies are calculated using Eq. (5.13):

$$\det |\omega^2 \mathbf{A}\mathbf{M} - \mathbf{I}| = 0$$

$$\det \left| 10^{-6}\omega^2 \begin{bmatrix} 1.68 & 1.16 \\ 1.16 & 1.68 \end{bmatrix} \begin{bmatrix} 500 & 0 \\ 0 & 375 \end{bmatrix} - \begin{bmatrix} 1 & 0 \\ 0 & 1 \end{bmatrix} \right| = 0$$

$$\begin{vmatrix} 8.4 \times 10^{-4}\omega^2 - 1 & 4.35 \times 10^{-4}\omega^2 \\ 5.8 \times 10^{-4}\omega^2 & 6.3 \times 10^{-4}\omega^2 - 1 \end{vmatrix} = 0$$

$$2.77 \times 10^{-7}\omega^4 - 1.47 \times 10^{-3}\omega^2 + 1 = 0$$

$$\omega = \left[\frac{1.47 \times 10^{-3} \pm \sqrt{(1.47 \times 10^{-3})^2 - 4(2.77 \times 10^{-7})(1)}}{2(2.77 \times 10^{-7})} \right]^{1/2}$$

$$\omega_1 = 28.3 \frac{\text{rad}}{\text{s}}, \qquad \omega_2 = 67.1 \frac{\text{rad}}{\text{s}}$$

5.27 The beam described in Problem 5.26 is made of a material of mass density 7800 kg/m³ and has a cross-sectional area of 4.36×10^{-3} m². Determine the natural frequencies of the system when inertia effects of the beam are approximated by adding particles of appropriate mass at the nodes.

The total mass of the beam is

$$m_b = \rho A L = (7800 \text{ kg})(4.36 \times 10^{-3} \text{ m}^2)(6 \text{ m}) = 204.0 \text{ kg}$$

The mass added at each node represents the mass of a segment of the beam. The boundary between segments for adjacent nodes is midway between the nodes. The boundary of a segment for a node adjacent to a support is midway between the node and the support. The inertia of particles near the supports is neglected. If included equally with other particles, the inertia of the beam would be overapproximated. Thus for this model, the mass added to each node is $m_3/3 = 68.0$ kg. Thus the mass matrix becomes

$$\mathbf{M} = \begin{bmatrix} 568 & 0 \\ 0 & 443 \end{bmatrix}$$

A procedure similar to that used in Problem 5.26 is followed, leading to

$$\omega_1 = 26.4 \frac{\text{rad}}{\text{s}}, \qquad \omega_2 = 62.3 \frac{\text{rad}}{\text{s}}$$

5.28 Determine the mode shape vectors for the system of Problem 5.25 using x and θ as generalized coordinates.

The normal mode solution implies that the ratios of the values of the generalized coordinates are constant for each mode. Let X be the displacement of the mass center of the bar at an arbitrary instant for either mode, and let Θ be the angular rotation of the bar at this instant. The mode shapes are calculated using the results of Problem 5.25:

$$\begin{bmatrix} 4 \times 10^5 - 5\omega^2 & 4 \times 10^4 \\ 4 \times 10^4 & 4 \times 10^4 \times 0.5\omega^2 \end{bmatrix} \begin{bmatrix} X \\ \Theta \end{bmatrix} = \begin{bmatrix} 0 \\ 0 \end{bmatrix}$$

The two equations represented by the previous matrix system are dependent. From the top equation,

$$(4 \times 10^5 - 5\omega^2)X + 4 \times 10^4 \Theta = 0$$

$$\Theta = -\frac{4 \times 10^5 - 5\omega^2}{4 \times 10^4} X$$

Substituting $\omega = 233.9$ rad/s leads to $\Theta = -3.16X$. Substituting $\omega = 324.5$ rad/s leads to $\Theta = 3.16X$. Arbitrarily setting $X = 1$, the mode shape vectors are

$$\mathbf{X}_1 = \begin{bmatrix} 1 \\ -3.16 \end{bmatrix}, \qquad \mathbf{X}_2 = \begin{bmatrix} 1 \\ 3.16 \end{bmatrix}$$

5.29   Both ends of the bar of Problems 5.25 and 5.28 are given a 1.8-mm displacement from equilibrium, and the bar released from rest. Determine the time history of the resulting motion.

Use of the normal mode solution leads to four linearly independent solutions to the homogeneous set of differential equations. The most general solution is a linear combination of all homogeneous solutions. To this end,

$$\mathbf{x}(t) = \tilde{C}_1\mathbf{X}_1 e^{i\omega_1 t} + \tilde{C}_2\mathbf{X}_1 e^{-i\omega_1 t} + \tilde{C}_3\mathbf{X}_2 e^{i\omega_2 t} + \tilde{C}_4\mathbf{X}_2 e^{-i\omega_2 t}$$

Euler's identity is used to replace the complex exponentials by trigonometric functions:

$$\mathbf{x}(t) = C_1\mathbf{X}_1 \cos \omega_1 t + C_2\mathbf{X}_1 \sin \omega_1 t + C_3\mathbf{X}_2 \cos \omega_2 t + C_4\mathbf{X}_2 \sin \omega_2 t$$

where C_1, C_2, C_3, and C_4 are constants of integration. The initial conditions are

$$\mathbf{x}(0) = \begin{bmatrix} 0.0018 \\ 0 \end{bmatrix}, \qquad \dot{\mathbf{x}}(0) = \begin{bmatrix} 0 \\ 0 \end{bmatrix}$$

whose application lead to

$$x(0) = 0.0018 = C_1 + C_3$$
$$\theta(0) = 0 = -3.16 C_1 + 3.16 C_3$$
$$\dot{x}(0) = 0 = \omega_1 C_2 + \omega_2 C_4$$
$$\dot{\theta}(0) = 0 = -3.16\omega_1 C_2 + 3.16\omega_2 C_4$$

whose solution is $C_1 = C_3 = 0.0009$, $C_2 = C_4 = 0$, leading to

$$\begin{bmatrix} x_1(t) \\ x_2(t) \end{bmatrix} = \begin{bmatrix} 0.0009 \\ -0.00284 \end{bmatrix} \cos 233.9t + \begin{bmatrix} 0.0009 \\ 0.00284 \end{bmatrix} \cos 324.5t$$

5.30   Determine the natural frequencies of the system of Fig. 5-1.

The natural frequencies are the square roots of the eigenvalues of $\mathbf{M}^{-1}\mathbf{K}$. To this end,

$$\mathbf{M}^{-1}\mathbf{K} = \frac{k}{m}\begin{bmatrix} 1 & 0 & 0 \\ 0 & \frac{1}{2} & 0 \\ 0 & 0 & \frac{1}{2} \end{bmatrix}\begin{bmatrix} 3 & -2 & 0 \\ -2 & 3 & -1 \\ 0 & -1 & 1 \end{bmatrix} = \frac{k}{m}\begin{bmatrix} 3 & -2 & 0 \\ -1 & \frac{3}{2} & -\frac{1}{2} \\ 0 & -\frac{1}{2} & \frac{1}{2} \end{bmatrix}$$

The eigenvalues are calculated from

$$\det |\mathbf{M}^{-1}\mathbf{K} - \lambda\mathbf{I}| = 0$$

$$\begin{vmatrix} 3\dfrac{k}{m} - \lambda & -2\dfrac{k}{m} & 0 \\[2mm] -\dfrac{k}{m} & \dfrac{3}{2}\dfrac{k}{m} - \lambda & -\dfrac{1}{2}\dfrac{k}{m} \\[2mm] 0 & -\dfrac{1}{2}\dfrac{k}{m} & \dfrac{1}{2}\dfrac{k}{m} - \lambda \end{vmatrix} = 0$$

$$-\beta^3 + 5\beta^2 - \tfrac{9}{2}\beta + \tfrac{1}{2} = 0, \qquad \beta = \lambda\frac{m}{k}$$

The roots of the cubic equations are 0.129, 1, 3.870. Thus the natural frequencies are

$$\omega_1 = 0.359\sqrt{\frac{k}{m}}, \qquad \omega_2 = \sqrt{\frac{k}{m}}, \qquad \omega_3 = 1.97\sqrt{\frac{k}{m}}$$

5.31   Determine the mode shape vectors for the system of Fig. 5-1 and Problem 5.30.

Let $\mathbf{X}_i = [X_{i1} \quad X_{i2} \quad X_{i3}]^T$ be the mode shape vector corresponding to ω_i. The equations from which the mode shape vectors are determined are

$$\begin{bmatrix} 3\phi - \lambda_i & -2\phi & 0 \\ -2\phi & \frac{3}{2}\phi - \lambda_i & -\frac{1}{2}\phi \\ 0 & -\frac{1}{2}\phi & \frac{1}{2}\phi - \lambda_i \end{bmatrix} \begin{bmatrix} X_{i1} \\ X_{i2} \\ X_{i3} \end{bmatrix} = \begin{bmatrix} 0 \\ 0 \\ 0 \end{bmatrix}$$

where $\phi = k/m$. Since the previous equations are dependent, only two must be used in determining the mode shapes. To this end, arbitrarily choose $X_{i2} = 1$. Then

$$X_{i1} = \frac{2\phi}{3\phi - \lambda_i}, \qquad X_{i3} = \frac{\phi}{2(\frac{1}{2}\phi - \lambda_i)}$$

Substituting calculated values of λ_i from Problem 5.30 leads to

$$\mathbf{X}_1 = \begin{bmatrix} 0.697 \\ 1 \\ 1.347 \end{bmatrix}, \qquad \mathbf{X}_2 = \begin{bmatrix} 1 \\ 1 \\ -1 \end{bmatrix}, \qquad \mathbf{X}_3 = \begin{bmatrix} -2.298 \\ 1 \\ -0.1484 \end{bmatrix}$$

5.32 Use a 3-degree-of-freedom model to approximate the lowest natural frequency of a simply supported beam.

The inertia of the beam is approximated by placing particles at equally spaced nodes along the length of the beam, as shown in Fig. 5-23. The magnitude of the particle masses are obtained as in Problem 5.27. If m is the mass of the beam,

$$\mathbf{M} = \frac{m}{4} \begin{bmatrix} 1 & 0 & 0 \\ 0 & 1 & 0 \\ 0 & 0 & 1 \end{bmatrix}$$

The natural frequencies are the reciprocals of the square roots of the eigenvalues of $\mathbf{AM}$. To this end, using the flexibility matrix of Problem 5.19,

$$\det |\mathbf{AM} - \lambda \mathbf{I}| = 0$$

$$\frac{mL^3}{4(768)EI} \det \left| \begin{bmatrix} 9 & 11 & 7 \\ 11 & 16 & 11 \\ 7 & 11 & 9 \end{bmatrix} - \lambda \begin{bmatrix} 1 & 0 & 0 \\ 0 & 1 & 0 \\ 0 & 0 & 1 \end{bmatrix} \right| = 0$$

$$\begin{vmatrix} 9\phi - \lambda & 11\phi & 7\phi \\ 11\phi & 16\phi - \lambda & 11\phi \\ 7\phi & 11\phi & 9\phi - \lambda \end{vmatrix} = 0, \qquad \phi = \frac{mL^3}{3072EI}$$

$$-\beta^3 + 34\beta^2 - 78\beta + 28 = 0, \qquad \beta = \frac{\lambda}{\phi}$$

The roots of the above equation are $\beta = 0.444$, 2, and 31.556. The lowest natural frequency is

$$\omega_1 = \frac{1}{\sqrt{\lambda_3}} = 9.866 \sqrt{\frac{EI}{mL^3}}$$

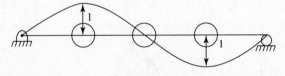

Fig. 5-23

5.33 Determine and graphically illustrate the second mode of the beam of Problem 5.32.

The mode shape corresponding to $\lambda = 2\phi$ is determined from

$$\begin{bmatrix} 9\phi - 2\phi & 11\phi & 7\phi \\ 11\phi & 16\phi - 2\phi & 11\phi \\ 7\phi & 11\phi & 9\phi - 2\phi \end{bmatrix} \begin{bmatrix} X_{21} \\ X_{22} \\ X_{23} \end{bmatrix} = \begin{bmatrix} 0 \\ 0 \\ 0 \end{bmatrix}$$

Arbitrarily setting $X_{21} = 1$ and using the first two of the previous equations leads to

$$11X_{22} + 7X_{23} = -7$$
$$14X_{22} + 11X_{23} = -11$$

whose solution is $X_{22} = 0$, $X_{23} = -1$, leading to $\mathbf{X}_2 = [1 \quad 0 \quad -1]^T$. The mode shape is illustrated in Fig. 5-23.

5.34 The coupling of three identical railroad cars of mass m is shown in Fig. 5-24. The stiffness in the coupling between each car is k. Describe the time history of motion of the three cars after coupling.

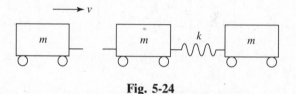

Fig. 5-24

The differential equations governing the motion of the system is

$$\begin{bmatrix} m & 0 & 0 \\ 0 & m & 0 \\ 0 & 0 & m \end{bmatrix} \begin{bmatrix} \ddot{x}_1 \\ \ddot{x}_2 \\ \ddot{x}_3 \end{bmatrix} = \begin{bmatrix} k & -k & 0 \\ -k & 2k & -k \\ 0 & -k & k \end{bmatrix} \begin{bmatrix} x_1 \\ x_2 \\ x_3 \end{bmatrix} = \begin{bmatrix} 0 \\ 0 \\ 0 \end{bmatrix}$$

where x_1, x_2, and x_3 are the displacements of the railroad cars. The system's initial conditions are

$$\mathbf{x}(0) = \begin{bmatrix} 0 \\ 0 \\ 0 \end{bmatrix}, \qquad \dot{\mathbf{x}}(0) = \begin{bmatrix} v \\ 0 \\ 0 \end{bmatrix}$$

The natural frequencies are calculated as

$$\det |\mathbf{K} - \omega^2 \mathbf{M}| = 0$$

$$\begin{vmatrix} k - m\omega^2 & -k & 0 \\ -k & 2k - m\omega^2 & -k \\ 0 & -k & k - m\omega^2 \end{vmatrix} = 0$$

$$\omega_1 = 0, \qquad \omega_2 = \sqrt{\frac{k}{m}}, \qquad \omega_3 = \sqrt{3\frac{k}{m}}$$

The mode shapes are determined as

$$\mathbf{X}_1 = \begin{bmatrix} 1 \\ 1 \\ 1 \end{bmatrix}, \qquad \mathbf{X}_2 = \begin{bmatrix} 1 \\ 0 \\ -1 \end{bmatrix}, \qquad \mathbf{X}_3 = \begin{bmatrix} -1 \\ 2 \\ -1 \end{bmatrix}$$

The general solution is

$$\begin{bmatrix} x_1 \\ x_2 \\ x_3 \end{bmatrix} = \begin{bmatrix} 1 \\ 1 \\ 1 \end{bmatrix}(C_1 + C_2 t) + \begin{bmatrix} 1 \\ 0 \\ -1 \end{bmatrix}\left(C_3 \cos \sqrt{\frac{k}{m}}t + C_4 \sin \sqrt{\frac{k}{m}}t\right)$$

$$+ \begin{bmatrix} -1 \\ 2 \\ -1 \end{bmatrix}\left(C_5 \cos \sqrt{3\frac{k}{m}}t + C_6 \sin \sqrt{3\frac{k}{m}}t\right)$$

Application of the initial conditions leads to

$$x_1(0) = 0 = C_1 + C_2 - C_5$$

$$x_2(0) = 0 = C_1 + 2C_5$$

$$x_3(0) = 0 = C_1 - C_3 - C_5$$

$$\dot{x}_1(0) = v = C_2 + \sqrt{\frac{k}{m}}\,C_4 + \sqrt{3\frac{k}{m}}\,C_6$$

$$\dot{x}_2(0) = 0 = C_2 + 2\sqrt{3\frac{k}{m}}\,C_6$$

$$\dot{x}_3(0) = 0 = C_2 - \sqrt{\frac{k}{m}}\,C_4 - \sqrt{3\frac{k}{m}}\,C_6$$

whose solution is $C_1 = C_3 = C_5 = 0$,

$$C_2 = \frac{v}{3}, \qquad C_4 = \frac{v}{2}\sqrt{\frac{m}{k}}, \qquad C_6 = -\frac{v}{6}\sqrt{\frac{m}{3k}}$$

Thus

$$x_1(t) = \frac{v}{3}t + \frac{v}{2}\sqrt{\frac{m}{k}}\sin\sqrt{\frac{k}{m}}t - \frac{v}{6}\sqrt{\frac{m}{3k}}\sin\sqrt{3\frac{k}{m}}t$$

$$x_2(t) = \frac{v}{3}t - \frac{v}{3}\sqrt{\frac{m}{3k}}\sin\sqrt{3\frac{k}{m}}t$$

$$x_3(t) = \frac{v}{3}t - \frac{v}{2}\sqrt{\frac{m}{k}}\sin\sqrt{\frac{k}{m}}t - \frac{v}{6}\sqrt{\frac{m}{3k}}\sin\sqrt{3\frac{k}{m}}t$$

5.35 Determine the natural frequencies of the torsional system of Fig. 5-22.

The natural frequencies are calculated using the stiffness matrix derived in Problem 5.23:

$$\det|\mathbf{K} - \omega^2\mathbf{M}| = 0$$

$$\begin{vmatrix} 1.70 \times 10^5 - 2.5\omega^2 & -1.70 \times 10^5 & 0 \\ -1.70 \times 10^5 & 5.72 \times 10^5 - 4.5\omega^2 & -4.02 \times 10^5 \\ 0 & -4.02 \times 10^5 & 4.02 \times 10^5 - 2.8\omega^2 \end{vmatrix} = 0$$

$$-31.5\omega^6 + 1.067 \times 10^7\omega^4 - 6.696 \times 10^{11}\omega^2 = 0$$

$$\omega_1 = 0, \qquad \omega_2 = 288.5\,\frac{\text{rad}}{\text{s}}, \qquad \omega_3 = 505.4\,\frac{\text{rad}}{\text{s}}$$

5.36 Demonstrate orthogonality of the mode shapes of the system of Problem 5.28.

$$\mathbf{X}_1^T\mathbf{M}\mathbf{X}_2 = \begin{bmatrix} 1 & -3.16 \end{bmatrix}\begin{bmatrix} 5 & 0 \\ 0 & -0.5 \end{bmatrix}\begin{bmatrix} 1 \\ 3.16 \end{bmatrix}$$

$$= \begin{bmatrix} 1 & -3.16 \end{bmatrix}\begin{bmatrix} 5 \\ 1.58 \end{bmatrix} = (1)(5) + (-3.16)(1.58) \approx 0$$

5.37 Demonstrate orthogonality of the mode shapes of the system of Problem 5.30.

$$\mathbf{X}_1^T\mathbf{M}\mathbf{X}_2 = m\begin{bmatrix} 0.697 & 1 & 1.347 \end{bmatrix}\begin{bmatrix} 1 & 0 & 0 \\ 0 & 2 & 0 \\ 0 & 0 & 2 \end{bmatrix}\begin{bmatrix} 1 \\ 1 \\ -1 \end{bmatrix}$$

$$= m\begin{bmatrix} 0.697 & 1 & 1.347 \end{bmatrix}\begin{bmatrix} 1 \\ 2 \\ -2 \end{bmatrix} = m[(0.697)(1) + (1)(2) + (1.347)(-2)] \approx 0$$

$$\mathbf{X}_1{}^T\mathbf{M}\mathbf{X}_3 = m[0.697 \quad 1 \quad 1.347]\begin{bmatrix} 1 & 0 & 0 \\ 0 & 2 & 0 \\ 0 & 0 & 2 \end{bmatrix}\begin{bmatrix} -2.298 \\ 1 \\ -0.1484 \end{bmatrix}$$

$$= m[0.697 \quad 1 \quad 1.347]\begin{bmatrix} -2.298 \\ 2 \\ -0.2968 \end{bmatrix}$$

$$= m[(0.697)(-2.298) + (1)(2) + (1.347)(-0.2968)] \approx 0$$

$$\mathbf{X}_3{}^T\mathbf{M}\mathbf{X}_2 = m[-2.298 \quad 1 \quad -0.1484]\begin{bmatrix} 1 & 0 & 0 \\ 0 & 2 & 0 \\ 0 & 0 & 2 \end{bmatrix}\begin{bmatrix} 1 \\ 1 \\ -1 \end{bmatrix}$$

$$= m[-2.298 \quad 1 \quad -0.1484]\begin{bmatrix} 1 \\ 2 \\ -2 \end{bmatrix}$$

$$= m[(-2.298)(1) + (1)(2) + (-0.1484)(-2)] \approx 0$$

5.38 Normalize the mode shape vectors for the system of Problem 5.30.

Normalization of a mode shape vector $\mathbf{X}$ is achieved by dividing every component of the vector by $[\mathbf{X}^T\mathbf{M}\mathbf{X}]^{1/2}$. To this end, for the mode shape vectors of Problem 5.30,

$$\mathbf{X}_1{}^T\mathbf{M}\mathbf{X}_1 = [0.697 \quad 1 \quad 1.347]\begin{bmatrix} m & 0 & 0 \\ 0 & 2m & 0 \\ 0 & 0 & 2m \end{bmatrix}\begin{bmatrix} 0.697 \\ 1 \\ 1.347 \end{bmatrix}$$

$$= [0.697 \quad 1 \quad 1.347]\begin{bmatrix} 0.697m \\ 2m \\ 2.694m \end{bmatrix}$$

$$(0.697)(0.697m) + (1)(2m) + (1.347)(2.694m) = 6.115m$$

$$\mathbf{X}_2{}^T\mathbf{M}\mathbf{X}_2 = [1 \quad 1 \quad -1]\begin{bmatrix} m & 0 & 0 \\ 0 & 2m & 0 \\ 0 & 0 & 2m \end{bmatrix}\begin{bmatrix} 1 \\ 1 \\ -1 \end{bmatrix}$$

$$= [1 \quad 1 \quad -1]\begin{bmatrix} m \\ 2m \\ -2m \end{bmatrix} = (1)(m) + (1)(2m) + (-1)(-2m) = 5m$$

$$\mathbf{X}_3{}^T\mathbf{M}\mathbf{X}_3 = [-2.298 \quad 1 \quad -0.1484]\begin{bmatrix} m & 0 & 0 \\ 0 & 2m & 0 \\ 0 & 0 & 2m \end{bmatrix}\begin{bmatrix} -2.298 \\ 1 \\ -0.1484 \end{bmatrix}$$

$$= [2.298 \quad 1 \quad -0.1484]\begin{bmatrix} -2.298m \\ 2m \\ -0.2968m \end{bmatrix}$$

$$= (-2.298)(-2.298m) + (1)(2m) + (-0.1484)(-0.2968m) = 7.325m$$

The normalized mode shapes are

$$\mathbf{X}_1 = \frac{1}{\sqrt{6.115m}}\begin{bmatrix} 0.697 \\ 1 \\ 1.347 \end{bmatrix} = \frac{1}{\sqrt{m}}\begin{bmatrix} 0.2819 \\ 0.4044 \\ 0.5447 \end{bmatrix}, \qquad \mathbf{X}_2 = \frac{1}{\sqrt{5m}}\begin{bmatrix} 1 \\ 1 \\ -1 \end{bmatrix} = \frac{1}{\sqrt{m}}\begin{bmatrix} 0.4472 \\ 0.4472 \\ -0.4472 \end{bmatrix},$$

$$\mathbf{X}_3 = \frac{1}{\sqrt{7.325m}}\begin{bmatrix} -2.298 \\ 1 \\ -0.1484 \end{bmatrix} = \frac{1}{\sqrt{m}}\begin{bmatrix} -0.8491 \\ 0.3695 \\ -0.0548 \end{bmatrix}$$

5.39 A 2-degree-of-freedom system whose mass matrix is

$$\mathbf{M} = \begin{bmatrix} 100 & 40 \\ 40 & 150 \end{bmatrix}$$

has a normalized mode shape of $X_1 = [0.0341 \quad 0.0682]^T$. Determine the normalized mode shape for the second mode.

Assume the second mode shape as $[1 \quad a]^T$. Imposing orthogonality,

$$\mathbf{X}_1{}^T \mathbf{M} \mathbf{X}_2 = 0 = [0.0341 \quad 0.0682] \begin{bmatrix} 100 & 40 \\ 40 & 150 \end{bmatrix} \begin{bmatrix} 1 \\ a \end{bmatrix}$$

$$= [0.0341 \quad 0.0682] \begin{bmatrix} 100 + 40a \\ 40 + 150a \end{bmatrix}$$

$$= (0.0341)(100 + 40a) + (0.0682)(40 + 150a) = 6.138 + 11.594a$$

$$a = -0.529$$

Normalizing the mode shape

$$\mathbf{X}_2{}^T \mathbf{M} \mathbf{X}_2 = [1 \quad -0.529] \begin{bmatrix} 100 & 40 \\ 40 & 150 \end{bmatrix} \begin{bmatrix} 1 \\ -0.529 \end{bmatrix} = 99.66$$

The normalized mode shape is

$$\mathbf{X}_2 = \frac{1}{\sqrt{99.66}} \begin{bmatrix} 1 \\ -0.529 \end{bmatrix} = \begin{bmatrix} 0.0100 \\ -0.0530 \end{bmatrix}$$

5.40 Use matrix iteration with the trial vector $\mathbf{u}_0 = [1 \quad 0 \quad 0]^T$ to approximate the lowest natural frequency and its mode shape for the system of Problem 5.32.

Using the matrix $\mathbf{AM}$ calculated in Problem 5.32, matrix iteration is used as shown:

$$\mathbf{u}_1 = \mathbf{AM}\mathbf{u}_0 = \begin{bmatrix} 9\phi & 11\phi & 7\phi \\ 11\phi & 16\phi & 11\phi \\ 7\phi & 11\phi & 9\phi \end{bmatrix} \begin{bmatrix} 1 \\ 0 \\ 0 \end{bmatrix} = \begin{bmatrix} 9\phi \\ 11\phi \\ 7\phi \end{bmatrix}, \qquad \tilde{\mathbf{u}}_1 = \begin{bmatrix} 0.8182 \\ 1 \\ 0.6364 \end{bmatrix}$$

$$\mathbf{u}_2 = \mathbf{AM}\tilde{\mathbf{u}}_1 = \begin{bmatrix} 9\phi & 11\phi & 7\phi \\ 11\phi & 16\phi & 11\phi \\ 7\phi & 11\phi & 9\phi \end{bmatrix} \begin{bmatrix} 0.8182 \\ 1 \\ 0.6364 \end{bmatrix} = \begin{bmatrix} 22.82\phi \\ 32\phi \\ 22.46\phi \end{bmatrix}, \qquad \tilde{\mathbf{u}}_2 = \begin{bmatrix} 0.7131 \\ 1 \\ 0.7019 \end{bmatrix}$$

$$\mathbf{u}_3 = \mathbf{AM}\tilde{\mathbf{u}}_2 = \begin{bmatrix} 9\phi & 11\phi & 7\phi \\ 11\phi & 16\phi & 11\phi \\ 7\phi & 11\phi & 9\phi \end{bmatrix} \begin{bmatrix} 0.7131 \\ 1 \\ 0.7019 \end{bmatrix} = \begin{bmatrix} 23.33\phi \\ 31.57\phi \\ 23.31\phi \end{bmatrix}, \qquad \tilde{\mathbf{u}}_3 = \begin{bmatrix} 0.7389 \\ 1 \\ 0.7389 \end{bmatrix}$$

$$\mathbf{u}_4 = \mathbf{AM}\tilde{\mathbf{u}}_3 = \begin{bmatrix} 9\phi & 11\phi & 7\phi \\ 11\phi & 16\phi & 11\phi \\ 7\phi & 11\phi & 9\phi \end{bmatrix} \begin{bmatrix} 0.7389 \\ 1 \\ 0.7389 \end{bmatrix} = \begin{bmatrix} 22.82\phi \\ 32.22\phi \\ 22.82\phi \end{bmatrix}, \qquad \tilde{\mathbf{u}}_4 = \begin{bmatrix} 0.7073 \\ 1 \\ 0.7073 \end{bmatrix}$$

$$\mathbf{u}_5 = \mathbf{AM}\tilde{\mathbf{u}}_4 = \begin{bmatrix} 9\phi & 11\phi & 7\phi \\ 11\phi & 16\phi & 11\phi \\ 7\phi & 11\phi & 9\phi \end{bmatrix} \begin{bmatrix} 0.7073 \\ 1 \\ 0.7073 \end{bmatrix} = \begin{bmatrix} 22.32\phi \\ 31.56\phi \\ 22.32\phi \end{bmatrix}, \qquad \tilde{\mathbf{u}}_5 = \begin{bmatrix} 0.7073 \\ 1 \\ 0.7073 \end{bmatrix}$$

$$\mathbf{u}_6 = \mathbf{AM}\tilde{\mathbf{u}}_5 = \begin{bmatrix} 9\phi & 11\phi & 7\phi \\ 11\phi & 16\phi & 11\phi \\ 7\phi & 11\phi & 9\phi \end{bmatrix} \begin{bmatrix} 0.7071 \\ 1 \\ 0.7071 \end{bmatrix} = \begin{bmatrix} 22.31\phi \\ 31.56\phi \\ 22.31\phi \end{bmatrix}, \qquad \tilde{\mathbf{u}}_5 = \begin{bmatrix} 0.7071 \\ 1 \\ 0.7071 \end{bmatrix}$$

Hence the iteration has converged to $\lambda = 31.56\phi$ and

$$\mathbf{X}_1 = [0.7071 \quad 1 \quad 0.7071]^T \text{ which leads to}$$

$$\omega_1 = \frac{1}{\sqrt{31.56\phi}} = \frac{1}{\sqrt{31.56\left(\dfrac{mL^3}{3072EI}\right)}} = 9.866\sqrt{\frac{EI}{mL^3}}$$

5.41 Use matrix iteration to determine the second natural frequency and mode shape of the system of Problem 5.32.

If $\mathbf{X}_2 = [A \quad B \quad C]^T$, then orthogonality requires

$$\mathbf{X}_2^T \mathbf{M} \mathbf{X}_1 = 0 = [A \quad B \quad C] \begin{bmatrix} \dfrac{m}{4} & 0 & 0 \\ 0 & \dfrac{m}{4} & 0 \\ 0 & 0 & \dfrac{m}{4} \end{bmatrix} \begin{bmatrix} 0.7071 \\ 1 \\ 0.7071 \end{bmatrix}$$

$$\frac{m}{4}(0.7071A + B + 0.7071C) = 0$$

$$A = -1.414B - C$$

Orthogonality to the first mode is imposed by defining

$$\mathbf{Q} = \begin{bmatrix} 9\phi & 11\phi & 7\phi \\ 11\phi & 16\phi & 11\phi \\ 7\phi & 11\phi & 9\phi \end{bmatrix} \begin{bmatrix} 0 & -1.414 & -1 \\ 0 & 1 & 0 \\ 0 & 0 & 1 \end{bmatrix}$$

$$= \begin{bmatrix} 0 & -1.726\phi & -2\phi \\ 0 & 0.4460\phi & 0 \\ 0 & 1.102\phi & 2\phi \end{bmatrix}$$

Matrix iteration, when used with the matrix $\mathbf{Q}$, imposes orthogonality of the iterate to the first mode, and thus the iteration converges to the mode shape for the second mode and yields the second natural frequency. To this end select $\mathbf{u}_0 = [0 \quad 0 \quad 1]^T$. Then

$$\mathbf{u}_1 = \mathbf{Q}\mathbf{u}_0 = \begin{bmatrix} 0 & -1.726\phi & -2\phi \\ 0 & 0.4460\phi & 0 \\ 0 & 1.102\phi & 2\phi \end{bmatrix} \begin{bmatrix} 0 \\ 0 \\ 1 \end{bmatrix} = \begin{bmatrix} -2\phi \\ 0 \\ 2\phi \end{bmatrix}, \quad \tilde{\mathbf{u}}_1 = \begin{bmatrix} -1 \\ 0 \\ 1 \end{bmatrix}$$

$$\mathbf{u}_2 = \mathbf{Q}\tilde{\mathbf{u}}_1 = \begin{bmatrix} 0 & -1.726\phi & -2\phi \\ 0 & 0.446\phi & 0 \\ 0 & 1.102\phi & 2\phi \end{bmatrix} \begin{bmatrix} -1 \\ 0 \\ 1 \end{bmatrix} = \begin{bmatrix} 2\phi \\ 0 \\ -2\phi \end{bmatrix}$$

Hence it is clear that $\mathbf{X}_2 = [1 \quad 0 \quad -1]^T$ and

$$\omega_2 = \frac{1}{\sqrt{2\phi}} = \frac{1}{\sqrt{2\left(\dfrac{mL^3}{3072EI}\right)}} = 39.19\sqrt{\frac{EI}{mL^3}}$$

5.42 Use the results of Problems 5.40 and 5.41 to determine the highest natural frequency of the system of Problem 5.32.

If $\mathbf{X}_3 = [D \quad E \quad F]^T$, then orthogonality with $\mathbf{X}_1$ requires $D = -1.414E - F$. Orthogonality with $\mathbf{X}_2$ requires

$$\mathbf{X}_3^T \mathbf{M} \mathbf{X}_2 = 0 = [D \quad E \quad F] \begin{bmatrix} \dfrac{m}{4} & 0 & 0 \\ 0 & \dfrac{m}{4} & 0 \\ 0 & 0 & \dfrac{m}{4} \end{bmatrix} \begin{bmatrix} 1 \\ 0 \\ -1 \end{bmatrix}$$

$$\frac{m}{4}(D - F) = 0 = 0 \;\rightarrow\; D = F$$

Arbitrarily setting $F = 1$ leads to $D = 1$ and $E = -1.414$. The third eigenvalue of $\mathbf{AM}$ is obtained by noting that $\mathbf{AMX}_3 = \lambda_3 \mathbf{X}_3$. Thus

$$\begin{bmatrix} 9\phi & 11\phi & 7\phi \\ 11\phi & 16\phi & 11\phi \\ 7\phi & 11\phi & 9\phi \end{bmatrix} \begin{bmatrix} 1 \\ -1.414 \\ 1 \end{bmatrix} = \begin{bmatrix} 0.446\phi \\ -0.624\phi \\ 0.446\phi \end{bmatrix}$$

Hence $\lambda_3 = 0.446$ and

$$\omega_3 = \frac{1}{\sqrt{0.446\phi}} = \frac{1}{\sqrt{0.446\left(\dfrac{mL^3}{3072EI}\right)}} = 82.99\sqrt{\frac{EI}{mL^3}}$$

5.43 Use matrix iteration to determine the highest natural frequency and its corresponding mode shape for the system of Problem 5.30.

Matrix iteration converges to the highest natural frequency when the matrix $\mathbf{M}^{-1}\mathbf{K}$ is used in the iteration procedure. Using $\mathbf{u}_0 = [1 \quad 0 \quad 0]^T$ and $\mathbf{M}^{-1}\mathbf{K}$ from Problem 5.30 with $\phi = k/m$,

$$\mathbf{u}_1 = \mathbf{M}^{-1}\mathbf{K}\mathbf{u}_0 = \begin{bmatrix} 3\phi & -2\phi & 0 \\ -\phi & \frac{3}{2}\phi & -\frac{1}{2}\phi \\ 0 & -\frac{1}{2}\phi & \frac{1}{2}\phi \end{bmatrix} \begin{bmatrix} 1 \\ 0 \\ 0 \end{bmatrix} = \begin{bmatrix} 3\phi \\ -\phi \\ 0 \end{bmatrix}, \qquad \tilde{\mathbf{u}}_1 = \begin{bmatrix} 1 \\ -0.3333 \\ 0 \end{bmatrix}$$

$$\mathbf{u}_2 = \mathbf{M}^{-1}\mathbf{K}\tilde{\mathbf{u}}_1 = \begin{bmatrix} 3\phi & -2\phi & 0 \\ -\phi & \frac{3}{2}\phi & -\frac{1}{2}\phi \\ 0 & -\frac{1}{2}\phi & \frac{1}{2}\phi \end{bmatrix} \begin{bmatrix} 1 \\ -0.3333 \\ 0 \end{bmatrix} = \begin{bmatrix} 3.667\phi \\ -1.5\phi \\ 0.1667\phi \end{bmatrix}, \qquad \tilde{\mathbf{u}}_2 = \begin{bmatrix} 1 \\ -0.4091 \\ 0.0455 \end{bmatrix}$$

$$\mathbf{u}_3 = \mathbf{M}^{-1}\mathbf{K}\tilde{\mathbf{u}}_2 = \begin{bmatrix} 3\phi & -2\phi & 0 \\ -\phi & \frac{3}{2}\phi & -\frac{1}{2}\phi \\ 0 & -\frac{1}{2}\phi & \frac{1}{2}\phi \end{bmatrix} \begin{bmatrix} 1 \\ -0.4091 \\ 0.0455 \end{bmatrix} = \begin{bmatrix} 3.818\phi \\ -1.636\phi \\ 0.2273\phi \end{bmatrix}, \qquad \tilde{\mathbf{u}}_3 = \begin{bmatrix} 1 \\ -0.4285 \\ 0.0595 \end{bmatrix}$$

$$\mathbf{u}_4 = \mathbf{M}^{-1}\mathbf{K}\tilde{\mathbf{u}}_3 = \begin{bmatrix} 3\phi & -2\phi & 0 \\ -\phi & \frac{3}{2}\phi & -\frac{1}{2}\phi \\ 0 & -\frac{1}{2}\phi & \frac{1}{2}\phi \end{bmatrix} \begin{bmatrix} 1 \\ -0.4295 \\ 0.0595 \end{bmatrix} = \begin{bmatrix} 3.859\phi \\ -1.674\phi \\ 0.2445\phi \end{bmatrix}, \qquad \tilde{\mathbf{u}}_4 = \begin{bmatrix} 1 \\ -0.4338 \\ 0.0636 \end{bmatrix}$$

$$\mathbf{u}_5 = \mathbf{M}^{-1}\mathbf{K}\tilde{\mathbf{u}}_4 = \begin{bmatrix} 3\phi & -2\phi & 0 \\ -\phi & \frac{3}{2}\phi & -\frac{1}{2}\phi \\ 0 & -\frac{1}{2}\phi & \frac{1}{2}\phi \end{bmatrix} \begin{bmatrix} 1 \\ -0.4338 \\ 0.0636 \end{bmatrix} = \begin{bmatrix} 3.868\phi \\ -1.683\phi \\ 0.2487\phi \end{bmatrix}, \qquad \tilde{\mathbf{u}}_5 = \begin{bmatrix} 1 \\ 0.4351 \\ 0.0643 \end{bmatrix}$$

$$\mathbf{u}_6 = \mathbf{M}^{-1}\mathbf{K}\tilde{\mathbf{u}}_5 = \begin{bmatrix} 3\phi & -2\phi & 0 \\ -\phi & \frac{3}{2}\phi & -\frac{1}{2}\phi \\ 0 & -\frac{1}{2}\phi & \frac{1}{2}\phi \end{bmatrix} \begin{bmatrix} 1 \\ -0.4351 \\ 0.0643 \end{bmatrix} = \begin{bmatrix} 3.870\phi \\ -1.685\phi \\ 0.2497\phi \end{bmatrix}, \qquad \tilde{\mathbf{u}}_6 = \begin{bmatrix} 1 \\ 0.4354 \\ 0.06452 \end{bmatrix}$$

$$\mathbf{u}_7 = \mathbf{M}^{-1}\mathbf{K}\tilde{\mathbf{u}}_6 = \begin{bmatrix} 3\phi & -2\phi & 0 \\ -\phi & \frac{3}{2}\phi & -\frac{1}{2}\phi \\ 0 & -\frac{1}{2}\phi & \frac{1}{2}\phi \end{bmatrix} \begin{bmatrix} 1 \\ 0.4354 \\ 0.0652 \end{bmatrix} = \begin{bmatrix} 3.871\phi \\ -1.685\phi \\ 0.2503\phi \end{bmatrix}, \qquad \tilde{\mathbf{u}}_7 = \begin{bmatrix} 1 \\ -0.4353 \\ 0.0647 \end{bmatrix}$$

The iteration has converged to $\mathbf{X}_3 = [1 \quad -0.4353 \quad 0.0647]^T$ and $\lambda_3 = 3.871\phi$, leading to

$$\omega_3 = \sqrt{3.871\phi} = 1.967\sqrt{\frac{k}{m}}$$

5.44 For what values of c will both modes of the system of Fig. 5-25 be underdamped?

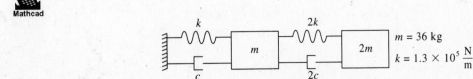

Fig. 5-25

The differential equations governing the motion of the system of Fig. 5-25 are

$$\begin{bmatrix} m & 0 \\ 0 & 2m \end{bmatrix}\begin{bmatrix} \ddot{x}_1 \\ \ddot{x}_2 \end{bmatrix} + \begin{bmatrix} 3c & -2c \\ -2c & 2c \end{bmatrix}\begin{bmatrix} \dot{x}_1 \\ \dot{x}_2 \end{bmatrix} + \begin{bmatrix} 3k & -2k \\ -2k & 2k \end{bmatrix}\begin{bmatrix} x_1 \\ x_2 \end{bmatrix} = \begin{bmatrix} 0 \\ 0 \end{bmatrix}$$

The system has viscous damping which is proportional with $\alpha = c/k$ and $\beta = 0$. The undamped natural frequencies are determined from

$$\det |\mathbf{K} - \omega^2\mathbf{M}| = 0$$

$$\begin{vmatrix} 3k - m\omega^2 & -2k \\ -2k & 2k - 2m\omega^2 \end{vmatrix} = 0$$

$$\omega_1 = 0.5177\sqrt{\frac{k}{m}}, \qquad \omega_2 = 1.932\sqrt{\frac{k}{m}}$$

Then from Eq. (5.21), for $\beta = 0$, the mode with the highest natural frequency has the highest damping ratio. Thus for $\zeta_2 < 1$,

$$\tfrac{1}{2}\alpha\omega_2 < 1$$

$$\frac{1}{2}\frac{c}{k}\left(1.932\sqrt{\frac{k}{m}}\right) < 1$$

$$c < 1.035\sqrt{mk} = 1.035\sqrt{\left(1.3 \times 10^5\,\frac{N}{m}\right)(36\ \text{kg})} = 2.24 \times 10^3\,\frac{\text{N-s}}{\text{m}}$$

5.45   Determine the general free vibration response of the system of Fig. 5-26.

Mathcad

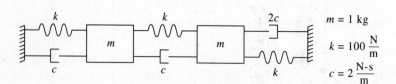

$m = 1$ kg

$k = 100\,\dfrac{\text{N}}{\text{m}}$

$c = 2\,\dfrac{\text{N-s}}{\text{m}}$

Fig. 5-26

The differential equations governing the motion of the system of Fig. 5-26 are

$$\begin{bmatrix} m & 0 \\ 0 & m \end{bmatrix}\begin{bmatrix} \ddot{x}_1 \\ \ddot{x}_2 \end{bmatrix} + \begin{bmatrix} 2c & -c \\ -c & 3c \end{bmatrix}\begin{bmatrix} \dot{x}_1 \\ \dot{x}_2 \end{bmatrix} + \begin{bmatrix} 2k & -k \\ -k & 2k \end{bmatrix}\begin{bmatrix} x_1 \\ x_2 \end{bmatrix} = \begin{bmatrix} 0 \\ 0 \end{bmatrix}$$

The $2n \times 2n$ partitioned matrices of Eq. (5.23) are

$$\tilde{\mathbf{M}} = \begin{bmatrix} 0 & 0 & m & 0 \\ 0 & 0 & 0 & m \\ m & 0 & 2c & -c \\ 0 & m & -c & 3c \end{bmatrix} \qquad \tilde{\mathbf{K}} = \begin{bmatrix} -m & 0 & 0 & 0 \\ 0 & -m & 0 & 0 \\ 0 & 0 & 2k & -k \\ 0 & 0 & -k & 2k \end{bmatrix}$$

The values of γ are obtained as eigenvalues of

$$\tilde{\mathbf{M}}^{-1}\tilde{\mathbf{K}} = \begin{bmatrix} 4 & -2 & 200 & -100 \\ -2 & 6 & -100 & 200 \\ -1 & 0 & 0 & 0 \\ 0 & -1 & 0 & 0 \end{bmatrix}$$

which are determined from

$$\begin{bmatrix} 4-\gamma & -2 & 200 & -100 \\ -2 & 6-\gamma & -100 & 200 \\ -1 & 0 & -\gamma & 0 \\ 0 & -1 & 0 & -\gamma \end{bmatrix} = 0$$

The determinant can be evaluated using row expansion by the third row, leading to the evaluation of two 3×3 determinants. The resulting equation is

$$\gamma^4 - 10\gamma^3 + 420\gamma^2 - 1600\gamma + 30,000 = 0$$

whose roots are

$$\gamma = 1.502 \pm 9.912i, \qquad 3.497 \pm 16.918i$$

If $\mathbf{X} = [X_1 \ X_2 \ X_3 \ X_4]^T$ is an eigenvector for an eigenvalue, $\gamma = \gamma_r + i\gamma_i$, of $\mathbf{M}^{-1}\mathbf{K}$, then $\mathbf{X} = [X_3 \ X_4]^T$ is the mode shape vector that is of the form $\mathbf{X}_r + i\mathbf{X}_i$, and then the general solution corresponding to γ is

$$e^{-\gamma_r t}[C_1(\mathbf{X}_r - \mathbf{X}_i) \cos \gamma_i t + C_2(\mathbf{X}_r + \mathbf{X}_i) \sin \gamma_i t]$$

After performing the necessary calculations, the solution is obtained as

$$\mathbf{x} = e^{-1.502t}\left(C_1 \begin{bmatrix} 0.940 \\ 1 \end{bmatrix} \cos 9.912t + C_2 \begin{bmatrix} 1.172 \\ 1 \end{bmatrix} \sin 9.912t \right)$$

$$+ e^{-3.497t}\left(C_3 \begin{bmatrix} -1.071 \\ 1 \end{bmatrix} \cos 16.918t + C_4 \begin{bmatrix} -0.704 \\ 1 \end{bmatrix} \sin 16.918t \right)$$

Supplementary Problems

5.46 Use the free body diagram method to derive the differential equations governing the motion of the system of Fig. 5-27 using x_1, x_2, and x_3 as generalized coordinates.

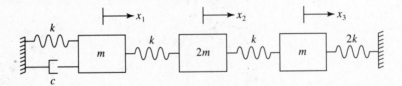

Fig. 5-27

Ans.

$$m\ddot{x}_1 + c\dot{x}_1 + 2kx_1 - kx_2 = 0$$
$$2m\ddot{x}_2 - kx_1 + 2kx_2 - kx_3 = 0$$
$$m\ddot{x}_3 - kx_2 + 3kx_3 = 0$$

5.47 Use the free body diagram method to derive the differential equations governing the motion of the system of Fig. 5-28 using θ_1 and θ_2 as generalized coordinates.

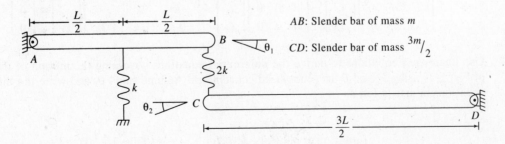

Fig. 5-28

Ans.

$$\tfrac{1}{3}mL^2\ddot{\theta}_1 + \tfrac{9}{4}kL^2\theta_1 - 3kL^2\theta_2 = 0$$

$$\tfrac{9}{8}mL^2\ddot{\theta}_2 - 3kL^2\theta_1 + \tfrac{9}{2}kL^2\theta_2 = 0$$

5.48 Use the free body diagram method to derive the differential equations governing the motion of the system of Fig. 5-29 using θ, x_1, and x_2 as generalized coordinates.

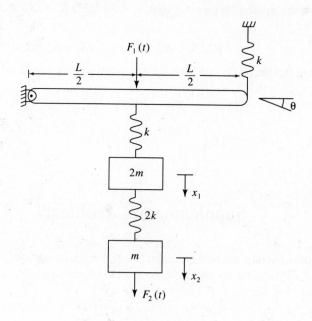

Fig. 5-29

Ans.

$$\tfrac{1}{3}mL^2\ddot{\theta} + \tfrac{5}{4}kL^2\theta - \tfrac{1}{2}kLx_1 = \tfrac{1}{2}F_1(t)L$$

$$2m\ddot{x}_1 - \tfrac{1}{2}kL\theta + 3kx_1 - 2kx_2 = 0$$

$$m\ddot{x}_2 - 2kx_1 + 2kx_2 = F_2(t)$$

5.49 Use Lagrange's equations to derive the differential equations governing the motion of the system of Fig. 5-27 using x_1, x_2, and x_3 as generalized coordinates.

Ans.

$$\begin{bmatrix} m & 0 & 0 \\ 0 & 2m & 0 \\ 0 & 0 & m \end{bmatrix}\begin{bmatrix} \ddot{x}_1 \\ \ddot{x}_2 \\ \ddot{x}_3 \end{bmatrix} + \begin{bmatrix} c & 0 & 0 \\ 0 & 0 & 0 \\ 0 & 0 & 0 \end{bmatrix}\begin{bmatrix} \dot{x}_1 \\ \dot{x}_2 \\ \dot{x}_3 \end{bmatrix} = \begin{bmatrix} 2k & -k & 0 \\ -k & 2k & -k \\ 0 & -k & 3k \end{bmatrix}\begin{bmatrix} x_1 \\ x_2 \\ x_3 \end{bmatrix} = \begin{bmatrix} 0 \\ 0 \\ 0 \end{bmatrix}$$

5.50 Use Lagrange's equations to derive the differential equations governing the motion of the system of Fig. 5-28 using θ_1 and θ_2 as generalized coordinates. Assume small θ, and write the differential equations in matrix form.

Ans.

$$\begin{bmatrix} \tfrac{1}{3}mL^2 & 0 \\ 0 & \tfrac{9}{8}mL^2 \end{bmatrix}\begin{bmatrix} \ddot{\theta}_1 \\ \ddot{\theta}_2 \end{bmatrix} + \begin{bmatrix} \tfrac{9}{4}kL^2 & -3kL^2 \\ -3kL^2 & \tfrac{9}{2}kL^2 \end{bmatrix}\begin{bmatrix} \theta_1 \\ \theta_2 \end{bmatrix} = \begin{bmatrix} 0 \\ 0 \end{bmatrix}$$

5.51 Use Lagrange's equations to derive the differential equations governing the motion of the system of Fig. 5-29 using θ, x_1, and x_2 as generalized coordinates.

Ans.

$$\begin{bmatrix} \frac{1}{3}mL^2 & 0 & 0 \\ 0 & 2m & 0 \\ 0 & 0 & m \end{bmatrix}\begin{bmatrix} \ddot{\theta} \\ \ddot{x}_1 \\ \ddot{x}_2 \end{bmatrix} + \begin{bmatrix} \frac{5}{4}kL^2 & -\frac{1}{2}kL & 0 \\ -\frac{1}{2}kL & 3k & -2k \\ 0 & -2k & 2k \end{bmatrix}\begin{bmatrix} \theta \\ x_1 \\ x_2 \end{bmatrix} = \begin{bmatrix} \frac{1}{2}F_1(t)L \\ 0 \\ F_2(t) \end{bmatrix}$$

5.52 Use Lagrange's equations to derive the differential equations governing the motion of the system of Fig. 5-30 using θ_1 and θ_2 as generalized coordinates. Assume small θ_1 and θ_2.

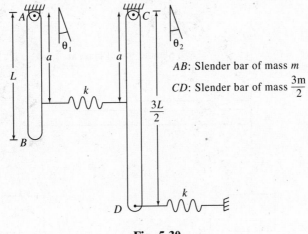

AB: Slender bar of mass m

CD: Slender bar of mass $\dfrac{3m}{2}$

Fig. 5-30

Ans.

$$\begin{bmatrix} \frac{1}{3}mL^2 & 0 \\ 0 & \frac{9}{8}mL^2 \end{bmatrix}\begin{bmatrix} \ddot{\theta}_1 \\ \ddot{\theta}_2 \end{bmatrix} + \begin{bmatrix} ka^2 + \frac{1}{2}mgL & -ka^2 \\ -ka^2 & k(a^2 + \frac{9}{4}L^2) + \frac{3}{4}mgL \end{bmatrix}\begin{bmatrix} \theta_1 \\ \theta_2 \end{bmatrix} = \begin{bmatrix} 0 \\ 0 \end{bmatrix}$$

5.53 Use Lagrange's equations to derive the differential equations governing the motion of the system of Fig. 5-31 using x and θ as generalized coordinates. Assume small θ.

AB: Bar of mass $2m$ and centroidal moment of inertia I

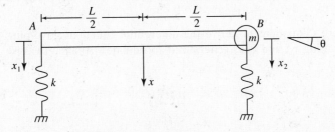

Fig. 5-31

Ans.

$$\begin{bmatrix} 3m & \frac{1}{2}mL \\ \frac{1}{2}mL & I + \frac{1}{4}mL^2 \end{bmatrix}\begin{bmatrix} \ddot{x} \\ \ddot{\theta} \end{bmatrix} + \begin{bmatrix} 2k & 0 \\ 0 & \frac{1}{2}kL^2 \end{bmatrix} = \begin{bmatrix} 0 \\ 0 \end{bmatrix}$$

5.54   Use Lagrange's equations to derive the differential equations governing the motion of the system of Fig. 5-31 using x_1 and x_2 as generalized coordinates.

Ans.

$$\begin{bmatrix} \dfrac{m}{2}+\dfrac{I}{L^2} & \dfrac{m}{2}-\dfrac{I}{L^2} \\[2ex] \dfrac{m}{2}-\dfrac{I}{L^2} & \dfrac{3}{2}m+\dfrac{I}{L^2} \end{bmatrix}\begin{bmatrix} \ddot{x}_1 \\ \ddot{x}_2 \end{bmatrix}+\begin{bmatrix} k & 0 \\ 0 & k \end{bmatrix}\begin{bmatrix} x_1 \\ x_2 \end{bmatrix}=\begin{bmatrix} 0 \\ 0 \end{bmatrix}$$

5.55   Use Lagrange's equations to derive the motion of the system of Fig. 5-32 using x_1, x_2, and θ as generalized coordinates.

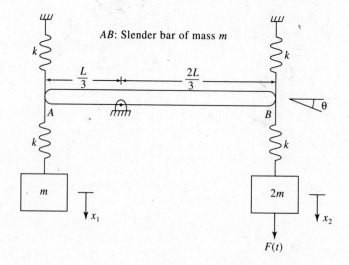

AB: Slender bar of mass m

Fig. 5-32

Ans.

$$\begin{bmatrix} \frac{1}{9}mL^2 & 0 & 0 \\ 0 & m & 0 \\ 0 & 0 & 2m \end{bmatrix}\begin{bmatrix} \ddot{\theta} \\ \ddot{x}_1 \\ \ddot{x}_2 \end{bmatrix}+\begin{bmatrix} \frac{10}{9}kL^2 & \frac{1}{3}kL & -\frac{2}{3}kL \\ \frac{1}{3}kL & k & 0 \\ -\frac{2}{3}kL & 0 & k \end{bmatrix}\begin{bmatrix} \theta \\ x_1 \\ x_2 \end{bmatrix}=\begin{bmatrix} 0 \\ 0 \\ F(t) \end{bmatrix}$$

5.56   Use Lagrange's equations to derive the differential equations governing the motion of the system of Fig. 5-33 using x_1, and x_2 as generalized coordinates. Assume the disk rolls without slip.

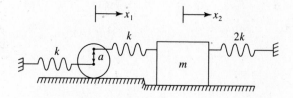

Fig. 5-33

Ans.

$$\begin{bmatrix} \dfrac{3}{2}m & 0 \\ 0 & m \end{bmatrix}\begin{bmatrix} \ddot{x}_1 \\ \ddot{x}_2 \end{bmatrix} + \begin{bmatrix} k\left(2+2\dfrac{a}{r}+\dfrac{a^2}{r^2}\right) & -k\left(1+\dfrac{a}{r}\right) \\ -k\left(1+\dfrac{a}{r}\right) & 3k \end{bmatrix}\begin{bmatrix} x_1 \\ x_2 \end{bmatrix} = \begin{bmatrix} 0 \\ 0 \end{bmatrix}$$

5.57   Use Lagrange's equations to derive the differential equations governing the motion of the system of Fig. 5-34 using x_1, x_2, and θ as generalized coordinates.

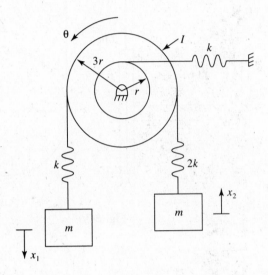

Fig. 5-34

Ans.

$$\begin{bmatrix} m & 0 & 0 \\ 0 & m & 0 \\ 0 & 0 & I \end{bmatrix}\begin{bmatrix} \ddot{x}_1 \\ \ddot{x}_2 \\ \ddot{\theta} \end{bmatrix} + \begin{bmatrix} k & 0 & -3kr \\ 0 & 2k & -6kr \\ -3kr & -6kr & 28kr^2 \end{bmatrix}\begin{bmatrix} x_1 \\ x_2 \\ \theta \end{bmatrix} = \begin{bmatrix} 0 \\ 0 \\ 0 \end{bmatrix}$$

5.58   Derive the differential equations governing the motion of the system of Fig. 5-35 using θ_1 and θ_2, as generalized coordinates.

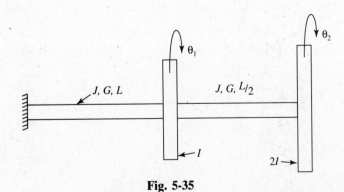

Fig. 5-35

Ans.

$$\begin{bmatrix} I & 0 \\ 0 & 2I \end{bmatrix}\begin{bmatrix} \ddot{\theta}_1 \\ \ddot{\theta}_2 \end{bmatrix} + \frac{JG}{L}\begin{bmatrix} 3 & -2 \\ -2 & 2 \end{bmatrix}\begin{bmatrix} \theta_1 \\ \theta_2 \end{bmatrix} = \begin{bmatrix} 0 \\ 0 \end{bmatrix}$$

5.59   Derive the differential equations governing the motion of the system of Fig. 5-36 using θ_1, θ_2, and x as generalized coordinates.

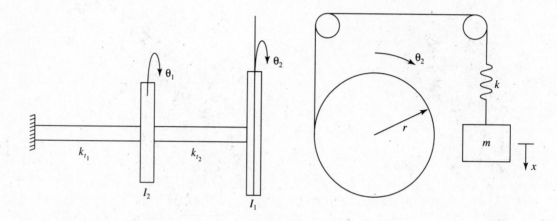

Fig. 5-36

Ans.

$$\begin{bmatrix} I_1 & 0 & 0 \\ 0 & I_2 & 0 \\ 0 & 0 & m \end{bmatrix}\begin{bmatrix} \ddot{\theta}_1 \\ \ddot{\theta}_2 \\ \ddot{x} \end{bmatrix} + \begin{bmatrix} k_{t_1} + k_{t_2} & -k_{t_1} & 0 \\ -k_{t_1} & k_{t_2} + kr^2 & -kr \\ 0 & -kr & k \end{bmatrix}\begin{bmatrix} \theta_1 \\ \theta_2 \\ x \end{bmatrix} = \begin{bmatrix} 0 \\ 0 \\ 0 \end{bmatrix}$$

5.60   Derive the differential equations governing the motion of an automobile suspension system using the 4-degree-of-freedom model of Fig. 5-37 using x_1, x_2, x_3, and x_4 as generalized coordinates.

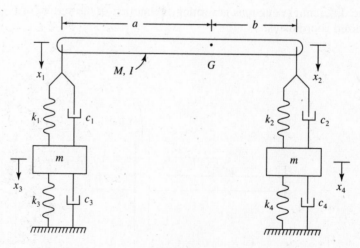

Fig. 5-37

Ans.

$$\begin{bmatrix} \dfrac{b^2M+I}{(a+b)^2} & \dfrac{abM-I}{(a+b)^2} & 0 & 0 \\[2mm] \dfrac{abM+I}{(a+b)^2} & \dfrac{a^2M+I}{(a+b)^2} & 0 & 0 \\[2mm] 0 & 0 & m & 0 \\[2mm] 0 & 0 & 0 & m \end{bmatrix} \begin{bmatrix} \ddot{x}_1 \\ \ddot{x}_2 \\ \ddot{x}_3 \\ \ddot{x}_4 \end{bmatrix} + \begin{bmatrix} c_1 & 0 & -c_1 & 0 \\ c_2 & 0 & -c_2 & 0 \\ -c_1 & 0 & c_1+c_3 & 0 \\ 0 & -c_2 & 0 & c_2+c_4 \end{bmatrix} \begin{bmatrix} \dot{x}_1 \\ \dot{x}_2 \\ \dot{x}_3 \\ \dot{x}_4 \end{bmatrix}$$

$$+ \begin{bmatrix} k_1 & 0 & -k_1 & 0 \\ 0 & k_2 & 0 & -k_2 \\ -k_1 & 0 & k_1+k_3 & 0 \\ 0 & -k_2 & 0 & k_2+k_4 \end{bmatrix} \begin{bmatrix} x_1 \\ x_2 \\ x_3 \\ x_4 \end{bmatrix} = \begin{bmatrix} 0 \\ 0 \\ 0 \\ 0 \end{bmatrix}$$

5.61 Use stiffness influence coefficients to derive the stiffness matrix for the system of Fig. 5-27 using x_1, x_2, and x_3 as generalized coordinates.

Ans.

$$\begin{bmatrix} 2k & -k & 0 \\ -k & 2k & -k \\ 0 & -k & 3k \end{bmatrix}$$

5.62 Use stiffness influence coefficients to derive the stiffness matrix for the system of Fig. 5-28 using θ_1 and θ_2 as generalized coordinates.

Ans.

$$\begin{bmatrix} \tfrac{9}{4}kL^2 & -3kL^2 \\ -3kL^2 & \tfrac{9}{2}kL^2 \end{bmatrix}$$

5.63 Use stiffness influence coefficients to derive the stiffness matrix for the system of Fig. 5-30 using θ_1 and θ_2 as generalized coordinates.

Ans.

$$\begin{bmatrix} ka^2 + \tfrac{1}{2}mgL & -ka^2 \\ -ka^2 & k(a^2 + \tfrac{9}{4}L^2) + \tfrac{3}{4}mgL \end{bmatrix}$$

5.64 Use stiffness influence coefficients to derive the stiffness matrix for the system of Fig. 5-32 using x_1, x_2, and θ as generalized coordinates.

Ans.

$$\begin{bmatrix} \tfrac{10}{9}kL^2 & \tfrac{1}{3}kL & -\tfrac{2}{3}kL \\ \tfrac{1}{3}kL & k & 0 \\ -\tfrac{2}{3}kL & 0 & k \end{bmatrix}$$

5.65 Use stiffness influence coefficients to derive the stiffness matrix for the system of Fig. 5-35 using x_1 and x_2 as generalized coordinates.

Ans.

$$\frac{JG}{L}\begin{bmatrix} 3 & -2 \\ -2 & 2 \end{bmatrix}$$

5.66 Use flexibility influence coefficients to derive the flexibility matrix for the system of Fig. 5-27 using x_1, x_2, and x_3 as generalized coordinates.

Ans.

$$\frac{1}{7k}\begin{bmatrix} 5 & 3 & 1 \\ 3 & 6 & 2 \\ 1 & 2 & 3 \end{bmatrix}$$

5.67 Use flexibility influence coefficients to derive the flexibility matrix for the system of Fig. 5-28 using θ_1 and θ_2 as generalized coordinates.

Ans.

$$\frac{1}{kL^2}\begin{bmatrix} 4 & \frac{8}{3} \\ \frac{8}{3} & 2 \end{bmatrix}$$

5.68 Two machines are equally spaced along the span of a simply supported beam of length L, elastic modulus E, and cross-sectional moment of inertia I. Determine the flexibility matrix for a 2-degree-of-freedom model of the system using the displacements of the machines as generalized coordinates.

Ans.

$$\frac{L^3}{EI}\begin{bmatrix} 0.01646 & 0.0144 \\ 0.0144 & 0.01646 \end{bmatrix}$$

5.69 Determine the flexibility matrix for a 4-degree-of-freedom model of a fixed-fixed beam of length L, elastic modulus E, and cross-sectional moment of inertia I.

Ans.

$$\frac{L^3}{EI}\begin{bmatrix} 1.365 & 2.016 & 1.451 & 0.5013 \\ 2.016 & 4.608 & 3.925 & 1.451 \\ 1.451 & 3.925 & 4.608 & 2.016 \\ 0.5013 & 1.451 & 2.016 & 1.365 \end{bmatrix}10^{-3}$$

5.70 Determine the flexibility matrix for the system of Fig. 5-38.

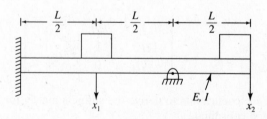

Fig. 5-38

Ans.

$$\frac{L^3}{EI}\begin{bmatrix} 9.116 & -15.54 \\ -15.54 & 104.2 \end{bmatrix}10^{-3}$$

5.71   Determine the flexibility matrix for the system of Fig. 5-39.

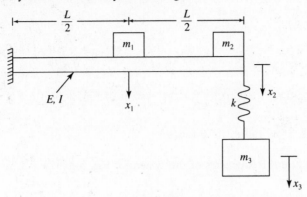

Fig. 5-39

Ans.

$$\frac{L^3}{EI} \begin{bmatrix} 0.0417 & 0.1042 & 0.1042 \\ 0.1042 & 0.3333 & 0.3333 \\ 0.1042 & 0.3333 & \frac{EI}{kL^3}+0.3333 \end{bmatrix}$$

5.72   Determine the natural frequencies for a 2-degree-of-freedom system whose governing differential equations are

$$\begin{bmatrix} 100 & 60 \\ 60 & 120 \end{bmatrix}\begin{bmatrix} \ddot{x}_1 \\ \ddot{x}_2 \end{bmatrix} + \begin{bmatrix} 30,000 & -10,000 \\ -10,000 & 20,000 \end{bmatrix}\begin{bmatrix} x_1 \\ x_2 \end{bmatrix} = \begin{bmatrix} 0 \\ 0 \end{bmatrix}$$

Ans.   9.044 rad/s, 26.98 rad/s.

5.73   Determine the mode shape vectors for the system of Problem 5.72.

Ans. $[0.04332 \quad 0.06342]^T$, $[0.111 \quad -0.08879]^T$

5.74   Demonstrate orthogonality of the mode shapes for the system of Problem 5.72.

5.75   Determine the natural frequencies of the system of Fig. 5-28.

Ans.

$$0.536\sqrt{\frac{k}{m}}, \qquad 3.23\sqrt{\frac{k}{m}}$$

5.76   Determine the natural frequencies of the system of Fig. 5-31 assuming $I = \frac{1}{12}mL^2$.

Ans.

$$0.760\sqrt{\frac{k}{m}}, \qquad 1.521\sqrt{\frac{k}{m}}$$

5.77   Determine the natural frequencies of the system of Fig. 5-35.

Ans.

$$0.5176\sqrt{\frac{JG}{IL}}, \qquad 1.932\sqrt{\frac{JG}{IL}}$$

5.78   Determine the mode shape vectors for the system of Fig. 5-35.

Ans. $[0.4597 \quad 0.6277]^T$, $[0.8881 \quad -0.3251]^T$

5.79   Determine the natural frequencies for the system of Problem 5.68 if the machines both have a mass m.

Ans.

$$5.692 \sqrt{\frac{EI}{mL^3}}, \qquad 22.03 \sqrt{\frac{EI}{mL^3}}$$

5.80   Determine the natural frequencies for the system of Fig. 5-29 if $k = 3 \times 10^5$ N/m, $m = 15$ kg, and $L = 1.6$ m.

Ans.   70.7 rad/s, 244.9 rad/s, 282.8 rad/s.

5.81   Determine the natural frequencies of the system of Fig. 5-32 if $k = 1.3 \times 10^5$ N/m, $m = 2.6$ kg, and $L = 1.0$ m.

Ans.   114.3 rad/s, 215.3 rad/s, 718.0 rad/s.

5.82   Determine the mode shape vectors for the system of Problem 5.81.

Ans. $[0.487 \quad -0.220 \quad 0.680]^T$, $[-0.212 \quad 0.970 \quad 0.165]^T$, $[2.953 \quad 0.106 \quad -0.100]^T$.

5.83   Demonstrate orthogonality of the mode shape vectors for Problem 5.82.

5.84   Determine the natural frequencies for the system of Fig. 5-39 if $L = 2$ m, $E = 200 \times 10^9$ N/m^2, $I = 1.5 \times 10^{-5}$ m^4, $k = 4 \times 10^5$ N/m, $m_1 = 60$ kg, $m_2 = 80$ kg, and $m_3 = 40$ kg. Assume the beam is massless.

Ans.   77.7 rad/s, 147.3 rad/s, 857.4 rad/s.

5.85   Use a 3-degree-of-freedom model to approximate the lowest natural frequencies of a fixed-fixed beam.

Ans.

$$22.3 \sqrt{\frac{EI}{mL^3}}, \qquad 59.26 \sqrt{\frac{EI}{mL^3}}, \qquad 97.4 \sqrt{\frac{EI}{mL^3}}$$

5.86   Use a 3-degree-of-freedom model to approximate the lowest natural frequencies of a fixed-free beam.

Ans.

$$3.346 \sqrt{\frac{EI}{mL^3}}, \qquad 18.86 \sqrt{\frac{EI}{mL^3}}, \qquad 46.77 \sqrt{\frac{EI}{mL^3}}$$

5.87   Determine the natural frequencies of the system of Fig. 5-37 if $c_1 = c_2 = c_3 = c_4 = 0$, $a = 3$ m, $b = 1$ m, $M = 200$ kg, $m = 30$ kg, $I = 200$ kg-m^2, $k_1 = k_2 = 4 \times 10^5$ N/m, and $k_3 = k_4 = 1 \times 10^5$ N/m.

Ans.   23.0 rad/s, 44.3 rad/s, 138.5 rad/s, 188.8 rad/s.

5.88   Use matrix iteration to determine the natural frequencies of the system of Problem 5.82.

5.89   Use matrix iteration to determine the natural frequencies of the system of Problem 5.85.

5.90   Use matrix iteration to determine the lowest natural frequency of a fixed-fixed beam using a 3-degree-of-freedom model.

Ans.

$$22.3 \sqrt{\frac{EI}{mL^3}}$$

5.91   Determine the general free vibration response of the system of Fig. 5-37 and Problem 5.87 if $c_1 = c_2 = 2000$ N-s/m and $c_3 = c_4 = 500$ N-s/m.

Ans.

$$e^{-1.323t}(C_1 \cos 23.0t + C_2 \sin 23.0t)$$
$$+ e^{-4.906t}(C_3 \cos 44.0t + C_4 \sin 44.0t)$$
$$+ e^{-47.96t}(C_5 \cos 129.9t + C_6 \sin 129.9t)$$
$$+ e^{-89.1t}(C_7 \cos 166.4t + C_8 \sin 166.4t)$$

5.92   Determine the general free vibration response of the system of Fig. 5-40.

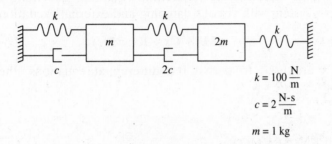

$$k = 100 \frac{N}{m}$$
$$c = 2 \frac{N\text{-}s}{m}$$
$$m = 1 \text{ kg}$$

Fig. 5-40

Ans.

$$e^{-0.268t}\left\{ C_1 \begin{bmatrix} 0.732 \\ 1 \end{bmatrix} \cos 7.96t + C_2 \begin{bmatrix} 0.732 \\ 1 \end{bmatrix} \sin 7.96t \right\}$$
$$+ e^{-3.73t}\left\{ C_3 \begin{bmatrix} -2.73 \\ 1 \end{bmatrix} \cos 14.92t + C_4 \begin{bmatrix} -2.73 \\ 1 \end{bmatrix} \sin 14.92t \right\}$$

5.93   Show that if a mode shape vector $\mathbf{X}$ is normalized according to Eq. (5.16), then Eq. (5.17) follows.

5.94   Let $\mathbf{X}_i$ be the mode shape vector corresponding to a natural frequency ω_i of a n-degree-of-freedom system. Then

$$\mathbf{KX}_i = \omega_i^2 \mathbf{MX}_i$$

Note that if $\mathbf{M}$ is symmetric, then for any two vectors $\mathbf{u}$ and $\mathbf{v}$:

$$\mathbf{u}^T \mathbf{Mv} = \mathbf{v}^T \mathbf{Mu}$$

Use the above to derive the orthogonality relation, Eq. (5.14).

Chapter 6

Forced Vibrations of Multi-Degree-of-Freedom Systems

6.1 GENERAL SYSTEM

The standard matrix form for the differential equations governing the motion of a linear n-degree-of-freedom system with viscous damping and external excitation is

$$\mathbf{M\ddot{x}} + \mathbf{C\dot{x}} + \mathbf{Kx} = \mathbf{F}(t) \qquad (6.1)$$

If energy methods are used to derive the differential equations, then $\mathbf{M}$, $\mathbf{C}$, and $\mathbf{K}$ are symmetric.

6.2 HARMONIC EXCITATION

If

$$\mathbf{F}(t) = \mathbf{R} \sin \omega t + \mathbf{S} \cos \omega t \qquad (6.2)$$

then the steady-state solution of Eq. (6.1) is

$$\mathbf{x}(t) = \mathbf{U} \sin \omega t + \mathbf{V} \cos \omega t \qquad (6.3)$$

where $\mathbf{U}$ and $\mathbf{V}$ are solutions of

$$(-\omega^2 \mathbf{M} + \mathbf{K})\mathbf{U} - \omega \mathbf{C V} = \mathbf{R} \qquad (6.4)$$

and

$$\omega \mathbf{C U} + (-\omega^2 \mathbf{M} + \mathbf{K}) = \mathbf{S} \qquad (6.5)$$

If $x_i(t) = X_i \sin (\omega t - \phi_i)$ then

$$X_i = \sqrt{U_i^2 + V_i^2} \qquad (6.6)$$

6.3 LAPLACE TRANSFORM SOLUTIONS

Let $\bar{\mathbf{x}}(s)$ be the vector of Laplace transforms of the generalized coordinates and $\bar{\mathbf{F}}(s)$ is the Laplace transform of $\mathbf{F}(t)$. If $\mathbf{x}(0) = \mathbf{0}$ and $\dot{\mathbf{x}}(0) = \mathbf{0}$, then taking the Laplace transform of Eq. (6.1) and solving for $\bar{\mathbf{x}}(s)$ leads to

$$\mathbf{Z}(s)\bar{\mathbf{x}}(s) = \bar{\mathbf{F}}(s) \qquad (6.7)$$

where the *impedance matrix* $\mathbf{Z}(s)$ is defined by

$$\mathbf{Z}(s) = s^2 \mathbf{M} + s\mathbf{C} + \mathbf{K} \qquad (6.8)$$

Equations (6.7) can be solved for $\bar{\mathbf{x}}(s)$ and the result inverted to obtain $\mathbf{x}(t)$.

6.4 MODAL ANALYSIS FOR SYSTEMS WITH PROPORTIONAL DAMPING

Let $\omega_1, \omega_2, \ldots, \omega_n$ be the natural frequencies of an undamped n-degree-of-freedom system. Let $X_1, X_2, \ldots, X_n$ be their corresponding mode shapes. The *modal matrix* P is the matrix whose ith column is X_i. Orthogonality of the mode shapes implies

$$P^T M P = I \tag{6.9}$$

where I is the $n \times n$ identity matrix,

$$P^T K P = \Omega = \text{diag}\{\omega_1{}^2, \omega_2{}^2, \ldots, \omega_n{}^2\} \tag{6.10}$$

and if the viscous damping is proportional,

$$P^T C P = Z = \text{diag}\{2\zeta_1\omega_1, 2\zeta_2\omega_2, \ldots, 2\zeta_n\omega_n\} \tag{6.11}$$

The *principal coordinates* p are defined through the linear transformation

$$p = P^{-1}x \tag{6.12}$$

or
$$x = Pp \tag{6.13}$$

When C is of the form of Eq. (5.19), the principal coordinates are used as dependent variables and Eq. (6.1) is rewritten as

$$\ddot{p} + Z\dot{p} + \Omega p = G(t) \tag{6.14}$$

where
$$G(t) = P^T F \tag{6.15}$$

Differential equations represented by Eq. (6.14) are uncoupled and of the form

$$\ddot{p}_1 + 2\zeta_i\omega_i\dot{p}_i + \omega_i^2 p_i = G_i(t) \qquad i = 1, 2, \ldots, n \tag{6.16}$$

The procedure where the principal coordinates are used to uncouple the differential equations is referred to as *modal analysis*. The convolution integral is used to determine the solution for each principal coordinate as

$$p_i(t) = \frac{1}{\omega_i\sqrt{1 - \zeta_i^2}} \int_0^t e^{-\zeta_i\omega_i(t-\tau)} \sin \omega_i\sqrt{1 - \zeta_i^2}(t - \tau)G_i(\tau) \, d\tau \tag{6.17}$$

6.5 MODAL ANALYSIS FOR SYSTEMS WITH GENERAL DAMPING

If a n-degree-of-freedom system is subject to viscous damping, but the damping matrix is not of the form of Eq. (5.19), a more complicated form of modal analysis must be used. The definition of Eqs. (5.23) and (5.24) are used to rewrite Eq. (6.1) as

$$\tilde{M}\dot{y} + \tilde{K}y = \tilde{F} \tag{6.18}$$

where
$$\tilde{F} = \begin{bmatrix} 0 \\ F \end{bmatrix} \tag{6.19}$$

is a $2n \times 1$ column vector. Define P as a $2n \times 2n$ matrix whose ith column is the eigenvector Φ_i of $\tilde{M}^{-1}\tilde{K}$ normalized such that

$$\Phi_i{}^T\tilde{M}\Phi_i = 1 \tag{6.20}$$

Equation (6.18) is uncoupled when the coordinates

$$\tilde{p} = \tilde{P}^{-1}y \tag{6.21}$$

are used as dependent variables. The uncoupled differential equations are

$$\tilde{p}_i - \gamma_i \tilde{p}_i = \tilde{g}_i \tag{6.22}$$

where

$$\tilde{\mathbf{G}} = \tilde{\mathbf{P}}^T \tilde{\mathbf{F}} \tag{6.23}$$

The solution of Eq. (6.22) is

$$\tilde{p}_i = \int_0^t \tilde{g}_i(\tau) e^{-\gamma_i(t-\tau)} \, d\tau \tag{6.24}$$

Solved Problems

6.1   Determine the steady-state response of the system of Fig. 6-1.

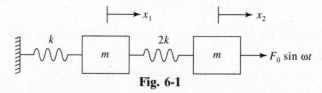

Fig. 6-1

The differential equations governing the motion of the system of Fig. 6-1 are

$$\begin{bmatrix} m & 0 \\ 0 & m \end{bmatrix} \begin{bmatrix} \ddot{x}_1 \\ \ddot{x}_2 \end{bmatrix} + \begin{bmatrix} 3k & -2k \\ -2k & 2k \end{bmatrix} \begin{bmatrix} x_1 \\ x_2 \end{bmatrix} = \begin{bmatrix} 0 \\ F_0 \end{bmatrix} \sin \omega t$$

Since the system is undamped and $\mathbf{S} = \mathbf{0}$, $\mathbf{V} = \mathbf{0}$. Then Eq. (6.4) reduces to

$$\begin{bmatrix} 3k - m\omega^2 & -2k \\ -2k & 2k - m\omega^2 \end{bmatrix} \begin{bmatrix} U_1 \\ U_2 \end{bmatrix} = \begin{bmatrix} 0 \\ F_0 \end{bmatrix}$$

Solving simultaneously leads to

$$U_1 = \frac{2kF_0}{(2k - m\omega^2)(3k - m\omega^2) - 4k^2}$$

$$U_2 = \frac{(3k - m\omega^2)F_0}{(2k - m\omega^2)(3k - m\omega^2) - 4k^2}$$

6.2   Determine the steady-state amplitudes of the blocks of the system of Fig. 6-2.

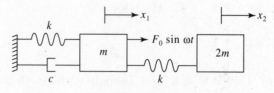

Fig. 6-2

The differential equations governing the motion of the system of Fig. 6-2 are

$$\begin{bmatrix} m & 0 \\ 0 & 2m \end{bmatrix} \begin{bmatrix} \ddot{x}_1 \\ \ddot{x}_2 \end{bmatrix} + \begin{bmatrix} c & 0 \\ 0 & 0 \end{bmatrix} \begin{bmatrix} \dot{x}_1 \\ \dot{x}_2 \end{bmatrix} + \begin{bmatrix} 2k & -k \\ -k & k \end{bmatrix} \begin{bmatrix} x_1 \\ x_2 \end{bmatrix} = \begin{bmatrix} F_0 \\ 0 \end{bmatrix} \sin \omega t$$

Equations (6.4) and (6.5) become

$$\begin{bmatrix} 2k - m\omega^2 & -k \\ -k & k - 2m\omega^2 \end{bmatrix} \begin{bmatrix} U_1 \\ U_2 \end{bmatrix} + \begin{bmatrix} -\omega c & 0 \\ 0 & 0 \end{bmatrix} \begin{bmatrix} V_1 \\ V_2 \end{bmatrix} = \begin{bmatrix} F_0 \\ 0 \end{bmatrix}$$

$$\begin{bmatrix} \omega c & 0 \\ 0 & 0 \end{bmatrix} \begin{bmatrix} U_1 \\ U_2 \end{bmatrix} + \begin{bmatrix} 2k - m\omega^2 & -k \\ -k & k - 2m\omega^2 \end{bmatrix} \begin{bmatrix} V_1 \\ V_2 \end{bmatrix} = \begin{bmatrix} 0 \\ 0 \end{bmatrix}$$

Solving simultaneously leads to

$$U_1 = \frac{(k - 2m\omega^2)(k^2 - 5km\omega^2 + 2m^2\omega^4)F_0}{D}$$

$$U_2 = \frac{k(k^2 - 5km\omega^2 + 2m^2\omega^4)F_0}{D}$$

$$V_1 = -\frac{\omega c(k - 2m\omega^2)^2 F_0}{D}$$

$$V_2 = -\frac{k\omega c(k - 2m\omega^2)F_0}{D}$$

where

$$D = k^4 + (k^2 c^2 - 10k^3 m)\omega^2 + (29k^2 m^2 - 4c^2 km)\omega^4$$
$$+ (4c^2 m^2 - 20km^3)\omega^6 + 4m^4\omega^8$$

The blocks' steady-state amplitudes are

$$X_1 = \sqrt{U_1^2 + V_1^2} = \frac{(k - 2m\omega^2)F_0}{\sqrt{D}}$$

$$X_2 = \sqrt{U_2^2 + V_2^2} = \frac{kF_0}{\sqrt{D}}$$

6.3   A 110-kg machine with a 0.45-kg-m rotating unbalance is placed at the end of a 1.5-m-long steel ($E = 200 \times 10^9$ N/m^2, $\rho = 7800$ kg/m^3) fixed-free beam of cross-sectional area 1.4×10^{-2} m^2, moment of inertia 3.5×10^{-6} m^4, and length 1.5 m. The machine operates at 200 Hz. Use a 3-degree-of-freedom model for the beam, and approximate the machine's steady-state amplitude.

The flexibility matrix for a 3-degree-of-freedom model of the fixed-free beam with equally spaced nodes is determined using the methods of Chap. 5 as

$$\mathbf{A} = 10^{-8} \begin{bmatrix} 5.95 & 14.9 & 23.8 \\ 14.9 & 47.6 & 83.3 \\ 23.8 & 83.3 & 160.7 \end{bmatrix} \frac{\text{m}}{\text{N}}$$

The beam's mass is

$$m_b = \rho A L = \left(7800 \ \frac{\text{kg}}{\text{m}^3}\right)(1.4 \times 10^{-2} \text{ m}^2)(1.5 \text{ m}) = 163.8 \text{ kg}$$

Lumping the mass of the beam at the three nodes, the mass matrix is

$$\begin{bmatrix} \dfrac{m_b}{3} & 0 & 0 \\ 0 & \dfrac{m_b}{3} & 0 \\ 0 & 0 & \dfrac{m_b}{6} + M \end{bmatrix} = \begin{bmatrix} 54.6 & 0 & 0 \\ 0 & 54.6 & 0 \\ 0 & 0 & 137.3 \end{bmatrix} \text{kg}$$

The magnitude of the harmonic excitation provided by the rotating unbalance is

$$F_0 = m_0 e \omega^2 = (0.45 \text{ kg-m})\left(200 \frac{\text{cycle}}{\text{s}} \frac{2\pi \text{ rad}}{\text{cycle}}\right)^2 = 7.11 \times 10^5 \text{ N}$$

The force vector is

$$\mathbf{F} = \begin{bmatrix} 0 \\ 0 \\ 7.11 \times 10^5 \sin 1257t \end{bmatrix}$$

The differential equations for this model are

$$\mathbf{AM\ddot{x}} + \mathbf{x} = \mathbf{AF}$$

$$10^{-8}\begin{bmatrix} 5.95 & 14.9 & 23.8 \\ 14.9 & 47.6 & 83.3 \\ 23.8 & 83.3 & 160.7 \end{bmatrix}\begin{bmatrix} 54.6 & 0 & 0 \\ 0 & 54.6 & 0 \\ 0 & 0 & 137.3 \end{bmatrix}\begin{bmatrix} \ddot{x}_1 \\ \ddot{x}_2 \\ \ddot{x}_3 \end{bmatrix} + \begin{bmatrix} x_1 \\ x_2 \\ x_3 \end{bmatrix}$$

$$= 10^{-8}\begin{bmatrix} 5.95 & 14.9 & 23.8 \\ 14.9 & 47.6 & 83.3 \\ 23.8 & 83.3 & 160.7 \end{bmatrix}\begin{bmatrix} 0 \\ 0 \\ 7.11 \times 10^5 \end{bmatrix}\sin 1257t$$

$$10^{-8}\begin{bmatrix} 324.9 & 813.5 & 3267.7 \\ 813.5 & 2599.0 & 11{,}437.1 \\ 1299.5 & 4548.2 & 22{,}064.1 \end{bmatrix}\begin{bmatrix} \ddot{x}_1 \\ \ddot{x}_2 \\ \ddot{x}_3 \end{bmatrix} + \begin{bmatrix} x_1 \\ x_2 \\ x_3 \end{bmatrix} = 10^{-8}\begin{bmatrix} 1.692 \times 10^7 \\ 5.922 \times 10^7 \\ 1.143 \times 10^8 \end{bmatrix}$$

When the assumed steady-state solution $\mathbf{x} = [U_1 \quad U_2 \quad U_4]^T \sin 1257t$ is substituted into the above equations, the following algebraic equations are obtained:

$$\begin{bmatrix} -4.13 & -12.85 & -51.63 \\ -12.85 & -40.06 & -180.7 \\ -20.53 & -71.86 & -347.6 \end{bmatrix}\begin{bmatrix} U_1 \\ U_2 \\ U_3 \end{bmatrix} = \begin{bmatrix} 0.169 \\ 0.592 \\ 1.143 \end{bmatrix}$$

whose solution is

$$U_1 = 1.32 \times 10^{-3}, \qquad U_2 = -4.30 \times 10^{-4}, \qquad U_3 = -3.28 \times 10^{-3}$$

Hence the machine's steady-state amplitude is 3.28 mm.

6.4 An auxiliary system consisting of a block of mass m_2 is connected to the primary system of Fig. 6-3 by a spring of stiffness k_2. The auxiliary system can be used as a vibration absorber if the parameters k_2 and m_2 are chosen correctly. Show that if

$$\sqrt{\frac{k_2}{m_2}} = \omega$$

then the steady-state amplitude of the primary system is zero.

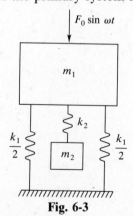

Fig. 6-3

The differential equations governing the motion of the system are

$$\begin{bmatrix} m_1 & 0 \\ 0 & m_2 \end{bmatrix}\begin{bmatrix} \ddot{x}_1 \\ \ddot{x}_2 \end{bmatrix} + \begin{bmatrix} k_1 + k_2 & -k_2 \\ -k_2 & k_2 \end{bmatrix}\begin{bmatrix} x_1 \\ x_2 \end{bmatrix} = \begin{bmatrix} F_0 \sin \omega t \\ 0 \end{bmatrix}$$

Application of Eq. (6.4) leads to

$$\begin{bmatrix} -\omega^2 m_1 + k_1 + k_2 & -k_2 \\ -k_2 & -\omega^2 m_2 + k_2 \end{bmatrix}\begin{bmatrix} U_1 \\ U_2 \end{bmatrix} = \begin{bmatrix} F_0 \\ 0 \end{bmatrix}$$

The solution for U_1 is

$$U_1 = \frac{F_0(k_2 - m_2\omega^2)}{(k_1 + k_2 - m_1\omega^2)(k_2 - m_2\omega^2) - k_2^2}$$

Hence $U_1 = 0$ if $k_2/m_2 = \omega^2$.

6.5   Use of the Laplace transform method to solve Problem 6.2 assuming that the system is at rest in equilibrium at $t = 0$, $m = 1$ kg, $k = 100$ N/m, $c = 2$ N-s/m, and $\omega = 10$ rad/s.

The impedance matrix for the system of Problem 6.2 is

$$\mathbf{Z}(s) = \begin{bmatrix} s^2 + 2s + 200 & -100 \\ -100 & 2s^2 + 100 \end{bmatrix}$$

and its inverse is

$$\mathbf{Z}^{-1}(s) = \frac{1}{D(s)}\begin{bmatrix} s^2 + 50 & 50 \\ 50 & \frac{1}{2}(s^2 + 2s + 200) \end{bmatrix}$$

where

$$D(s) = s^4 + 2s^3 + 250s^2 + 100s + 5000$$
$$= [(s + 0.136)^2 + 21.96][(s + 0.864)^2 + 226.8]$$

The Laplace transform of the force vector is

$$\bar{\mathbf{F}}(s) = \begin{bmatrix} \dfrac{10F_0}{s^2 + 100} \\ 0 \end{bmatrix}$$

The Laplace transform of the displacement vector is

$$\bar{\mathbf{x}} = \mathbf{Z}^{-1}\bar{\mathbf{F}}(s) = \frac{F_0}{(s^2 + 100)D(s)}\begin{bmatrix} 10(s^2 + 50) \\ 500 \end{bmatrix}$$

Partial fraction decomposition leads to

$$\bar{x}_1(s) = F_0\Bigg[\frac{-4.95 \times 10^{-4}s + 4.95 \times 10^{-2}}{s^2 + 100} + \frac{-2.33 \times 10^{-4}s + 1.74 \times 10^{-2}}{(s + 0.136)^2 + 21.96}$$
$$+ \frac{7.28 \times 10^{-4}s - 6.57 \times 10^{-2}}{(s + 0.864)^2 + 226.8}\Bigg]$$

$$\bar{x}_2(s) = F_0\Bigg[\frac{4.95 \times 10^{-4}s - 4.95 \times 10^{-2}}{s^2 + 100} + \frac{-1.11 \times 10^{-4}s + 3.11 \times 10^{-2}}{(s + 0.136)^2 + 21.96}$$
$$+ \frac{-3.83 \times 10^{-4}s + 1.77 \times 10^{-2}}{(s + 0.864)^2 + 226.8}\Bigg]$$

Inversion of the transforms gives

$$x_1(t) = F_0[-4.95 \times 10^{-4}\cos 10t + 4.95 \times 10^{-3}\sin 10t$$
$$+ e^{-0.136t}(-2.33 \times 10^{-4}\cos 4.69t + 3.72 \times 10^{-3}\sin 4.69t)$$
$$+ e^{-0.864t}(7.28 \times 10^{-4}\cos 15.07t - 4.36 \times 10^{-3}\sin 15.07t)]$$

$$x_2(t) = F_0[4.95 \times 10^{-4}\cos 10t - 4.95 \times 10^{-3}\sin 10t$$
$$+ e^{-0.136t}(-1.11 \times 10^{-4}\cos 4.69t + 6.63 \times 10^{-3}\sin 4.69t)$$
$$+ e^{-0.864t}(-3.83 \times 10^{-4}\cos 15.07t + 1.18 \times 10^{-3}\sin 15.07t)]$$

6.6   Use the Laplace transform method to determine $x_1(t)$ for the system of Fig. 6-4 assuming that the system is at rest in equilibrium at $t = 0$.

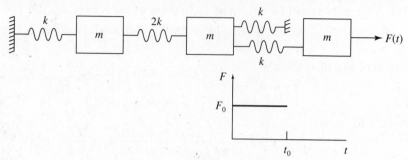

Fig. 6-4

The differential equations governing the motion of the system of Fig. 6-4 are

$$\begin{bmatrix} m & 0 & 0 \\ 0 & m & 0 \\ 0 & 0 & m \end{bmatrix}\begin{bmatrix} \ddot{x}_1 \\ \ddot{x}_2 \\ \ddot{x}_3 \end{bmatrix} + \begin{bmatrix} 3k & -2k & 0 \\ -2k & 4k & -k \\ 0 & -k & k \end{bmatrix}\begin{bmatrix} x_1 \\ x_2 \\ x_3 \end{bmatrix} = \begin{bmatrix} 0 \\ 0 \\ F_0[u(t) - u(t - t_0)] \end{bmatrix}$$

The impedance matrix for this system is

$$\mathbf{Z}(s) = \begin{bmatrix} ms^2 + 3k & -2k & 0 \\ -2k & ms^2 + 4k & -k \\ 0 & -k & ms^2 + k \end{bmatrix}$$

and its inverse is

$$\mathbf{Z}^{-1}(s) = \frac{1}{D(S)}\begin{bmatrix} m^2s^4 + 5ms^2k + 3k^2 & 2k(ms^2 + k) & 2k^2 \\ 2k(ms^2 + k) & (ms^2 + 3k)(ms^2 + k) & k(ms^2 + 3k) \\ 2k^2 & k(ms^2 + 3k) & m^2s^4 + 7ms^2 + 8k^2 \end{bmatrix}$$

where

$$D(s) = m^3s^6 + 8m^2s^4k + 14ms^2k^2 + 5k^3$$

The Laplace transform of the force vector is

$$\bar{\mathbf{F}}(s) = \begin{bmatrix} 0 \\ 0 \\ \dfrac{F_0}{s}(1 - e^{-st_0}) \end{bmatrix}$$

Then

$$\bar{\mathbf{x}}(s) = \mathbf{Z}^{-1}\bar{\mathbf{F}}(s) = \frac{1}{sD(S)}\begin{bmatrix} 2k^2 \\ k(ms^2 + 3k) \\ m^2s^4 + 7ms^2 + 8k^2 \end{bmatrix}F_0(1 - e^{-st_0})$$

Partial fraction decomposition leads to

$$\bar{x}_1(s) = 2\frac{k^2}{m^3}\left[\frac{0.2}{s} - \frac{0.297s}{s^2 + 0.4818\dfrac{k}{m}} + \frac{0.106s}{s^2 + 1.820\dfrac{k}{m}} + \frac{0.00874s}{s^2 + 5.698\dfrac{k}{m}}\right]F_0(1 - e^{-st_0})$$

which when inverted, using in part the second shifting theorem, leads to

$$x_1(t) = 2F_0\frac{k^2}{m^3}\left\{\left[0.2 - 0.297\cos 0.694\sqrt{\frac{k}{m}}t + 0.106\cos 1.395\sqrt{\frac{k}{m}}t\right.\right.$$

$$\left. + 0.00874\cos 2.387\sqrt{\frac{k}{m}}t\right]u(t) - \left[0.2 - 0.297\cos 0.694\sqrt{\frac{k}{m}}(t - t_0)\right.$$

$$\left.\left. + 0.106\cos 1.395\sqrt{\frac{k}{m}}(t - t_0) + 0.00874\cos 2.387\sqrt{\frac{k}{m}}(t - t_0)\right]\right\}$$

6.7 Derive Eq. (6.14).

Equation (6.1) is rewritten with the principal coordinates as the dependent variables by substituting Eq. (6.13) into Eq. (6.1), leading to

$$\mathbf{MP\ddot{p} + CP\dot{p} + KPp = F}$$

Premultiplying the previous equation by $\mathbf{P}^T$ leads to

$$\mathbf{P}^T\mathbf{MP\ddot{p}} + \mathbf{P}^T\mathbf{CP\dot{p}} + \mathbf{P}^T\mathbf{KPp} = \mathbf{P}^T\mathbf{F}$$

When we use Eqs. (6.9), (6.10), (6.11), and (6.15), the previous equation becomes

$$\mathbf{\ddot{p} + Z\dot{p} + \Omega p = G}$$

6.8 Determine the relationship between the generalized coordinates and the principal coordinates for the system of Fig. 5-3 and Problem 5.4 if $I = mL^2/12$.

The natural frequencies of the system are determined from

$$\begin{vmatrix} 2k - m\omega^2 & \frac{1}{4}kL \\ \frac{1}{4}kL & \frac{5}{16}kL^2 - \frac{1}{12}mL^2\omega^2 \end{vmatrix} = 0$$

$$\omega_1 = 1.28\sqrt{\frac{k}{m}}, \qquad \omega_2 = 2.03\sqrt{\frac{k}{m}}$$

The mode shapes are determined from

$$\begin{bmatrix} 2k - m\omega^2 & \frac{1}{4}kL \\ \frac{1}{4}kL & \frac{5}{16}kL^2 - \frac{1}{12}mL^2\omega^2 \end{bmatrix}\begin{bmatrix} X \\ \Theta \end{bmatrix} = \begin{bmatrix} 0 \\ 0 \end{bmatrix}$$

from which the top equation yields

$$\Theta = \frac{-4(2k - m\omega^2)}{kL}$$

which leads to

$$\mathbf{X}_1 = C_1\begin{bmatrix} 1 \\ \dfrac{-1.42}{L} \end{bmatrix}, \qquad \mathbf{X}_2 = C_2\begin{bmatrix} 1 \\ \dfrac{8.42}{L} \end{bmatrix}$$

The mode shapes are normalized by

$$\mathbf{X}_1{}^T\mathbf{MX}_1 = 1 = C_1{}^2\begin{bmatrix} 1 & \dfrac{-1.42}{L} \end{bmatrix}\begin{bmatrix} m & 0 \\ 0 & \dfrac{1}{12}mL^2 \end{bmatrix}\begin{bmatrix} 1 \\ \dfrac{1.42}{L} \end{bmatrix} \rightarrow C_1 = \frac{0.925}{\sqrt{m}}$$

$$\mathbf{X}_2{}^T\mathbf{MX}_2 = 1 = C_2{}^2\begin{bmatrix} 1 & \dfrac{8.42}{L} \end{bmatrix}\begin{bmatrix} m & 0 \\ 0 & \dfrac{1}{12}mL^2 \end{bmatrix}\begin{bmatrix} 1 \\ \dfrac{8.42}{L} \end{bmatrix} \rightarrow C_2 = \frac{0.380}{\sqrt{m}}$$

Thus the modal matrix and its inverse are

$$\mathbf{P} = \frac{1}{\sqrt{m}}\begin{bmatrix} 0.925 & 0.380 \\ -\dfrac{1.31}{L} & \dfrac{3.20}{L} \end{bmatrix}, \qquad \mathbf{P}^{-1} = \sqrt{m}\begin{bmatrix} 0.925 & -0.110L \\ 0.378 & 0.268L \end{bmatrix}$$

Hence the principal coordinates are related to the generalized coordinates by

$$\mathbf{p} = \mathbf{P}^{-1}\mathbf{x} = \sqrt{m}\begin{bmatrix} 0.925 & 0.110L \\ 0.378 & 0.268L \end{bmatrix}\begin{bmatrix} x \\ \theta \end{bmatrix}$$

6.9 Use modal analysis to determine the response of the system of Fig. 5-3 and Problems 5.4 and 6.8 if

$$M(t) = M_0[1 - u(t - t_0)]$$

The vector $\mathbf{G}(t)$ is determined as

$$\mathbf{G} = \mathbf{P}^T\mathbf{F} = \frac{1}{\sqrt{m}}\begin{bmatrix} 0.925 & \dfrac{1.31}{L} \\[2ex] 0.380 & -\dfrac{3.20}{L} \end{bmatrix}\begin{bmatrix} 0 \\[1ex] M(t) \end{bmatrix} = \frac{1}{\sqrt{m}}\begin{bmatrix} \dfrac{-1.31}{L}M(t) \\[2ex] \dfrac{3.20}{L}M(t) \end{bmatrix}$$

The differential equations for the principal coordinates become

$$\ddot{p}_1 + 1.644\frac{k}{m}p_1 = \frac{-1.31}{\sqrt{m}L}M_0[1 - u(t - t_0)]$$

$$\ddot{p}_2 + 4.104\frac{k}{m}p_2 = \frac{3.20}{\sqrt{m}L}M_0[1 - u(t - t_0)]$$

The convolution integral is used to solve for the principal coordinates as

$$p_1(t) = \frac{1}{1.28\sqrt{\dfrac{k}{m}}}\int_0^t \frac{-1.31}{\sqrt{m}L}M_0[1 - u(\tau - t_0)]\sin\left(1.28\sqrt{\frac{k}{m}}(t-\tau)\right)d\tau$$

$$= \frac{0.797\sqrt{m}}{kL}M_0\left\{\left[1 - \cos\left(1.282\sqrt{\frac{k}{m}}t\right)\right]u(t)\right.$$

$$\left. - \left[1 - \cos\left(1.282\sqrt{\frac{k}{m}}(t - t_0)\right)\right]u(t - t_0)\right\}$$

$$p_2(t) = \frac{1}{2.03\sqrt{\dfrac{k}{m}}}\int_0^t \frac{3.20}{\sqrt{m}L}M_0[1 - u(\tau - t_0)]\sin\left(2.03\sqrt{\frac{k}{m}}(t-\tau)\right)d\tau$$

$$= \frac{-0.777\sqrt{m}}{kL}M_0\left\{\left[1 - \cos\left(2.03\sqrt{\frac{k}{m}}t\right)\right]\right.$$

$$\left. - \left[1 - \cos\left(2.03\sqrt{\frac{k}{m}}(t - t_0)\right)\right]\right\}$$

The generalized coordinates are calculated from

$$x = \frac{1}{\sqrt{m}}[0.925p_1(t) + 0.380p_2(t)]$$

$$= \frac{-M_0}{kL}\left(\left[0.442 - 0.737\cos\left(1.28\sqrt{\frac{k}{m}}t\right) + 0.295\cos\left(2.03\sqrt{\frac{k}{m}}t\right)\right]u(t)\right.$$

$$- \left\{0.442 - 0.737\cos\left[1.28\sqrt{\frac{k}{m}}(t - t_0)\right]\right.$$

$$\left.\left. + 0.295\cos\left(2.03\sqrt{\frac{k}{m}}(t - t_0)\right)\right\}u(t - t_0)\right)$$

$$\theta = \frac{1}{\sqrt{m}}L[-1.31p_1(t) + 3.20p_2(t)]$$

$$= \frac{M_0}{kL^2}\left(\left[3.53 - 1.04\cos\left(1.28\sqrt{\frac{k}{m}}t\right)\right.\right.$$

$$\left. - 2.47\cos\left(2.03\sqrt{\frac{k}{m}}\right)\right]u(t) - \left[3.53 - 1.04\cos\left(1.28\sqrt{\frac{k}{m}}(t - t_0)\right)\right.$$

$$\left.\left. - 2.47\cos\left(2.03\sqrt{\frac{k}{m}}(t - t_0)\right)\right]u(t - t_0)\right)$$

6.10   The right-hand block of the system of Fig. 5-1 is subject to a constant force of magnitude

F_0 at $t = 0$. Use modal analysis to determine the resulting motion of the system.

From Problems 5.30, 5.31, and 5.38, the natural frequencies and modal matrix for the system are

$$\omega_1 = 0.359 \sqrt{\frac{k}{m}}, \qquad \omega_2 = \sqrt{\frac{k}{m}}, \qquad \omega_2 = 1.97 \sqrt{\frac{k}{m}}$$

$$\mathbf{P} = \frac{1}{\sqrt{m}} \begin{bmatrix} 0.282 & 0.447 & -0.849 \\ 0.404 & 0.447 & 0.370 \\ 0.545 & -0.447 & -0.055 \end{bmatrix}$$

The force vector is

$$\mathbf{F} = \begin{bmatrix} 0 \\ 0 \\ F_0 \end{bmatrix}$$

The vector $\mathbf{G}(t)$ is calculated as

$$\mathbf{G} = \mathbf{P}^T \mathbf{F} = \frac{1}{\sqrt{m}} \begin{bmatrix} 0.282 & 0.404 & 0.545 \\ 0.447 & 0.447 & -0.447 \\ -0.849 & 0.370 & -0.055 \end{bmatrix} \begin{bmatrix} 0 \\ 0 \\ F_0 \end{bmatrix} = \frac{1}{\sqrt{m}} \begin{bmatrix} 0.545F_0 \\ -0.447F_0 \\ -0.055F_0 \end{bmatrix}$$

The differential equations for the principal coordinates become

$$\ddot{p}_1 + 0.129 \frac{k}{m} p_1 = 0.545 \frac{F_0}{\sqrt{m}}$$

$$\ddot{p}_2 + \frac{k}{m} = -0.447 \frac{F_0}{\sqrt{m}}$$

$$\ddot{p}_3 + 3.88 \frac{k}{m} p_3 = -0.055 \frac{F_0}{\sqrt{m}}$$

The solutions for the principal coordinates are

$$p_1(t) = 4.23 \frac{\sqrt{m}}{k} F_0 \left(1 - \cos 0.359 \sqrt{\frac{k}{m}} t \right)$$

$$p_2(t) = -0.447 \frac{\sqrt{m}}{k} F_0 \left(1 - \cos \sqrt{\frac{k}{m}} t \right)$$

$$p_3(t) = -0.0142 \frac{\sqrt{m}}{k} F_0 \left(1 - \cos 1.97 \sqrt{\frac{k}{m}} \right)$$

The original generalized coordinates are calculated from

$$x_1(t) = \frac{1}{\sqrt{m}} [0.282 p_1(t) + 0.447 p_2(t) - 0.849 p_3(t)]$$

$$= \frac{F_0}{k} \left(1.005 - 1.193 \cos 0.359 \sqrt{\frac{k}{m}} t + 0.200 \cos \sqrt{\frac{k}{m}} t \right.$$

$$\left. - 0.0121 \cos 1.97 \sqrt{\frac{k}{m}} \right)$$

$$x_2(t) = \frac{1}{\sqrt{m}} [0.404 p_1(t) + 0.447 p_2(t) + 0.370 p_3(t)]$$

$$= \frac{F_0}{k} \left(1.504 - 1.708 \cos 0.359 \sqrt{\frac{k}{m}} t \right.$$

$$\left. + 0.200 \cos \sqrt{\frac{k}{m}} t + 0.00525 \cos 1.97 \sqrt{\frac{k}{m}} t \right)$$

$$x_3(t) = \frac{1}{\sqrt{m}}[0.545p_1(t) - 0.447p_2(t) - 0.055p_3(t)]$$

$$= \frac{F_0}{k}\left(2.506 - 2.305\cos 0.359\sqrt{\frac{k}{m}}t + 0.200\cos\sqrt{\frac{k}{m}}t\right.$$

$$\left. + 0.000781\cos 1.97\sqrt{\frac{k}{m}}t\right)$$

6.11   Repeat Problem 6.3 using modal analysis.

Mathcad

The flexibility matrix **A** and the mass matrix **M** are as given in Problem 6.3. The natural frequencies are the reciprocals of the square roots of the eigenvalues of **AM**. The columns of the modal matrix are the corresponding normalized eigenvectors. Calculations lead to

$$\omega_1 = 63.7\ \frac{rad}{s}, \qquad \omega_2 = 572.3\ \frac{rad}{s}, \qquad \omega_3 = 1660\ \frac{rad}{s}$$

$$\mathbf{P} = \begin{bmatrix} 0.0123 & 0.0768 & 0.111 \\ 0.0423 & 0.1034 & -0.0764 \\ 0.0807 & -0.0262 & 0.00923 \end{bmatrix}$$

$$\mathbf{G} = \mathbf{P}^T\mathbf{F} = \begin{bmatrix} 0.0123 & 0.0423 & 0.0807 \\ 0.0768 & 0.1034 & -0.0262 \\ 0.111 & -0.0764 & 0.00923 \end{bmatrix}\begin{bmatrix} 0 \\ 0 \\ 7.11 \times 10^5 \end{bmatrix}\sin 1257t$$

$$= \begin{bmatrix} 5.74 \times 10^4 \\ -1.86 \times 10^4 \\ 6.57 \times 10^3 \end{bmatrix}\sin 1257t$$

Thus the differential equations for the principal coordinates are

$$\ddot{p}_1 + 4.06 \times 10^3 p_1 = 5.74 \times 10^4 \sin 1257t$$

$$\ddot{p}_2 + 3.28 \times 10^5 p_2 = -1.86 \times 10^4 \sin 1257t$$

$$\ddot{p}_3 + 2.76 \times 10^6 p_3 = 6.57 \times 10^3 \sin 1257t$$

The steady-state responses are

$$p_1(t) = \frac{5.75 \times 10^4}{4.06 \times 10^3 - (1257)^2}\sin 1257t = -3.65 \times 10^{-2}\sin 1257t$$

$$p_2(t) = \frac{-1.86 \times 10^4}{3.58 \times 10^5 - (1257)^2}\sin 1257t = 1.52 \times 10^{-2}\sin 1257t$$

$$p_3(t) = \frac{6.57 \times 10^3}{2.76 \times 10^6 - (1257)^2}\sin 1257t = 5.57 \times 10^{-3}\sin 1257t$$

The steady-state response for x_3 is determined as

$$x_3 = 0.0807p_1 - 0.0262p_2 + 0.00923p_3 = -3.28 \times 10^{-3}\sin 1257t$$

6.12   Experiments indicate that the modal damping ratios for the first two modes of the system of Problems 6.3 and 6.11 are 0.04 and 0.15. Repeat Problem 6.11 assuming proportional damping.

Mathcad

If the damping is proportional, the damping ratios are of the form

$$\zeta_i = \frac{1}{2}\left(\alpha\omega_i + \frac{\beta}{\omega_i}\right)$$

Substituting the experimental damping ratios and the natural frequencies calculated in Problem 6.11 leads to

$$0.04 = \tfrac{1}{2}(63.7\alpha + 0.0157\beta)$$

$$0.15 = \tfrac{1}{2}(572.3\alpha + 0.00175\beta)$$

Solving simultaneously gives $\alpha = 5.15 \times \times 10^{-4}$ and $\beta = 2.91$. Then

$$\zeta_3 = \tfrac{1}{2}[(5.15 \times 10^{-4})(1660) + (2.91)(6.02 \times 10^{-4})] = 0.428$$

Using Eq. (6.16), the differential equations governing the principal coordinates are

$$\ddot{p}_1 + 5.10\dot{p}_1 + 4.06 \times 10^3 p_1 = 5.74 \times 10^4 \sin 1257t$$

$$\ddot{p}_2 + 171.7\dot{p}_2 + 3.28 \times 10^5 p_2 = -1.86 \times 10^4 \sin 1257t$$

$$\ddot{p}_3 + 1421\dot{p}_3 + 2.76 \times 10^6 p_3 = 6.57 \times 10^3 \sin 1257t$$

The steady-state solutions are of the form

$$p_i(t) = \frac{G_i}{\omega_i^2} M(r_i, \zeta_i) \sin(1257t - \phi_i)$$

where
$$r_i = \frac{1257}{\omega_i}, \qquad M(r_i, \zeta_i) = \frac{1}{\sqrt{(1 - r_i^2)^2 + (2\zeta_i r_i)^2}},$$

$$\phi_i = \tan^{-1}\left(\frac{2\zeta_i r_i}{1 - r_i^2}\right)$$

As an example,

$$r_1 = \frac{1257}{63.7} = 19.73$$

$$M(r_1, \zeta_1) = \frac{1}{\sqrt{(1 - 19.73^2)^2 + [2(0.04)(19.73)]^2}} = 2.58 \times 10^{-3}$$

$$\phi_1 = \tan^{-1}\left(\frac{2(0.04)(19.73)}{1 - (19.73)^2}\right) = 3.137$$

Further calculations lead to

$$r_2 = 2.197, \qquad M(2.197, 0.15) = 0.258, \qquad \phi_2 = 2.97$$

$$r_3 = 0.757, \qquad M(0.757, 0.428) = 1.289, \qquad \phi_3 = 0.988$$

Hence,

$$p_1(t) = \left(\frac{5.75 \times 10^4}{4.06 \times 10^3}\right)(2.58 \times 10^{-3}) \sin(1257t - 3.137)$$

$$= 3.65 \times 10^{-2} \sin(1257t - 3.137)$$

$$P_2(t) = \left(\frac{-1.86 \times 10^4}{3.28 \times 10^5}\right)(0.258) \sin(1257t - 2.97)$$

$$= -1.46 \times 10^{-2} \sin(1257t - 2.97)$$

$$p_2(t) = \left(\frac{6.57 \times 10^3}{2.76 \times 10^6}\right)(1.289) \sin(1257t - 0.988)$$

$$= 3.07 \times 10^{-3} \sin(1257t - 0.988)$$

Modal analysis is used to obtain

$$x_3(t) = 0.0807 p_1(t) - 0.0262 p_2(t) + 0.00924 p_3(t)$$

$$= 2.95 \times 10^{-3} \sin(1257t - 3.137) + 3.83 \times 10^{-4} \sin(1257t - 2.97)$$

$$+ 2.84 \times 10^{-5} \sin(1257t - 0.988)$$

Trigonometric identities are used to rewrite

$$x_3(t) = -3.31 \times 10^{-3} \sin 1257t + 1.03 \times 10^{-4} \cos 1257t$$

and the steady-state amplitude is

$$X_3 = \sqrt{(-3.31 \times 10^{-3})^2 + (1.03 \times 10^{-4})^2} = 3.31 \times 10^{-3} \text{ m}$$

6.13 The three railroad cars of Problem 5.34 are coupled and at rest in equilibrium when the left car is subjected to an impulse of magnitude I. Determine the resulting motion of the coupled cars.

The differential equations governing the motion of the system are

$$\begin{bmatrix} m & 0 & 0 \\ 0 & m & 0 \\ 0 & 0 & m \end{bmatrix} \begin{bmatrix} \ddot{x}_1 \\ \ddot{x}_2 \\ \ddot{x}_3 \end{bmatrix} + \begin{bmatrix} k & -k & 0 \\ -k & 2k & -k \\ 0 & -k & k \end{bmatrix} \begin{bmatrix} x_1 \\ x_2 \\ x_3 \end{bmatrix} = \begin{bmatrix} I\,\delta(t) \\ 0 \\ 0 \end{bmatrix}$$

The natural frequencies are calculated in Problem 5.34 as

$$\omega_1 = 0, \qquad \omega_2 = \sqrt{\frac{k}{m}}, \qquad \omega_3 = \sqrt{3\frac{k}{m}}$$

The mode shapes of Problem 5.34 are normalized, leading to the modal matrix of

$$\mathbf{P} = \frac{1}{\sqrt{6m}} \begin{bmatrix} \sqrt{2} & \sqrt{3} & -1 \\ \sqrt{2} & 0 & 2 \\ \sqrt{2} & -\sqrt{3} & -1 \end{bmatrix}$$

Then, $\qquad \mathbf{G} = \mathbf{P}^T \mathbf{F} = \dfrac{1}{\sqrt{6m}} \begin{bmatrix} \sqrt{2} & \sqrt{2} & \sqrt{2} \\ \sqrt{3} & 0 & -\sqrt{3} \\ -1 & 2 & -1 \end{bmatrix} \begin{bmatrix} I\,\delta(t) \\ 0 \\ 0 \end{bmatrix} = \begin{bmatrix} \sqrt{2} \\ \sqrt{3} \\ -1 \end{bmatrix} \dfrac{I}{\sqrt{6m}} \delta(t)$

The differential equations for the principal coordinates are

$$\ddot{p}_1 = \frac{I}{\sqrt{3m}} \delta(t)$$

$$\ddot{p}_2 + \frac{k}{m} p_2 = \frac{I}{\sqrt{2m}} \delta(t)$$

$$\ddot{p}_2 + 3\frac{k}{m} p_3 = -\frac{I}{\sqrt{6m}} \delta(t)$$

The solutions for the principal coordinates subject to initial conditions of $p_i(0) = 0$ and $\dot{p}_i(0) = 0$ are

$$p_1(t) = \frac{I}{\sqrt{3m}} t\, u(t)$$

$$p_2(t) = \frac{I}{\sqrt{2m}} \sin \sqrt{\frac{k}{m}}\, t\, u(t)$$

$$p_3(t) = -\frac{I}{\sqrt{6m}} \sin \sqrt{3\frac{k}{m}}\, t\, u(t)$$

The $x_i(t)$ coordinates are obtained using Eq. (6.13) as

$$x_1 = \frac{1}{\sqrt{3m}} p_1 + \frac{1}{\sqrt{2m}} p_2 - \frac{1}{\sqrt{6m}} p_3$$

$$= \left(\frac{I}{3m} t + \frac{I}{2m} \sin \sqrt{\frac{k}{m}} t + \frac{I}{6m} \sin \sqrt{3\frac{k}{m}} t \right) u(t)$$

$$x_2(t) = \frac{1}{\sqrt{3m}} p_1 + \frac{2}{\sqrt{6m}} p_3$$

$$= \left(\frac{I}{3m} t - \frac{2I}{6m} \sin \sqrt{3\frac{k}{m}} t \right) u(t)$$

$$x_3(t) = \frac{1}{\sqrt{3m}} p_1(t) - \frac{1}{\sqrt{2m}} p_2(t) - \frac{1}{\sqrt{6m}} p_3(t)$$

$$= \left(\frac{I}{3m} t - \frac{I}{2m} \sin \sqrt{\frac{k}{m}} + \frac{I}{6m} \sin \sqrt{3\frac{k}{m}} \right) u(t)$$

6.14 Repeat Problem 6.13 if each coupling is modeled as a spring of stiffness k in parallel with a viscous damper of damping coefficient c.

The differential equations governing the motion of this system are

$$\begin{bmatrix} m & 0 & 0 \\ 0 & m & 0 \\ 0 & 0 & m \end{bmatrix} \begin{bmatrix} \ddot{x}_1 \\ \ddot{x}_2 \\ \ddot{x}_3 \end{bmatrix} + \begin{bmatrix} c & -c & 0 \\ -c & 2c & -c \\ 0 & -c & c \end{bmatrix} \begin{bmatrix} \dot{x}_1 \\ \dot{x}_2 \\ \dot{x}_3 \end{bmatrix} + \begin{bmatrix} k & -k & 0 \\ -k & 2k & -k \\ 0 & -k & k \end{bmatrix} \begin{bmatrix} x_1 \\ x_2 \\ x_3 \end{bmatrix} = \begin{bmatrix} I\,\delta(t) \\ 0 \\ 0 \end{bmatrix}$$

The damping is proportional with $\alpha = c/k$ and $\beta = 0$. Thus

$$\zeta_1 = 0, \qquad \zeta_2 = \frac{c}{2k} \sqrt{\frac{k}{m}} = \frac{c}{2\sqrt{mk}}, \qquad \zeta_3 = \frac{c}{2k} \sqrt{3\frac{k}{m}} = \frac{\sqrt{3}c}{2\sqrt{mk}}$$

The differential equations governing the principal coordinates are

$$\ddot{p}_1 = \frac{I}{\sqrt{3m}} \delta(t)$$

$$\ddot{p}_2 + \frac{c}{m} \dot{p}_2 + \frac{k}{m} p_2 = \frac{I}{\sqrt{2m}} \delta(t)$$

$$\ddot{p}_3 + 3\frac{c}{m} \dot{p}_3 + 3\frac{k}{m} p_3 = -\frac{I}{\sqrt{6m}} \delta(t)$$

The solutions for the principal coordinates are

$$p_1(t) = \frac{I}{\sqrt{3m}} t\, u(t)$$

$$p_2(t) = \frac{I}{\sqrt{2m}\omega_2\sqrt{1 - \zeta_2^2}} e^{-(c/2m)t} \sin\left(\omega_2\sqrt{1 - \zeta_2^2}\,t\right) u(t)$$

$$p_3(t) = -\frac{I}{\sqrt{6m}\omega_3\sqrt{1 - \zeta_3^2}} e^{-(3c/2m)t} \sin\left(\omega_3\sqrt{1 - \zeta_3^2}\,t\right) u(t)$$

Modal analysis is used to obtain

$$x_1 = \frac{1}{\sqrt{3m}} p_1 + \frac{1}{\sqrt{2m}} p_2 - \frac{1}{\sqrt{6m}} p_3$$

$$x_2 = \frac{1}{\sqrt{3m}} p_1 + \frac{2}{\sqrt{6m}} p_3$$

$$x_3 = \frac{1}{\sqrt{3m}} p_1 - \frac{1}{\sqrt{2m}} p_2 - \frac{1}{\sqrt{6m}} p_3$$

6.15 A simplified 4-degree-of-freedom model of a suspension system is shown in Fig. 6-5. Model the transverse motion of a vehicle due to a bump in the road as the response due to an impulse of magnitude I applied to the front wheel at $t = 0$ and an impulse of magnitude I applied to the rear wheel at $t = 0.05$ s.

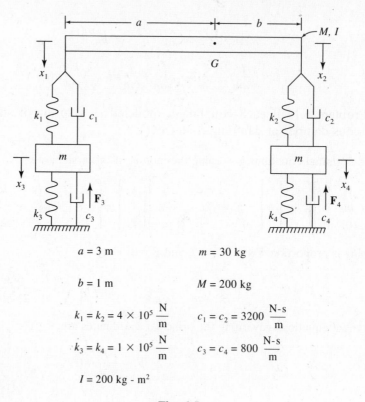

$a = 3$ m $\qquad\qquad$ $m = 30$ kg

$b = 1$ m $\qquad\qquad$ $M = 200$ kg

$k_1 = k_2 = 4 \times 10^5 \dfrac{\text{N}}{\text{m}} \qquad c_1 = c_2 = 3200 \dfrac{\text{N-s}}{\text{m}}$

$k_3 = k_4 = 1 \times 10^5 \dfrac{\text{N}}{\text{m}} \qquad c_3 = c_4 = 800 \dfrac{\text{N-s}}{\text{m}}$

$I = 200$ kg - m^2

Fig. 6-5

The kinetic energy of the system at an arbitrary instant is

$$T = \frac{1}{2} M \left(\frac{b\dot{x}_1 + a\dot{x}_2}{a+b} \right)^2 + \frac{1}{2} I \left(\frac{\dot{x}_2 - \dot{x}_1}{a+b} \right)^2$$

$$+ \frac{1}{2} m\dot{x}_3{}^2 + \frac{1}{2} m\dot{x}_4{}^2$$

The potential energy of the system at an arbitrary instant is

$$V = \tfrac{1}{2}k_1(x_3 - x_1)^2 + \tfrac{1}{2}k_2(x_4 - x_2)^2 + \tfrac{1}{2}k_3x_3{}^2 + \tfrac{1}{2}k_4x_4{}^2$$

The work done by the nonconservative forces as the system moves through virtual displacements is

$$\delta W = -I\,\delta(t)\,\delta x_3 - I\,\delta(t - 0.05)\,\delta x_4 - c_1(\dot{x}_3 - \dot{x}_1)\,\delta(x_3 - x_1)$$

$$- c_2(\dot{x}_4 - \dot{x}_2)\,\delta(x_4 - x_2) - c_3\dot{x}_3\,\delta x_3 - c_4\dot{x}_4\,\delta x_4$$

Lagrange's equations are applied to yield

$$\begin{bmatrix} \dfrac{Mb^2 + I}{(a+b)^2} & \dfrac{Mab - I}{(a+b)^2} & 0 & 0 \\[2mm] \dfrac{Mab - I}{(a+b)^2} & \dfrac{Ma^2 + I}{(a+b)^2} & 0 & 0 \\[2mm] 0 & 0 & m & 0 \\[1mm] 0 & 0 & 0 & m \end{bmatrix} \begin{bmatrix} \ddot{x}_1 \\ \ddot{x}_2 \\ \ddot{x}_3 \\ \ddot{x}_4 \end{bmatrix} + \begin{bmatrix} c_1 & 0 & -c_1 & 0 \\ 0 & c_2 & 0 & -c_2 \\ -c_1 & 0 & c_1 + c_3 & 0 \\ 0 & -c_2 & 0 & c_2 + c_4 \end{bmatrix} \begin{bmatrix} \dot{x}_1 \\ \dot{x}_2 \\ \dot{x}_3 \\ \dot{x}_4 \end{bmatrix}$$

$$+ \begin{bmatrix} k_1 & 0 & -k_1 & 0 \\ 0 & k_2 & 0 & -k_2 \\ -k_1 & 0 & k_1 + k_3 & 0 \\ 0 & -k_2 & 0 & k_2 + k_4 \end{bmatrix} \begin{bmatrix} x_1 \\ x_2 \\ x_3 \\ x_4 \end{bmatrix} = \begin{bmatrix} 0 \\ 0 \\ -I\,\delta(t) \\ -I\,\delta(t - 0.05) \end{bmatrix}$$

Substituting given values, the mass, damping, and stiffness matrices are

$$\mathbf{M} = \begin{bmatrix} 25 & 25 & 0 & 0 \\ 25 & 125 & 0 & 0 \\ 0 & 0 & 30 & 0 \\ 0 & 0 & 0 & 30 \end{bmatrix}, \quad \mathbf{C} = \begin{bmatrix} 3200 & 0 & -3200 & 0 \\ 0 & 3200 & 0 & -3200 \\ -3200 & 0 & 4000 & 0 \\ 0 & -3200 & 0 & 4000 \end{bmatrix}$$

$$\mathbf{K} = \begin{bmatrix} 4 \times 10^5 & 0 & -4 \times 10^5 & 0 \\ 0 & 4 \times 10^5 & 0 & -4 \times 10^5 \\ -4 \times 10^5 & 0 & 5 \times 10^5 & 0 \\ 0 & -4 \times 10^5 & 0 & 5 \times 10^5 \end{bmatrix}$$

The damping matrix is proportional to the stiffness matrix with $\alpha = 0.008$ s. Hence, the modal matrix for the undamped system is used to uncouple the differential equations. The natural frequencies are the square roots of the eigenvalues of $\mathbf{M}^{-1}\mathbf{K}$, and the columns of the modal matrix are the normalized mode shapes. The methods of Chap. 5 are applied leading to

$$\omega_1 = 23.0 \frac{\text{rad}}{\text{s}}, \qquad \omega_2 = 44.3 \frac{\text{rad}}{\text{s}},$$

$$\omega_2 = 138.5 \frac{\text{rad}}{\text{s}}, \qquad \omega_4 = 188.8 \frac{\text{rad}}{\text{s}}$$

$$\mathbf{P} = \begin{bmatrix} 0.0187 & 0.147 & -0.0074 & 0.167 \\ 0.0791 & -0.0347 & -0.0313 & -0.0395 \\ 0.0154 & 0.133 & 0.0390 & -0.117 \\ 0.0654 & -0.0315 & 0.165 & 0.0277 \end{bmatrix}$$

Then

$$\mathbf{G} = \mathbf{P}^T\mathbf{F} = -I \begin{bmatrix} 0.0154\,\delta(t) + 0.0654\,\delta(t-0.05) \\ 0.133\,\delta(t) - 0.0315\,\delta(t-0.05) \\ 0.0390\,\delta(t) + 0.165\,\delta(t-0.05) \\ -0.117\,\delta(t) + 0.0277\,\delta(t-0.05) \end{bmatrix}$$

The modal damping ratios are calculated as

$$\zeta_1 = \tfrac{1}{2}\alpha\omega_1 = 0.092, \qquad \zeta_2 = \tfrac{1}{2}\alpha\omega_2 = 0.177$$

$$\zeta_3 = \tfrac{1}{2}\alpha\omega_3 = 0.554, \qquad \zeta_4 = \tfrac{1}{2}\alpha\omega_4 = 0.755$$

The differential equations governing the principal coordinates are

$$\ddot{p}_1 + 4.23\dot{p}_1 + 5.29 \times 10^2 p_1 = -0.154I\,\delta(t) - 0.0654I\,\delta(t-0.05)$$

$$\ddot{p}_2 + 15.68\dot{p}_2 + 1.96 \times 10^3 p_2 = -0.133I\,\delta(t) + 0.0315I\,\delta(t-0.05)$$

$$\ddot{p}_3 + 153.5\dot{p}_3 + 1.92 \times 10^4 p_3 = 0.0390I\,\delta(t) - 0.165I\,\delta(t-0.05)$$

$$\ddot{p}_4 + 285.8\dot{p}_4 + 3.56 \times 10^4 p_4 = 0.117I\,\delta(t) - 0.0277I\,\delta(t-0.05)$$

The convolution integral, Eq. (6.17), is used to obtain

$$p_1(t) = -Ie^{-2.12t}[6.73 \times 10^{-3} \sin(22.9t)\,u(t)$$
$$+ 3.18 \times 10^{-3} \sin(22.9t - 1.145)\,u(t - 0.05)]$$

$$p_2(t) = -Ie^{-7.84t}[3.05 \times 10^{-3} \sin(43.6t)\,u(t)$$
$$- 1.07 \times 10^{-3} \sin(43.6t - 2.18)\,u(t - 0.05)]$$

$$p_3(t) = -Ie^{-76.7t}[3.38 \times 10^{-4} \sin(115.3t)\,u(t)$$
$$+ 6.62 \times 10^{-2} \sin(115.3t - 5.77)\,u(t - 0.05)]$$

$$p_4(t) = -Ie^{-142.5t}[9.45 \times 10^{-4} \sin(123.8t)\,u(t)$$
$$- 2.79 \times 10^{-1} \sin(123.8t - 6.19)\,u(t - 0.05)]$$

The generalized coordinates are calculated from

$$x_1(t) = 0.0187p_1(t) + 0.147p_2(t) - 0.0074p_3(t) + 0.167p_4(t)$$

$$x_2(t) = 0.0791p_1(t) - 0.0347p_2(t) - 0.0313p_3(t) - 0.0395p_4(t)$$

$$x_3(t) = 0.0154p_1(t) + 0.133p_2(t) + 0.0390p_3(t) - 0.117p_4(t)$$

$$x_4(t) = 0.00654p_1(t) - 0.0315p_2(t) + 0.165p_3(t) + 0.0277p_4(t)$$

6.16 Determine the time-dependent response of the system of Fig. 6-6 if the system is at rest in
 equilibrium at $t = 0$ when an impulse of magnitude 0.5 N-s is applied to the left block
followed by an impulse of 0.5 N-s at $t = 0.01$ s. The free vibration response of this system
is obtained in Problem 5.45.

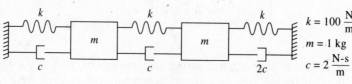

$$k = 100 \frac{N}{m}$$
$$m = 1 \text{ kg}$$
$$c = 2 \frac{N\text{-}s}{m}$$

Fig. 6-6

The differential equations governing the motion of the system are

$$\begin{bmatrix} 1 & 0 \\ 0 & 1 \end{bmatrix} \begin{bmatrix} \ddot{x}_1 \\ \ddot{x}_2 \end{bmatrix} + \begin{bmatrix} 4 & -2 \\ -2 & 6 \end{bmatrix} \begin{bmatrix} \dot{x}_1 \\ \dot{x}_2 \end{bmatrix} + \begin{bmatrix} 200 & -100 \\ -100 & 200 \end{bmatrix} \begin{bmatrix} x_1 \\ x_2 \end{bmatrix} = \begin{bmatrix} 0.5\,\delta(t) + 0.5\,\delta(t - 0.01) \\ 0 \end{bmatrix}$$

Since the viscous damping is not proportional, a general modal analysis is used. The eigenvalues of
$\tilde{\mathbf{M}}^{-1}\tilde{\mathbf{K}}$ are

$$\gamma_1 = 1.502 - 9.912i, \qquad \gamma_2 = 1.502 + 9.912i$$

$$\gamma_3 = 3.49 - 16.918i, \qquad \gamma_4 = 3.49 + 16.918i$$

The eigenvectors of $\tilde{\mathbf{M}}^{-1}\tilde{\mathbf{K}}$ are assumed of the form

$$\Phi_i = [-\gamma_i \quad -\gamma_i z_i \quad 1 \quad z_i]^T$$

Thus the problem to determine the eigenvector becomes

$$\begin{bmatrix} 4 - \gamma_i & -2 & 200 & -100 \\ -2 & 6 - \gamma_i & -100 & 200 \\ -1 & 0 & -\gamma_i & 0 \\ 0 & -1 & 0 & -\gamma_i \end{bmatrix} \begin{bmatrix} -\gamma_i \\ -\gamma_i z_i \\ 1 \\ z_i \end{bmatrix} = \begin{bmatrix} 0 \\ 0 \\ 0 \\ 0 \end{bmatrix}$$

from which it is determined that

$$z_i = \frac{200 - 4\gamma_i + \gamma_i^2}{100 - 2\gamma_i}$$

Substitution leads to

$$z_1 = 0.9898 - 0.100i, \qquad z_2 = 0.9898 + 0.100i$$

$$z_3 = -1.009 - 0.175i, \qquad z_4 = -1.009 + 0.175i$$

The mode shapes are normalized by requiring $\Phi_i^T \tilde{\mathbf{M}} \Phi_i = 1$. The columns of the modal matrix are
the normalized mode shapes. Calculations lead to

$$\tilde{\mathbf{P}} = \begin{bmatrix} 0.886 + 1.34i & 0.886 - 1.34i & 1.29 + 1.63i & 1.29 - 1.63i \\ 1.01 + 1.23i & 1.01 - 1.23i & -1.02 - 1.88i & -1.02 + 1.88i \\ 0.119 - 0.107i & 0.119 + 0107i & 0.0774 - 0.0925i & 0.0774 + 0.0925i \\ 0.107 - 0.118i & 0.107 + 0.118i & -0.0946 + 0.0797i & -0.0946 - 0.0797i \end{bmatrix}$$

Then
$$\tilde{\mathbf{G}} = \tilde{\mathbf{P}}^T \tilde{\mathbf{F}} = \begin{bmatrix} 0.119 - 0.107i \\ 0.119 + 0.107i \\ 0.0774 - 0.0925i \\ 0.0774 + 0.0925i \end{bmatrix} [0.5\,\delta(t) + 0.5\,\delta(t - 0.01)]$$

Application of Eq. (6.24) leads to

$$p_1(t) = 0.5(0.119 - 0.107i)[e^{-(1.502-9.912i)t} u(t)$$
$$+ e^{-(1.502-9.912i)(t-0.01)} u(t - 0.01)]$$

$$p_2(t) = 0.5(0.119 + 0.107i)[e^{-(1.502+9.912i)t} u(t)$$
$$+ e^{-(1.502+9.912i)(t-0.01)} u(t - 0.01)]$$

$$p_3(t) = 0.5(0.0774 - 0.0925i)[e^{-(3.49-16.918i)t} u(t)$$
$$+ e^{-(3.49-16.918i)(t-0.01)} u(t - 0.01)]$$

$$p_4(t) = 0.5(0.0774 + 0.0925i)[e^{-(3.49+16.918i)t} u(t)$$
$$+ e^{-(3.49+16.918i)(t-0.01)} u(t - 0.01)]$$

Noting that $x_1 = y_3$ and $x_2 = y_4$ leads to

$$x_1 = (0.119 - 0.107i)p_1 + (0.119 + 0.107i)p_2$$
$$+ (0.0074 - 0.0925i)p_3 + (0.0774 + 0.0925i)p_4$$

$$x_2 = (0.107 - 0.118i)p_1 + (0.107 + 0.118i)p_2$$
$$+ (-0.0946 + 0.0797i)p_3 + (-0.0946 - 0.0797i)p_4$$

Use of complex algebra and Euler's identity leads to

$$x_1(t) = e^{-1.502t}\{[0.0256 \cos (9.912t) - 0.0253 \sin (9.912t)] u(t)$$
$$+ (0.0260 \cos (9.912t - 0.099)$$
$$- 0.0257 \sin (9.912t - 0.099)] u(t - 0.01)\}$$
$$+ e^{-3.49t}(0.0145 \cos (16.918t) - 0.0143 \sin (16.918t)] u(t)$$
$$+ (0.0151 \cos (16.918t - 0.169)$$
$$- 0.0148 \sin (16.918t - 0.169)] u(t - 0.01)\}$$

$$x_2(t) = e^{-1.502t}\{[0.0254 \cos (9.912t) - 0.0252 \sin (9.912t)] u(t)$$
$$+ [0.0257 \cos (9.912t - 0.099)$$
$$- 0.0259 \sin (9.912t - 0.099)] u(t - 0.01)\}$$
$$+ e^{-3.49t}\{[0.0153 \cos (16.918t) - 0.0150 \sin (16.918t)] u(t)$$
$$+ [0.0158 \cos (16.918t - 0.169)$$
$$- 0.0155 \sin (16.918t - 0.169)] u(t - 0.01)\}$$

Supplementary Problems

6.17   Determine the steady-state amplitude of the 60-kg block of Fig. 6-7 if $F(t) = 250 \sin 40t$.

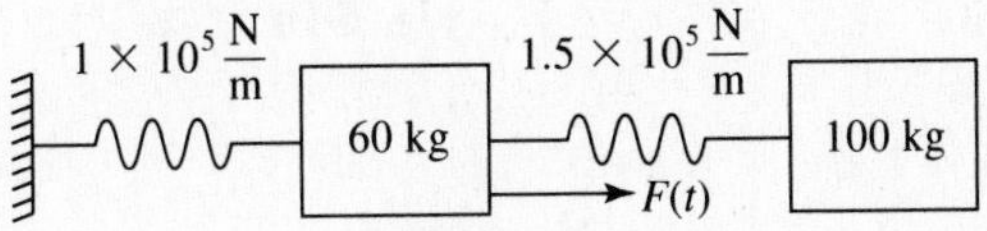

Fig. 6-7

Ans. 1.04×10^{-4} m

6.18   Determine the steady-state amplitude of the 100-kg block of Fig. 6-8 if $\omega = 80$ rad/s.

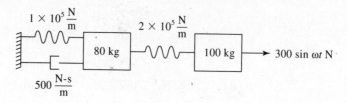

Fig. 6-8

Ans. 1.27×10^{-3} m

6.19   For what values of ω is the steady-state amplitude of the 100-kg block of Fig. 6-8 less than 1 mm?

Ans. $\omega > 81.0$ rad/s

6.20   Determine the steady-state amplitude of the 60-kg block of Fig. 6-9.

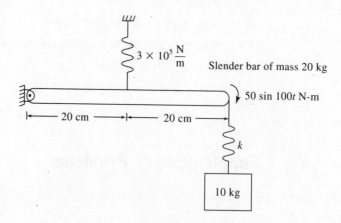

Fig. 6-9

Ans. 1.08×10^{-3} m

6.21   For what value of k is the steady-state amplitude of angular oscillation of the bar of Fig. 6-10 identically zero?

Fig. 6-10

Ans. 1×10^5 N/m

6.22   Use the Laplace transform method to solve Problem 6.17.

Ans. 1.04×10^{-4} m

6.23   Use the Laplace transform method to determine the displacement of the 60-kg block of Fig. 6-7 if $F(t) = 250[u(t) - u(t - 0.1)]$.

Ans.

$$[2.5 \times 10^{-3} - 1.87 \times 10^{-3} \cos(21.96t) - 6.29 \times 10^{-4} \cos(72.0t)]\, u(t)$$

$$- [2.5 \times 10^{-3} - 1.87 \times 10^{-3} \cos(21.96t - 2.196) - 6.29 \times 10^{-4} \cos(72.0t - 7.2)]$$

$$u(t - 0.1)$$

6.24   A moment $10e^{-5t}$ N-m is applied to the upper bar of the system of Fig. 5-28. Use the Laplace transform method to determine the time-dependent response of the lower bar if $L = 1$ m, $k = 100$ N/m, and $m = 10$ kg.

Ans.

$$7.38 \times 10^{-5} e^{-0.5t} + 2.78 \times 10^{-3} \sin(1.693t) - 9.40 \times 10^{-5} \cos 1.693t$$

$$- 9.88 \times 10^{-6} \sin(10.23t) + 2.02 \times 10^{-5} \cos(10.23t)$$

6.25   Solve Problem 6.24 using modal analysis.

6.26   Solve Problem 6.20 using modal analysis.

6.27   Use modal analysis to determine the steady-state response of the system of Fig. 5-1 when the leftmost block is subjected to a force $F_0 \sin 1.5\sqrt{k/m}\, t$.

Ans.

$$x_1 = 0.245 \frac{F_0}{k} \sin 1.5 \sqrt{\frac{k}{m}}\, t, \qquad x_2 = -0.406 \frac{F_0}{k} \sin 1.5 \sqrt{\frac{k}{m}}\, t,$$

$$x_3 = 0.116 \frac{F_0}{k} \sin 1.5 \sqrt{\frac{k}{m}}\, t$$

6.28   Use modal analysis to determine the steady-state amplitude of angular oscillation of the leftmost disk of the system of Fig. 6-11 when the middle disk is subjected to a torque of $50 \sin 30t$ N-m.

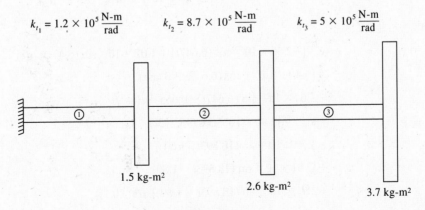

Fig. 6-11

Ans. 4.46×10^{-4} rad

6.29   Three 20-kg machines are equally spaced along the span of a 2-m simply supported beam of elastic modulus 200×10^9 N/m^2 and cross-sectional moment of inertia 1.35×10^{-6} m^4. The machine near the left support has a rotating unbalance of magnitude 0.5 kg-m and operates at 100 rad/s. Determine the steady-state amplitude of the machine at the midspan.

Ans.   0.0028 m

6.30   Repeat Problem 6.29 as if the system had proportional damping with the damping ratio for the lowest mode equal to 0.04. Assume the damping matrix is proportional to the stiffness matrix.

Ans.   0.0028 m

6.31   Use modal analysis to determine the steady-state amplitude of the 60-kg block of the system of Fig. 6-12.

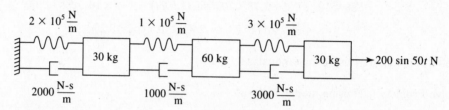

Fig. 6-12

Ans. 1.68×10^{-4} m

6.32   Use modal analysis to determine $x_1(t)$ for the system of Fig. 6-13 if the leftmost block is subjected to an impulse of magnitude 1 N-s at $t = 0$ and the rightmost block is subjected to an impulse of magnitude 1.5 N-s at $t = 0.1$ s.

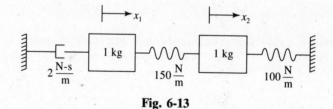

Fig. 6-13

Ans.

$$e^{-0.660t}\{[-2.56 \times 10^{-3} \cos (6.647t) + 1.02 \times 10^{-1} \sin (6.47t)] \, u(t)$$

$$+ [-4.11 \times 10^{-3} \cos (6.647t - 0.665)$$

$$+ 1.64 \times 10^{-1} \sin (6.647t - 0.665)] \, u(t - 0.1)\}$$

$$+ e^{-0.339t}\{[2.96 \times 10^{-3} \cos (18.893t)$$

$$- 1.79 \times 10^{-2} \sin (18.893t)] \, u(t)$$

$$+ [3.91 \times 10^{-2} \sin (18.893t - 1.89)$$

$$+ 3.91 \times 10^{-2} \sin (18.893t - 1.89)] \, u(t - 0.1)\}$$

Chapter 7

Vibrations of Continuous Systems

Continuous, or *distributed parameter,* systems are systems in which inertia is continuously distributed throughout the system. A continuous system's dependent kinematic properties are functions of spatial variables, as well as time. Vibrations of continuous systems are governed by partial differential equations.

7.1 WAVE EQUATION

Free vibrations of certain one-dimensional systems are governed by the *wave equation*

$$c^2 \frac{\partial^2 u}{\partial x^2} = \frac{\partial^2 u}{\partial t^2} \tag{7.1}$$

where x is a spatial coordinate, $u(x, t)$ is the displacement of a particle in the system whose equilibrium position is identified by x, and c is the *wave speed,* the velocity at which waves propagate in the system. Problems governed by the wave equation and the system's wave speed are given in Table 7-1.

7.2 WAVE SOLUTION

The general solution of Eq. (7.1) can be expressed as

$$u(x, t) = f(x - ct) + g(x + ct) \tag{7.2}$$

where f and g are arbitrary functions of a single variable.

7.3 NORMAL MODE SOLUTION

The *normal mode solution* of Eq. (7.1) is

$$u(x, t) = X(x)e^{i\omega t} \tag{7.3}$$

where ω is a natural frequency of the system and $X(x)$ is the *mode shape* corresponding to that natural frequency. Equation (7.3) is substituted into Eq. (7.1), leading to

$$c^2 \frac{d^2 X}{dx^2} + \omega^2 X = 0 \tag{7.4}$$

The general solution of Eq. (7.4) is

$$X(c) = C_1 \cos \frac{\omega}{c} x + C_2 \sin \frac{\omega}{c} x \tag{7.5}$$

where C_1 and C_2 are constants of integration. Boundary conditions for a specific system are used in Eq. (7.5) to develop a *characteristic equation* that is satisfied by the natural frequencies.

201

Table 7-1

Problem	Schematic	Nondimensional wave equation	Wave speed	
Torsional oscillations of circular cylinder		$\dfrac{\partial^2 \theta}{\partial x^2} = \dfrac{\partial^2 \theta}{\partial t^2}$	$c = \sqrt{\dfrac{G}{\rho}}$	G = shear modulus ρ = mass density
Longitudinal oscillations of bar	$w(x,t)$	$\dfrac{\partial^2 w}{\partial x^2} = \dfrac{\partial^2 w}{\partial t^2}$	$c = \sqrt{\dfrac{E}{\rho}}$	E = elastic modulus ρ = mass density
Transverse vibrations of taut string	x $y(x,t)$	$\dfrac{\partial^2 y}{\partial x^2} = \dfrac{\partial^2 y}{\partial t^2}$	$c = \sqrt{\dfrac{T}{\mu}}$	T = tension μ = linear density
Pressure waves in an ideal gas	$p(x,t)$	$\dfrac{\partial^2 p}{\partial x^2} = \dfrac{\partial^2 p}{\partial t^2}$	$c = \sqrt{kRT}$	k = ratio of specific heats R = gas constant T = temperature
Waterhammer waves in rigid pipe	$p(x,t)$	$\dfrac{\partial^2 p}{\partial x^2} = \dfrac{\partial^2 p}{\partial t^2}$	$c = \sqrt{\dfrac{k}{\rho}}$	k = bulk modulus of fluid ρ = mass density

Continuous systems have an infinite, but countable, number of natural frequencies. Application of the boundary conditions also leads to a relation between C_1 and C_2 and thus determination of a mode shape. Let $X_i(x)$ and $X_j(x)$ be mode shapes corresponding to distinct natural frequencies ω_i and ω_j, respectively. The mode shape satisfy an *orthogonality condition*, which for most systems has the form

$$\int_0^L X_i(x)\,X_j(x)\,dx = 0 \tag{7.6}$$

where L is the length of the continuous system. The mode shapes are *normalized* by requiring

$$\int_0^L X_i^2(x)\,dx = 1 \tag{7.7}$$

7.4 BEAM EQUATION

The partial differential equation governing the free transverse vibrations $w(x,t)$ of a uniform beam of elastic modulus E, mass density ρ, cross-sectional moment of inertia I, and cross-sectional area A is

$$EI\frac{\partial^4 w}{\partial x^4} + \rho A \frac{\partial^2 w}{\partial t^2} = 0 \tag{7.8}$$

The effects of axial loads, rotary inertia, and shear deformation are ignored in Eq. (7.8). A normal mode solution,

$$w(x, t) = X(x)e^{i\omega t} \qquad (7.9)$$

of Eq. (7.8) leads to

$$\frac{d^4X}{dx^4} - \frac{\rho A}{EI}\omega^2 X = 0 \qquad (7.10)$$

The solution of Eq. (7.10) is

$$X(x) = C_1 \cos \lambda x + C_2 \sin \lambda x + C_3 \cosh \lambda x + C_4 \sinh \lambda x \qquad (7.11)$$

where

$$\lambda = \left(\frac{\rho A}{EI}\omega^2\right)^{1/4} \qquad (7.12)$$

Boundary conditions are applied for specific end conditions, leading to a characteristic equation solved by an infinite, but countable, number of natural frequencies. The mode shapes satisfy an orthogonality condition similar to Eq. (7.6).

7.5 MODAL SUPERPOSITION

The *modal superposition method* is used to determine the response of a continuous system due to initial conditions or external excitations. It can be used for systems whose free vibrations are governed by the wave equation or the beam equation. If $u(x, t)$ is the time-dependent response of a system with natural frequencies $\omega_1, \omega_2, \ldots$ and corresponding normalized mode shapes $X_1(x), X_2(x), \ldots$, then the modal superposition formula is

$$u(x, t) = \sum_{k=1}^{\infty} p_k(t) X_k(x) \qquad (7.13)$$

where the $p_k(t)$ are to be determined. If $F(x, t)$ is a nonhomogeneous term appearing in the differential equation, it can be expanded as

$$F(x, t) = \sum_{k=1}^{\infty} C_k(t) X_k(x) \qquad (7.14)$$

where

$$C_k(t) = \int_0^L F(x, t) X_k(x) \, dx \qquad (7.15)$$

Equations (7.13) and (7.14) are substituted into the partial differential equation governing $u(x, t)$. The resulting equation is multiplied by $X_i(t)$, for an arbitrary i, and integrated between 0 and L. Orthogonality conditions are used to generate an uncoupled set of ordinary differential equations for the p_k's.

7.6 RAYLEIGH'S QUOTIENT

Let $f(x)$ be any continuous function satisfying the boundary conditions for a continuous system whose free vibrations are governed by the wave equation. *Rayleigh's quotient* is a functional defined as

$$R(f) = \frac{\displaystyle\int_0^L g(x)\left(\frac{df}{dx}\right)^2 dx}{\displaystyle\int_0^L m(x) f(x)^2\, dx + \sum_{i=1}^n m_i f(x_i)^2} \qquad (7.16)$$

where $g(x)$ and $m(x)$ are known functions of geometry, inertia properties, and elastic properties for the system. For longitudinal oscillations in a bar, $g(x) = EA(x)$ and $m(x) = \rho A(x)$. For torsional oscillations in a shaft, $g(x) = GJ(x)$ and $m(x) = \rho J(x)$. The summation in the denominator of Eq. (7.16) is taken over all discrete masses. Rayleigh's quotient is stationary if and only if $f(x)$ is a mode shape for the system. In this case

$$R[X_i(x)] = \omega_i^2 \qquad (7.17)$$

Thus the minimum value of $R(f)$ is ω_1^2.

Rayleigh's quotient for beam vibrations problems is

$$R(f) = \frac{\displaystyle\int_0^L EI\left(\frac{d^2f}{dx^2}\right)^2 dx}{\displaystyle\int_0^L \rho A\, f(x)^2\, dx + \sum_{i=1}^n m_i f(x_i)^2} \qquad (7.18)$$

7.7 RAYLEIGH-RITZ METHOD

The *Rayleigh-Ritz method* is an energy method that can be used to approximate natural frequencies, mode shapes, and forced responses of continuous systems. Let $\phi_1, \phi_2, \ldots, \phi_n$ be a set of n linearly independent functions satisfying at least the system's geometric boundary conditions (zeroth-order derivatives for the wave equation and zeroth- and first-order derivatives for the beam equation). For free vibrations problems an approximation to the mode shape is of the form

$$X(x) = \sum_{k=1}^n c_k \phi_k(x) \qquad (7.19)$$

When Eq. (7.19) is substituted into the wave equation or beam equation, the following system of equations is obtained:

$$\sum_{j=1}^n (\alpha_{ij} - \omega^2 \beta_{ij}) c_j = 0 \qquad (7.20)$$

where the forms of α_{ij} and β_{ij} are given in Table 7-2. The determinant of the coefficient matrix of the system of equations represented by Eq. (7.20) is set to zero, resulting in an nth-order polynomial to solve for ω^2.

Table 7-2

Situation	α_{ij}	β_{ij}
Longitudinal oscillations in a bar	$\displaystyle\int_0^L EA\left(\frac{d\phi_i}{dx}\right)\left(\frac{d\phi_j}{dx}\right)dx$ $\displaystyle+\sum_{\ell=1}^{n}k_\ell\phi_i(x_\ell)\phi_j(x_\ell)$	$\displaystyle\int_0^L \rho A\phi_i(x)\phi_j(x)\,dx$ $\displaystyle+\sum_{\ell=1}^{n_m}m_\ell\phi_i(x_\ell)\phi_j(x_\ell)$
Torsional oscillations in a shaft	$\displaystyle\int_0^L GJ\left(\frac{d\phi_i}{dx}\right)\left(\frac{d\phi_j}{dx}\right)dx$ $\displaystyle+\sum_{\ell=1}^{n}k_{t_\ell}\phi_i(x_\ell)\phi_j(x_\ell)$	$\displaystyle\int_0^L \rho J\phi_i(x)\phi_j(x)\,dx$ $\displaystyle+\sum_{\ell=1}^{n_m}m_\ell\phi_i(x_\ell)\phi_j(x_\ell)$
Transverse oscillations of a beam	$\displaystyle\int_0^L EI\left(\frac{d^2\phi_i}{dx^2}\right)\left(\frac{d^2\phi_j}{dx^2}\right)dx$ $\displaystyle+\sum_{i=1}^{n}k_i\phi_i(x_\ell)\phi_j(x_\ell)$	$\displaystyle\int_0^L \rho A\phi_i(x)\phi_j(x)\,dx$ $\displaystyle+\sum_{\ell=1}^{n_m}m_i\phi_i(x_\ell)\phi_j(x_\ell)$

Note: x_ℓ represents the locations where discrete stiffness or mass elements are attached.

Solved Problems

7.1 What is the speed of torsional waves in a solid steel ($G = 80 \times 10^9$ N/m^2, $\rho = 7800$ kg/m^3) shaft of 20 mm diameter?

The speed of torsional waves in a shaft is obtained from Table 7-1 as

$$c = \sqrt{\frac{G}{\rho}} = \sqrt{\frac{80 \times 10^9\ \dfrac{\text{N}}{\text{m}^2}}{7800\ \dfrac{\text{kg}}{\text{m}^3}}} = 3.20 \times 10^3\ \frac{\text{m}}{\text{s}}$$

7.2 Derive the partial differential equation governing free longitudinal vibrations of a uniform bar.

Consider a longitudinal element of the bar of thickness dx, as shown in Fig. 7-1. The force developed on each face of the element is the resultant of the normal stress distribution across the face. If the normal stress σ is uniform across the face, the resultant force is σA. A Taylor series expansion leads to

$$\sigma(x + dx)\,A(x + dx) = \sigma(x)\,A(x) + \frac{\partial}{\partial x}(\sigma A)\,dx + \frac{1}{2}\frac{\partial^2}{\partial x^2}(\sigma A)\,(dx)^2 + \cdots +$$

Thus application of Newton's law to the differential element leads to

$$\sigma(x)\,A(x) + \frac{\partial}{\partial x}(\sigma A) + \frac{1}{2}\frac{\partial^2}{\partial x^2}(\sigma A)\,(dx)^2 + \cdots + -\sigma(x)\,A(x)$$

$$= \rho(x)\,A(x)\frac{\partial^2 u}{\partial t^2}\,dx$$

Neglecting terms with higher-order differentials leads to

$$\frac{\partial}{\partial x}(\sigma A) = \rho A \frac{\partial^2 u}{\partial t^2} \qquad (7.21)$$

The normal stress is related to the normal strain through Hooke's law:

$$\sigma = E\epsilon$$

where the normal strain is defined in terms of the displacement as

$$\epsilon = \frac{\partial u}{\partial x}$$

Substituting into Eq. (7.21) leads to

$$\frac{\partial}{\partial x}\left(EA\,\frac{\partial u}{\partial x}\right) = \rho A \frac{\partial^2 u}{\partial t^2}$$

If the bar is uniform and homogeneous, the previous equation reduces to

$$\frac{E}{\rho}\frac{\partial^2 u}{\partial x^2} = \frac{\partial^2 u}{\partial t^2}$$

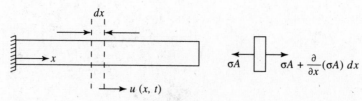

Fig. 7-1

7.3 Determine the natural fequencies and mode shapes of torsional oscillation of a uniform shaft of length L, mass density ρ, and cross-sectional polar moment of inertia J. The shaft is fixed at one end and free at the other end.

The torsional oscillations of the shaft $\theta(x, t)$ are governed by

$$\frac{G}{\rho}\frac{\partial^2 \theta}{\partial x^2} = \frac{\partial^2 \theta}{\partial t^2}$$

The fixed end is constrained against rotation; thus

$$\theta(0, t) = 0$$

The free end has no shear stress; thus

$$\frac{\partial \theta}{\partial x}(L, t) = 0$$

When used with the normal mode solution, $\theta(x, t) = X(x)e^{i\omega t}$, these conditions lead to

$$X(0) = 0, \qquad \frac{dX}{dx}(L) = 0$$

Application of the first boundary condition to Eq. (7.5) leads to

$$X(0) = 0 = C_1$$

Application of the second boundary condition to Eq. (7.5) then yields

$$\frac{dX}{dx}(L) = 0 = C_2 \frac{\omega}{c} \cos \frac{\omega}{c} L$$

If $C_2 = 0$ or $\omega = 0$, then $X(x) = 0$. Thus the system's natural frequencies are determined from

$$\cos \frac{\omega}{c} L = 0$$

$$\omega_n = \frac{(2n - 1)\pi}{2L} \sqrt{\frac{G}{\rho}} \qquad n = 1, 2, \ldots$$

The corresponding mode shapes are

$$X_n(x) = C_n \sin \left[\frac{(2n - 1)\pi x}{2L} \right]$$

for any nonzero C_n.

7.4 Determine the characteristic equation for longitudinal oscillation of a bar of length L, elastic modulus E, and mass density ρ that is fixed at one end and has a particle of mass m attached to the other end.

The equation governing the motion of the system is

$$\frac{E}{\rho} \frac{\partial^2 u}{\partial x^2} = \frac{\partial^2 u}{\partial t^2}$$

The end at $x = 0$ is fixed; thus

$$u(0, t) = 0$$

The boundary condition at $x = L$ is obtained by applying Newton's law to a free-body diagram of the particle, as shown in Fig. 7-2:

$$-EA \frac{\partial u}{\partial x}(L, t) = m \frac{\partial^2 u}{\partial t^2}(L, t)$$

Application of the normal mode solution, $u(x, t) = X(x)e^{i\omega t}$ to the boundary conditions leads to

$$X(0) = 0, \qquad EA \frac{dX}{dx}(L) = m\omega^2 X(L)$$

Application of the boundary condition at $x = 0$ to Eq. (7.5) leads to $C_1 = 0$. Application of the boundary condition at $x = L$ to Eq. (7.5) with $C_1 = 0$ leads to

$$EA\omega\sqrt{\frac{\rho}{E}}\cos\left(\omega\sqrt{\frac{\rho}{E}}L\right) = m\omega^2\sin\left(\omega\sqrt{\frac{\rho}{E}}L\right)$$

$$\sqrt{\rho E}\frac{A}{m} = \omega\tan\left(\omega\sqrt{\frac{\rho}{E}}L\right)$$

The solutions of the above characteristic equation are the system's natural frequencies.

$$EA\frac{\partial u}{\partial x}(L, t) \longleftarrow \boxed{} = \boxed{} \longrightarrow m\frac{\partial^2 u}{\partial t^2}(L, t)$$

Fig. 7-2

7.5   A ship's propeller is a 20-m steel ($E = 210 \times 10^9$ N/m, $\rho = 7800$ kg/m³) shaft of diameter 10 cm. The shaft is fixed at one end with a 500-kg propeller attached to the other end. What are the three lowest natural frequencies of longitudinal vibration of the propeller-shaft system?

The propeller-shaft system is modeled by the system of Problem 7.4. The transcendental equation governing the natural frequencies is

$$\frac{A}{m}\sqrt{\rho E} = \omega\tan\left(\omega L\sqrt{\frac{\rho}{E}}\right)$$

$$\frac{\rho A L}{m} = \phi\tan\phi$$

where

$$\phi = \omega L\sqrt{\frac{\rho}{E}} = \omega(20\text{ m})\sqrt{\frac{7800\frac{\text{kg}}{\text{m}^3}}{210\times10^9\frac{\text{N}}{\text{m}^2}}} = 3.85\times10^{-3}\omega$$

and

$$\frac{\rho A L}{m} = \frac{\left(7800\frac{\text{kg}}{\text{m}^3}\right)\pi(0.05\text{ m})^2(20\text{ m})}{500\text{ kg}} = 2.45$$

The three smallest solutions of the transcendental equation are

$$\phi_1 = 1.137, \qquad \phi_2 = 3.725, \qquad \phi_3 = 6.637$$

leading to

$$\omega_1 = 295.3\frac{\text{rad}}{\text{s}}, \qquad \omega_2 = 967.4\frac{\text{rad}}{\text{s}}, \qquad \omega_3 = 1724.0\frac{\text{rad}}{\text{s}}$$

7.6 What is the required tension in a transmission line of length 15 m and linear density of 5 kg/m such that the transmission line's lowest natural frequency for transverse vibrations is 100 rad/s? Assume the line is simply supported.

The differential equation governing the transverse vibration $u(x, t)$ of the line is

$$\frac{T}{\mu}\frac{\partial^2 u}{\partial x^2} = \frac{\partial^2 u}{\partial t^2}$$

Since the line is simply supported, both ends are constrained from transverse motion; thus

$$u(0, t) = 0, \qquad u(L, t) = 0$$

When the normal mode solution $u(x, t) = X(x)e^{i\omega t}$ is assumed, the boundary conditions lead to

$$X(0) = 0, \qquad X(L) = 0$$

Application of $X(0) = 0$ in Eq. (7.5) leads to $C_1 = 0$. Application of $X(L) = 0$ then leads to

$$C_2 \sin\left(\omega L \sqrt{\frac{\mu}{T}}\right) = 0$$

The smallest value of ω for which a nontrivial solution occurs is

$$\omega_1 = \frac{\pi}{L}\sqrt{\frac{T}{\mu}}$$

which is rearranged as

$$T = \frac{\omega_1^2 L^2 \mu}{\pi^2}$$

Substituting given values, requiring $\omega_1 = 100$ rad/s leads to

$$T = \frac{\left(100\,\frac{\text{rad}}{\text{s}}\right)^2 (15\text{ m})^2 \left(5\,\frac{\text{kg}}{\text{m}}\right)}{\pi^2} = 1.14 \times 10^6 \text{ N}$$

7.7 Determine the characteristic equation for natural frequencies of the system of Fig. 7-3.

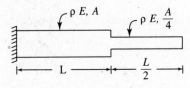

Fig. 7-3

Let $u_1(x, t)$ be the displacement in the left bar, and let $u_2(x, t)$ be the displacement in the right bar. The partial differential equations governing $u_1(x, t)$ and $u_2(x, t)$ are

$$\frac{E}{\rho}\frac{\partial^2 u_1}{\partial x^2} = \frac{\partial^2 u_1}{\partial t^2}$$

$$\frac{E}{\rho}\frac{\partial^2 u_2}{\partial x^2} = \frac{\partial^2 u_2}{\partial t^2}$$

The boundary conditions are

$$u_1(0, t) = 0, \qquad \frac{\partial u_2}{\partial x}\left(\frac{3}{2}L, t\right) = 0$$

The displacement must be continuous at the junction between the two bars:

$$u_1(L, t) = u_2(L, t)$$

The resultant force due to the normal stress distribution must be the same in each bar at their junction:

$$EA \frac{\partial u_1}{\partial x}(L, t) = E \frac{A}{4} \frac{\partial u_2}{\partial x}(L, t)$$

Use of the normal mode solution $u_1(x, t) = X_1(x)e^{i\omega t}$, $u_2(x, t) = X_2 e^{i\omega t}$ leads to

$$X_1(x) = C_1 \cos\left(\omega \sqrt{\frac{\rho}{E}} x\right) + C_2 \sin\left(\omega \sqrt{\frac{\rho}{E}} x\right)$$

$$X_2(x) = C_3 \cos\left(\omega \sqrt{\frac{\rho}{E}} x\right) + C_4 \sin\left(\omega \sqrt{\frac{\rho}{E}} x\right)$$

and

$$X_1(0) = 0, \qquad \frac{dX_2}{dx}\left(\frac{3}{2}L\right) = 0$$

$$X_1(L) = X_2(L), \qquad \frac{dX_1}{dx}(L) = \frac{1}{4}\frac{dX_2}{dx}(L)$$

Application of the boundary conditions leads to

$$X_1(0) = 0 \rightarrow C_1 = 0$$

$$\frac{dX_2}{dx}\left(\frac{3}{2}L\right) = 0 \rightarrow C_4 = C_3 \tan\left(\frac{3}{2}\phi\right), \qquad \phi = L\omega \sqrt{\frac{\rho}{E}}$$

$$X_1(L) = X_2(L) \rightarrow C_2 = C_3(\cot\phi + \tan\tfrac{3}{2}\phi)$$

$$\frac{dX_1}{dx}(L) = \frac{1}{4}\frac{dX_2}{dx}(L) \rightarrow 4\cos\phi(\cot\phi + \tan\tfrac{3}{2}\phi) = -\sin\phi + \cos\phi \tan\tfrac{3}{2}\phi$$

7.8   Determine the lowest natural frequency of longitudinal motion for the system of Fig. 7-4.

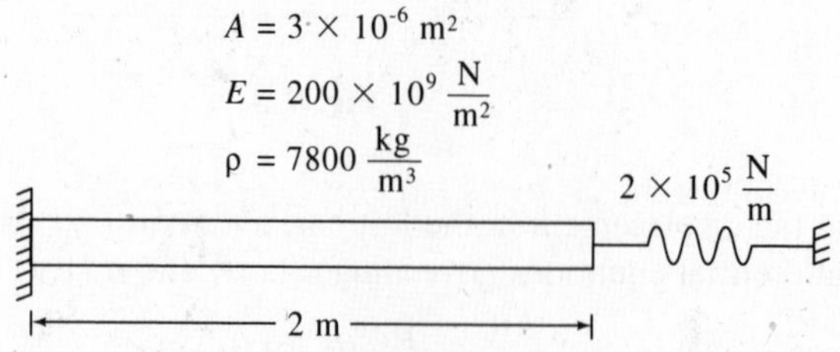

$$A = 3 \times 10^{-6} \text{ m}^2$$
$$E = 200 \times 10^9 \frac{\text{N}}{\text{m}^2}$$
$$\rho = 7800 \frac{\text{kg}}{\text{m}^3}$$

Fig. 7-4

The differential equation governing $u(x, t)$, the longitudinal displacement of the bar, is

$$\frac{E}{\rho} = \frac{\partial^2 u}{\partial x^2} = \frac{\partial^2 u}{\partial t^2}$$

Since the end at $x = 0$ is fixed,

$$u(0, t) = 0$$

The resultant of the normal stress at the right end of the bar must equal the force in the spring at any instant; thus

$$EA \frac{\partial u}{\partial x}(L, t) = -ku(L, t)$$

Application of the normal mode solution $u(x, t) = X(x)e^{i\omega t}$ to the boundary conditions leads to

$$X(0) = 0, \qquad EA \frac{dX}{dx}(L) = -kX(L)$$

Using $X(0) = 0$ in Eq. (7.5) leads to $C_1 = 0$. Application of the second boundary condition to Eq. (7.5) leads to

$$EA\omega \sqrt{\frac{\rho}{E}} \cos\left(\omega \sqrt{\frac{\rho}{E}} L\right) = -k \sin\left(\omega \sqrt{\frac{\rho}{E}} L\right)$$

$$\frac{EA}{k} \sqrt{\frac{\rho}{E}} \omega = -\tan\left(\omega L \sqrt{\frac{\rho}{E}}\right)$$

$$\frac{EA}{kL} \phi = -\tan \phi, \qquad \phi = \omega L \sqrt{\frac{\rho}{E}}$$

Substituting given values leads to

$$1.5\phi = -\tan \phi$$

The smallest solution of the transcendental equation is $\phi = 1.907$, which leads to

$$\omega_1 = \frac{\phi_1}{L} \sqrt{\frac{E}{\rho}} = \frac{1.907}{2 \text{ m}} \sqrt{\frac{200 \times 10^9 \frac{\text{N}}{\text{m}^2}}{7800 \frac{\text{kg}}{\text{m}^2}}} = 4.83 \times 10^3 \frac{\text{rad}}{\text{s}}$$

7.9   Show that the mode shapes of Problem 7.3 satisfy an orthogonality condition of the form of Eq. (7.6).

Let ω_i and ω_j be distinct natural frequencies of the system of Problem 7.3 with corresponding mode shapes $X_i(x)$ and $X_j(x)$. These mode shapes and natural frequencies satisfy the following problems:

$$\frac{d^2 X_i}{dx^2} + \frac{\omega_i^2}{c^2} X_i = 0 \qquad (7.22)$$

$$X_i(0) = 0, \qquad \frac{dX_i}{dx}(L) = 0$$

$$\frac{d^2 X_j}{dx^2} + \frac{\omega_j^2}{c^2} X_j = 0 \qquad (7.23)$$

$$X_j(0) = 0, \qquad \frac{dX_j}{dx}(L) = 0$$

Multiplying Eq. (7.22) by X_j and integrating from 0 to L leads to

$$\int_0^L X_j(x) \frac{d^2 X_i}{dx^2} dx + \frac{\omega_i^2}{c^2} \int_0^L X_i(x) X_j(X) dx = 0$$

Applying integration by parts twice to the first integral leads to

$$X_j(0) \frac{dX_i}{dx}(0) - X_j(L) \frac{dX_i}{dx}(L) - X_i(0) \frac{dX_j}{dx}(0) + X_i(L) \frac{dX_j}{dx}(L)$$

$$+ \int_0^L X_i(x) \frac{d^2 X_j}{dx^2} dx + \frac{\omega_i^2}{c^2} \int_0^L X_i(x) X_j(x) dx = 0$$

Application of the boundary conditions leads to

$$\int_0^L X_i \frac{d^2 X_j}{dx^2} dx + \frac{\omega_i^2}{c^2} \int_0^L X_i(x) X_j(x) = 0 \qquad (7.24)$$

From Eq. (7.23),

$$\frac{d^2 X_j}{dx^2} = - \frac{\omega_j^2}{c^2} X_j$$

which when substituted into Eq. (7.24) leads to

$$\frac{1}{c^2} (\omega_i^2 - \omega_j^2) \int_0^L X_i(x) X_j(x) dx = 0$$

and since $\omega_i \neq \omega_j$, Eq. (7.6) is satisfied.

7.10 Develop an orthogonality condition satisfied by the mode shapes of Problem 7.4.

Let ω_i and ω_j be distinct natural frequencies of Problem 7.4 with corresponding mode shapes X_i and X_j. The problems satisfied by these natural frequencies and mode shapes are

$$\frac{d^2 X_i}{dx^2} + \frac{\rho}{E} \omega_i^2 X_i = 0 \qquad (7.25)$$

$$X_i(0) = 0, \qquad EA \frac{dX_i}{dx}(L) = m\omega_i^2 X_i(L)$$

$$\frac{d^2 X_j}{dx^2} + \frac{\rho}{E} \omega_j^2 X_j = 0 \qquad (7.26)$$

$$X_j(0) = 0, \qquad EA \frac{dX_j}{dx}(L) = m\omega_j^2 X_j(L)$$

Multiplying Eq. (7.25) by X_j and integrating between 0 and L leads to

$$\int_0^L X_j(x) \frac{d^2 X_i}{dx^2} dx + \omega_i^2 \int_0^L X_i(x) X_j(x) dx = 0$$

Using integration by parts twice on the first integral leads to

$$X_j(L)\frac{dX_i}{dx}(L) - X_j(0)\frac{dX_i}{dx}(0) - X_i(L)\frac{dX_j}{dx}(L) + X_i(0)\frac{dX_j}{dx}(0)$$

$$+ \int_0^L X_i(x)\frac{d^2X_j}{dx^2}dx + \frac{\rho}{E}\omega_i^2 \int_0^L X_i(x)X_j(x)\,dx = 0$$

Using the boundary conditions in the previous equation leads to

$$\frac{m}{EA}\omega_i^2 X_i(L)X_j(L) - \frac{m}{EA}\omega_j^2 X_i(L)X_j(L)$$

$$+ \int_0^L X_i\frac{d^2X_j}{dx^2}dx + \frac{\rho}{E}\omega_i^2 \int_0^L X_i(x)X_j(x)\,dx = 0$$

Using Eq. (7.26) in the previous equation and rearranging leads to

$$(\omega_i^2 - \omega_j^2)\left[\frac{m}{EA}X_i(L)X_j(L) + \frac{\rho}{E}\int_0^L X_i(x)X_j(x)\,dx\right] = 0$$

Since $\omega_i \neq \omega_j$, the appropriate orthogonality condition is

$$\frac{m}{\rho A}X_i(L)X_j(L) + \int_0^L X_i(x)X_j(x)\,dx = 0$$

7.11 Determine the natural frequencies of a uniform simply supported beam of length L, elastic modulus E, mass density ρ, area A, and moment of inertia I.

The free transverse vibrations $w(x, t)$ a simply supported beam are governed by Eq. (7.8) subject to

$$w(0, t) = 0 \qquad \frac{\partial^2 w}{\partial x^2}(0, t) = 0$$

$$w(L, t) = 0 \qquad \frac{\partial^2 w}{\partial x^2}(L, t) = 0$$

Application of the normal mode solution $w(x, t) = X(x)e^{i\omega t}$ to the boundary conditions leads to

$$X(0) = 0 \qquad \frac{d^2X}{dx^2}(0) = 0$$

$$X(L) = 0 \qquad \frac{d^2X}{dx^2}(L) = 0$$

Application of the boundary conditions at $x = 0$ to Eq. (7.11) leads to

$$X(0) = 0 \;\rightarrow\; C_1 + C_3 = 0$$

$$\frac{d^2X}{dx^2}(0) = 0 \;\rightarrow\; -\lambda^2 C_1 + \lambda^2 C_3 = 0$$

from which it is determined that $C_1 = C_3 = 0$. Application of the boundary conditions at $x = L$ to Eq. (7.11) leads to

$$X(L) = 0 \rightarrow C_2 \sin \lambda L + C_4 \sinh \lambda L = 0$$

$$\frac{d^2 X}{dx^2}(L) = 0 \rightarrow -\lambda^2 C_2 \sin \lambda L + \lambda^2 C_4 \sinh \lambda L = 0$$

Nontrivial solutions of the above equations are obtained if and only if

$$\sin \lambda L = 0, \qquad C_4 = 0$$

Hence
$$\lambda = \frac{n\pi}{L} \qquad n = 1, 2, 3, \ldots$$

Then using Eq. (7.12), the natural frequencies are

$$\omega_n = (n\pi)^2 \sqrt{\frac{EI}{\rho A L^4}} \qquad n = 1, 2, 3, \ldots$$

7.12   Determine the characteristic equation for a beam pinned at one end free at its other end.

The problem governing the free transverse vibrations of a pinned-free beam pinned at $x = 0$ and free at $x = L$ is Eq. (7.8) subject to

$$w(0, t) = 0, \qquad \frac{\partial^2 w}{\partial t^2}(0, t) = 0$$

$$\frac{\partial^2 w}{\partial x^2}(L, t) = 0, \qquad \frac{\partial^3 w}{\partial x^3}(L, t) = 0$$

Application of the normal mode solution $w(x, t) = X(x)e^{i\omega t}$ to the boundary condition leads to

$$X(0) = 0 \qquad \frac{d^2 X}{dx^2}(0) = 0$$

$$\frac{d^2 X}{dx^2}(L) = 0 \qquad \frac{d^3 X}{dx^3}(L) = 0$$

Application of the boundary conditions at $x = 0$ to Eq. (7.11) leads to

$$X(0) = 0 \rightarrow C_1 + C_3 = 0$$

$$\frac{d^2 X}{dX^2}(0) = 0 \rightarrow -\lambda^2 C_1 + \lambda^2 C_3 = 0$$

from which it is determined that $C_1 = C_3 = 0$. Application of the boundary conditions at $x = L$ leads to

$$\frac{d^2 X}{dx^2}(L) = 0 \rightarrow -\lambda^2 C_2 \sin \lambda L + \lambda^2 \sinh \lambda L = 0$$

$$\frac{d^3 X}{dx^3}(L) = 0 \rightarrow -\lambda^3 C_2 \cos \lambda L + \lambda^3 C_4 \cosh \lambda L = 0$$

A nontrivial solution of the above equations exists if and only if the determinant of the coefficient matrix is zero,

$$-\sin \lambda L \cosh \lambda L + \sinh \lambda L \cos \lambda L = 0$$

leading to

$$\tan \lambda L = \tanh \lambda L$$

The solutions of the previous transcendental equation are used with Eq. (7.12) to determine the system's natural frequencies. The smallest solution is $\lambda = 0$, for which a nontrivial mode shape exists.

7.13 Determine the three lowest natural frequencies for the system of Fig. 7-5.

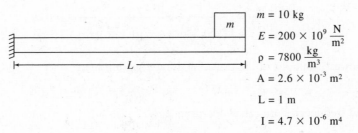

$$m = 10 \text{ kg}$$
$$E = 200 \times 10^9 \, \frac{\text{N}}{\text{m}^2}$$
$$\rho = 7800 \, \frac{\text{kg}}{\text{m}^3}$$
$$A = 2.6 \times 10^{-3} \, \text{m}^2$$
$$L = 1 \text{ m}$$
$$I = 4.7 \times 10^{-6} \, \text{m}^4$$

Fig. 7-5

The free transverse vibrations of the beam of Fig. 7-5 are governed by Eq. (7.8). Since the beam is fixed at $x = 0$,

$$w(0, t) = 0, \qquad \frac{\partial w}{\partial x}(0, t) = 0$$

The boundary conditions at $x = L$ are determined by applying Newton's law to the free body diagram of the block, Fig. 7-6:

$$\frac{\partial^2 w}{\partial x^2}(L, t) = 0 \qquad EI \frac{\partial^3 w}{\partial x^3}(L, t) = m \frac{\partial^2 w}{\partial t^2}(L, t)$$

Application of the normal mode solution $w(x, t) = X(x)e^{i\omega t}$ to the boundary conditions leads to

$$X(0) = 0, \qquad \frac{dX}{dx}(0) = 0$$

$$\frac{d^2 X}{dx^2}(L) = 0, \qquad EI \frac{d^3 X}{dx^3} = -m\omega^2 X$$

Application of the boundary conditions to Eq. (7.11) leads to

$$X(0) = 0 \rightarrow C_1 + C_3 = 0$$

$$\frac{dX}{dx}(0) = \lambda C_2 + \lambda C_4 = 0$$

$$\frac{d^2 X}{dx^2}(L) = 0 \rightarrow -\lambda^2 \cos \lambda L C_1 - \lambda^2 \sin \lambda L C_2 + \lambda^2 \cosh \lambda L C_3 + \lambda^2 \sinh \lambda L C_4 = 0$$

$$\frac{EI}{m} \frac{d^3 X}{dx^3}(L) = -\omega^2 X(L) \rightarrow$$

$$\left(\frac{\rho A}{m} \sin \lambda L + \lambda \cos \lambda L \right) C_1 + \left(-\frac{\rho A}{m} \cos \lambda L + \lambda \sin \lambda L \right) C_2$$

$$+ \left(\frac{\rho A}{m} \sinh \lambda L + \lambda \cosh \lambda L \right) C_3 + \left(\frac{\rho A}{m} \cosh \lambda L + \lambda \sinh \lambda L \right) C_4 = 0$$

where ω^2 has been replaced using Eq. (7.12). The previous equations represent a system of four homogeneous linear simultaneous equations for C_1, C_2, C_3, and C_4. A nontrivial solution exists if and only if the determinant of the system's coefficient matrix is zero. Setting the determinant to zero and simplifying leads to

$$(1 + \cos \phi \cosh \phi) + \frac{m\phi}{\rho AL}(\cos \phi \sinh \phi - \cosh \phi \sin \phi) = 0 \qquad \phi = \lambda L$$

Noting that

$$\frac{m}{\rho AL} = \frac{10 \text{ kg}}{\left(7800 \frac{\text{kg}}{\text{m}^2}\right)(2.6 \times 10^{-3} \text{ m}^2)(1 \text{ m})} = 0.493$$

the three lowest solutions of the transcendental equation are

$$\phi_1 = 1.423 \qquad \phi_2 = 4.113 \qquad \phi_3 = 7.192$$

The natural frequencies are calculated using Eq. (7.12):

$$\omega_i = \phi_1^2 \sqrt{\frac{EI}{\rho AL^4}} = \phi_i^2 \sqrt{\frac{\left(200 \times 10^9 \frac{\text{N}}{\text{m}^2}\right)(4.7 \times 10^{-6} \text{ m}^4)}{\left(7800 \frac{\text{kg}}{\text{m}^3}\right)(2.6 \times 10^{-3} \text{ m}^2)(1 \text{ m})^4}} = 215.3\phi_i^2$$

$$\omega_1 = 486.1 \frac{\text{rad}}{\text{s}} \qquad \omega_2 = 3.642 \times 10^3 \frac{\text{rad}}{\text{s}} \qquad \omega_3 = 1.114 \times 10^4 \frac{\text{rad}}{\text{s}}$$

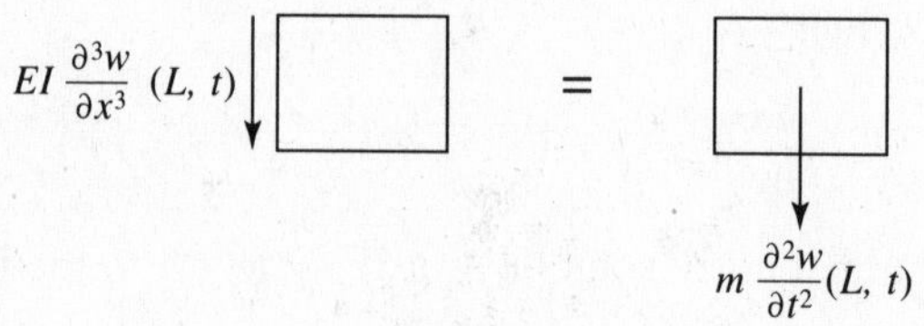

$$EI \frac{\partial^3 w}{\partial x^3}(L, t) \Big| \qquad\qquad = \qquad\qquad$$

$$m \frac{\partial^2 w}{\partial t^2}(L, t)$$

Fig. 7-6

7.14 Demonstrate that the mode shapes of a fixed-free beam satisfy an orthogonality condition of the form of Eq. (7.6).

Let ω_i and ω_j be distinct natural frequencies of a fixed-free beam with corresponding mode shapes X_i and X_j. The problems satisfied by these natural frequencies and mode shapes are

$$\frac{d^4X_i}{dx^4} - \frac{\rho A}{EI}\omega_i^2 X_i = 0 \qquad\qquad (7.27)$$

$$X_i(0) = 0, \qquad \frac{dX_i}{dx}(0) = 0$$

$$\frac{d^2X_i}{dx^2}(L) = 0, \qquad \frac{d^3X_i}{dx^3}(L) = 0$$

$$\frac{d^4X_j}{dx^4} - \frac{\rho A}{EI}\omega_j^2 X_j = 0 \qquad\qquad (7.28)$$

$$X_j(0) = 0 \qquad \frac{dX_j}{dx}(0) = 0$$

$$\frac{d^2X_j}{dx^2}(L) = 0 \qquad \frac{d^3X_j}{dx^3}(L) = 0$$

Multiplying Eq. (7.27) by X_j and integrating from 0 to L leads to

$$\int_0^L X_j \frac{d^4 X_i}{dx^4} dx - \frac{\rho A}{EI} \omega_i^2 \int_0^L X_i(x) X_j(x) dx = 0$$

Using integration by parts four times on the first integral leads to

$$X_j(L) \frac{d^3 X_i}{dx^3}(L) - X_j(0) \frac{d^3 X_i}{dx^3}(0) - \frac{dX_j}{dx}(L) \frac{d^2 X_i}{dx^2}(L)$$

$$+ \frac{dX_j}{dx}(0) \frac{d^2 X_i}{dx^2}(0) + \frac{d^2 X_j}{dx^2}(L) \frac{dX_i}{dx}(L) - \frac{d^2 X_j}{dx^2}(0) \frac{dX_i}{dx}(0)$$

$$- \frac{d^3 X_j}{dx^3}(L) X_i(L) + \frac{d^3 X_j}{dx^3}(0) X_i(0) + \int_0^L X_i \frac{d^4 X_j}{dx^4} dx$$

$$- \frac{\rho A}{EI} \omega_i^2 \int_0^L X_i(x) X_j(X) dx$$

which after application of boundary conditions reduces to

$$\int_0^L X_i \frac{d^4 X_j}{dx^4} dx - \frac{\rho A}{EI} \omega_i^2 \int_0^L X_i(x) X_j(x) dx$$

Using Eq. (7.28) in the previous equation and rearranging leads to

$$\frac{\rho A}{EI} (\omega_j^2 - \omega_i^2) \int_0^L X_i(X) X_j(x) dx = 0$$

Since $\omega_i \neq \omega_j$, the orthogonality condition is verified.

7.15 Determine the steady-state amplitude of the end of the shaft of Fig. 7-7.

Fig. 7-7

The problem governing the motion of the system of Fig. 7-7 is

$$\frac{G}{\rho} \frac{\partial^2 \theta}{\partial x^2} = \frac{\partial^2 \theta}{\partial t^2}$$

$$\theta(0, t) = 0$$

$$JG \frac{\partial \theta}{\partial x}(L, t) = T_0 \sin \omega t$$

The steady-state response is obtained by assuming

$$\theta(x, t) = Q(x) \sin \omega t$$

which when substituted into the partial differential equation and its boundary conditions leads to the following problem for $Q(x)$:

$$\frac{G}{\rho} \frac{d^2 Q}{dx^2} + \omega^2 Q = 0$$

$$Q(0) = 0$$

$$JG \frac{dQ}{dx}(L) = T_0$$

The solution of the differential equation is

$$Q(x) = C_1 \cos\left(\omega \sqrt{\frac{\rho}{G}} x\right) + C_2 \sin\left(\omega \sqrt{\frac{\rho}{G}} x\right)$$

Application of the boundary conditions leads to

$$Q(0) = 0 \; \rightarrow \; C_1 = 0$$

$$JG \frac{dQ}{dx}(L) = T_0 \; \rightarrow \; C_2 = \frac{T_0}{\omega J \sqrt{\rho G} \cos\left(\omega \sqrt{\frac{\rho}{G}} L\right)}$$

Hence the steady-state amplitude of the end of the shaft is

$$Q(L) = C_2 \sin\left(\omega \sqrt{\frac{\rho}{G}} L\right) = \frac{T_0}{\omega J \sqrt{\rho G}} \tan\left(\omega \sqrt{\frac{\rho}{G}} L\right)$$

7.16 The block at the end of the beam of Fig. 7-5 is a small reciprocating machine that operates at 100 rad/s. Determine the machine's steady-state amplitude if it has a rotating unbalance of 0.15 kg-m.

Mathcad

The mathematical problem goverening the response of the system is the same as that of Problem 7.13 except for the second boundary condition at $x = L$. This boundary condition is determined by applying Newton's law to the free body diagram of the machine, as shown in Fig. 7-8,

$$EI \frac{\partial^3 w}{\partial x^3}(L, t) = m \frac{\partial^2 w}{\partial t^2}(L, t) + m_0 e \omega^2 \sin \omega t$$

A steady-state solution is assumed as

$$w(x, t) = Q(x) \sin \omega t$$

which when substituted into the partial differential equation and the boundary conditions leads to

$$EI \frac{d^4 Q}{dx^4} - \omega^2 \rho A Q = 0$$

$$Q(0) = 0 \qquad \frac{dQ}{dx}(0) = 0$$

$$\frac{d^2 Q}{dx^2}(L) = 0 \qquad EI \frac{d^3 Q}{dx^3}(L) = -m \omega^2 Q(L) + m_0 e \omega^2$$

The solution of the differential equation is

$$Q(x) = C_1 \cos \beta x + C_2 \sin \beta x + C_3 \cosh \beta x + C_4 \sinh \beta x$$

where $\qquad \beta = \left(\dfrac{\omega^2 \rho A}{EI}\right)^{1/4} = \left[\dfrac{\left(100 \dfrac{\text{rad}}{\text{s}}\right)^2 \left(7800 \dfrac{\text{kg}}{\text{m}^3}\right)(2.6 \times 10^{-3} \text{ m}^2)}{\left(200 \times 10^9 \dfrac{\text{N}}{\text{m}^2}\right)(4.7 \times 10^{-6} \text{ m}^4)}\right]^{1/4} = 0.682$

Application of the boundary conditions leads to

$$C_1 + C_3 = 0$$

$$C_2 + C_4 = 0$$

$$-\cos \beta L\, C_1 - \sin \beta L\, C_2 + \cosh \beta L\, C_3 + \sinh \beta L\, C_4 = 0$$

$$\left(\frac{\beta m}{\rho A} \cos \beta L + \sin \beta L\right) C_1 + \left(\frac{\beta m}{\rho A} \sin \beta L - \cos \beta L\right) C_2$$

$$+ \left(\frac{\beta m}{\rho A} \cosh \beta L + \sinh \beta L\right) C_3 + \left(\frac{\beta m}{\rho A} \sinh \beta L + \cosh \beta L\right) = \frac{m_0 e \beta}{\rho A}$$

When numerical values are substituted, the previous two equations become

$$-0.776 C_1 - 0.630 C_2 + 1.242 C_3 + 0.736 C_4 = 0$$

$$0.891 C_1 - 0.564 C_2 + 1.153 C_3 + 1.489 C_4 = 5.04 \times 10^{-3}$$

The equations are solved simultaneously, leading to

$$C_1 = 1.82 \times 10^{-3} \qquad C_2 = -2.69 \times 10^{-3} \qquad C_3 = -1.82 \times 10^{-3} \qquad C_4 = 2.69 \times 10^{-3}$$

The steady-state amplitude of the machine is

$$Q(L) = C_1 \cos \beta L + C_2 \sin \beta L + C_3 \cosh \beta L + C_4 \sinh \beta L = 5.60 \times 10^{-4} \text{ m}$$

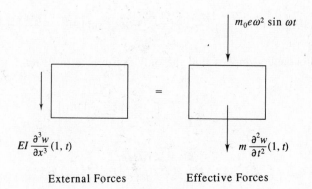

External Forces   Effective Forces

Fig. 7-8

7.17 A torque T is applied to the end of the shaft of Problem 7.3 and suddenly removed. Describe the resulting torsional oscillations of the shaft.

Removal of the torque induces torsional oscillations of the shaft. The initial angular displacement of a particle along the axis of the shaft is the static displacement due to a torque T applied at the end of the shaft. Thus, the initial conditions are

$$\theta(x, 0) = \frac{Tx}{JG}, \qquad \frac{\partial\theta}{\partial t}(x, 0) = 0$$

The natural frequencies and mode shapes are

$$\omega_i = \frac{(2i - 1)\pi}{2L} \sqrt{\frac{G}{\rho}}$$

$$X_i(x) = C_i \sin\left[\frac{(2i - 1)\pi x}{2L}\right]$$

The mode shapes are normalized according to Eq. (7.7) by

$$\int_0^L X_i^2(x)\, dx = 1 = \int_0^L C_i^2 \sin^2\left[\frac{(2i - 1)\pi x}{2L}\right] dx$$

$$C_i = \sqrt{\frac{2}{L}}$$

The general solution for the torsional oscillations is

$$u(x, t) = \sum_{i=1}^{\infty} \sqrt{\frac{2}{L}} \sin\left[\frac{(2i - 1)\pi x}{2L}\right](A_i \cos \omega_i t + B_i \sin \omega_i t)$$

Application of the initial velocity condition leads to $B_i = 0$. Application of the initial displacement condition leads to

$$\frac{Tx}{JG} = \sum_{i=1}^{\infty} \sqrt{\frac{2}{L}} \sin\left[\frac{(2i - 1)\pi x}{2L}\right] A_i$$

Multiplying the previous equation by $X_j(x)$ for an arbitrary j and integrating from 0 to L leads to

$$\frac{T}{JG} \sqrt{\frac{2}{L}} \int_0^L x \sin\left[\frac{(2j - 1)\pi x}{2L}\right] dx = \sum_{i=1}^{\infty} \frac{2}{L} A_i \int_0^L \sin\left[\frac{(2i - 1)\pi x}{2L}\right] \sin\left[\frac{(2j - 1)\pi x}{2L}\right] dx$$

Mode shape orthogonality implies that the only nonzero term in the infinite sum corresponds to $i = j$. Thus

$$A_j = \sqrt{\frac{2}{L}} \frac{T}{JG} \int_0^L x \sin\left[\frac{(2j - 1)\pi x}{2L}\right] dx$$

$$= \sqrt{\frac{2}{L}} \frac{TL^2(-1)^{j+1}}{\pi^2 JG(2j - 1)^2}$$

Thus the torsional response is

$$\theta(x, t) = \frac{8TL}{\pi^2 JG} \sum_{i=1}^{\infty} \frac{(-1)^{i+1}}{2i - 1} \sin\left[\frac{(2i - 1)\pi x}{2L}\right] \sin\left[\frac{(2i - 1)\pi}{2L} \sqrt{\frac{G}{\rho}} t\right]$$

7.18   A time-dependent torque of the form of Fig. 7-9 is applied to the midspan of a circular shaft of length L, shear modulus G, mass density ρ, and polar moment of inertia J. The shaft is fixed at one end and free at the other end. Use modal superposition to determine the time-dependent torsional response of the shaft.

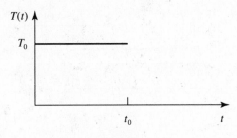

Fig. 7-9

The partial differential equation governing the torsional oscillation is

$$GJ\frac{\partial^2\theta}{\partial x^2} + T_0[u(t) - u(t - t_0)]\,\delta\!\left(x - \frac{L}{2}\right) = \rho J\frac{\partial^2\theta}{\partial t^2} \qquad (7.29)$$

From Problem 7.3 the natural frequencies and normalized mode shapes for the shaft are

$$\omega_n = \frac{(2n-1)\pi}{2L}\sqrt{\frac{G}{\rho}}, \qquad n = 1, 2, \ldots$$

$$X_n(x) = \sqrt{\frac{2}{L}}\sin\left[\frac{(2n-1)\pi x}{2L}\right]$$

The excitation is expanded in a series of mode shapes using Eq. (7.14) with

$$C_k = \int_0^L T_0[u(t) - u(t - t_0)]\,\delta\!\left(x - \frac{L}{2}\right)\sqrt{\frac{2}{L}}\sin\left[\frac{(2k-1)\pi x}{2L}\right]dx$$

$$= \sqrt{\frac{2}{L}}\,T_0\sin\left[\frac{(2k-1)\pi}{4}\right][u(t) - u(t - t_0)] \qquad (7.30)$$

$\theta(x, t)$ is expanded in a series of mode shapes of the form of Eq. (7.13). Substituting this expansion in Eq. (7.29) using Eq. (7.30) leads to

$$GJ\sum_{k=1}^{\infty} p_k(t)\frac{d^2X_k}{dx^2} + \sum_{k=1}^{\infty} C_k(t)X_k(x) = \rho J\sum_{k=1}^{\infty} \ddot{p}_k(t)X_k(x)$$

However,

$$\frac{d^2X_k}{dx^2} = -\omega_k^2\frac{\rho}{G}X_k$$

Thus

$$\sum_{k=1}^{\infty} (\rho J\ddot{p}_k + \rho J\omega_k^2 p_k - C_k)X_k = 0$$

Multiplying the above equation by $X_j(x)$ for an arbitrary j, integrating from 0 to L, and using mode shape orthogonality lead to

$$\ddot{p}_k + \omega_k^2 p_k = \sqrt{\frac{2}{L}}\frac{T_0}{\rho J}\sin\left[\frac{(2k-1)\pi}{4}\right][u(t) - u(t - t_0)]$$

The solution of the previous equation is obtained using the convolution integral as

$$p_k(t) = \sqrt{\frac{2}{L}}\frac{T_0}{\rho J\omega_k^2}\sin\left[\frac{(2k-1)\pi}{4}\right]\{(1 - \cos\omega_k t)\,u(t)$$

$$- [1 - \cos\omega_k(t - t_0)]\,u(t - t_0)\}$$

7.19 Use modal superposition to determine the time-dependent response of the system of Fig. 7-10.

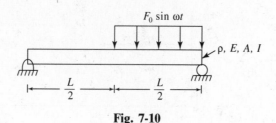

$F_0 \sin \omega t$

ρ, E, A, I

$$\frac{L}{2}$$

$$\frac{L}{2}$$

Fig. 7-10

The natural frequencies and normalized mode shapes for a simply supported beam are

$$\omega_n = (n\pi)^2 \sqrt{\frac{EI}{\rho A L^4}}$$

$$X_n(x) = \sqrt{\frac{2}{L}} \sin \frac{n\pi x}{L}$$

The differential equation governing the motion of the beam is

$$EI \frac{\partial^4 w}{\partial x^4} + \rho A \frac{\partial w^2}{\partial t^2} = F_0 \sin(\omega t)\, u\!\left(x - \frac{L}{2}\right) \qquad (7.31)$$

The excitation is expanded in a series of mode shapes according to Eq. (7.14) with

$$C_k = \int_0^L F_0 \sin(\omega t)\, u\!\left(x - \frac{L}{2}\right) \sqrt{\frac{2}{L}} \sin \frac{k\pi x}{L}\, dx$$

$$= \sqrt{2L}\, \frac{F_0}{k\pi} \sin \omega t \left[\cos\left(\frac{k\pi}{2}\right) - \cos(k\pi)\right]$$

$$= B_k \sin \omega t$$

$$B_k = \sqrt{2L}\, \frac{F_0}{k\pi} \begin{cases} 1 & k = 1, 3, 5, \ldots \\ -2 & k = 2, 6, 10, \ldots \\ 0 & k = 4, 8, 12, \ldots \end{cases}$$

The transverse deflection is expanded as $\sum p_k(t)\, X_k(x)$ and substituted into Eq. (7.31) leading to

$$EI \sum_{k=1}^{\infty} p_k(t) \frac{d^4 X_k}{dx^4} + \rho A \sum_{k=1}^{\infty} \ddot{p}_k(t)\, X_k(x) = \sum_{k=1}^{\infty} C_k(t)\, X_k(x)$$

Noting that

$$\frac{d^4 X_k}{dx^4} = \omega_k^2 \frac{\rho A}{EI} X_k$$

and rearranging lead to

$$\sum_{k=1}^{\infty} [\rho A(\ddot{p}_k + \omega_k^2 p_k) - C_k] X_k(x) = 0$$

Multiplying the previous equation by $X_j(x)$, integrating from 0 to L, and using mode shape orthogonality lead to

$$\ddot{p}_k + \omega_k^2 p_k = B_k \sin \omega t$$

The solution of this equation subject to $p_k(0) = 0$ and $\dot{p}_k(0) = 0$ and $\omega \neq \omega_k$ is

$$p_k(t) = \frac{B_k}{\omega_k^2 - \omega^2} \left(\sin \omega t - \frac{\omega}{\omega_k} \sin \omega_k t\right)$$

7.20   Use modal superposition to determine the response of the system of Fig. 7-11.

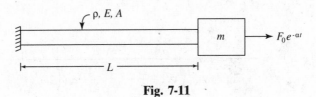

Fig. 7-11

The problem governing the motion of the system of Fig. 7-11 is

$$\frac{E}{\rho}\frac{\partial^2 u}{\partial x^2} = \frac{\partial^2 u}{\partial t^2}$$

subject to

$$u(0, t) = 0$$

$$-EA\frac{\partial u}{\partial x}(L, t) + F_0 e^{-\alpha t} = m\frac{\partial^2 u}{\partial t^2}(L, t)$$

The natural frequencies and mode shapes for this system are determined in Problem 7.4. Substituting the modal superposition, Eq. (7.13), into the differential equation and non-homogeneous boundary condition leads to

$$\frac{E}{\rho}\sum_{k=1}^{\infty} p_k \frac{d^2 X_k}{dx^2} = \sum_{k=1}^{\infty} \ddot{p}_k X_k$$

$$-EA\sum_{k=1}^{\infty}\frac{dX_k}{dx}(L)\,p_k(t) + F_0 e^{-\alpha t} = m\sum_{k=1}^{\infty}\ddot{p}_k X_k(L)$$

Noting that

$$\frac{d^2 X_k}{dx^2} = -\omega_k^2 \frac{\rho}{E} X_k$$

$$\frac{dX_k}{dx}(L) = \frac{m}{EA}\omega_k^2 X_k(L)$$

leads to

$$\sum_{k=1}^{\infty}(\ddot{p}_k + \omega_k^2 p_k)X_k = 0$$

$$\sum_{k=1}^{\infty}(\ddot{p}_k + \omega_k^2 p_k)\,X_k(L) = \frac{F_0}{m}e^{-\alpha t}$$

The first equation is multiplied by $X_j(x)$ for an arbitrary j and integrated from 0 to L. The second equation is multiplied by $mX_j(L)/(\rho A)$. The two equations are then added leading to

$$\sum_{k=1}^{\infty}(\ddot{p}_k + \omega_k^2 p_k)\left[\frac{m}{\rho A}X_j(L)\,X_k(L) + \int_0^L X_j(x)\,X_k(x)\,dx\right] = \frac{X_j(L)}{\rho A}F_0 e^{-\alpha t}$$

Using the mode shape orthogonality condition for this problem, derived in Problem 7.10, the only nonzero term in the sum corresponds to $k = j$. Thus

$$\ddot{p}_j + \omega_j^2 p_j = C_j F_0 e^{-\alpha t}$$

where
$$C_j = \dfrac{\dfrac{1}{\rho A} X_j(L)}{\dfrac{m}{\rho A} X_j^2(L) + \displaystyle\int_0^L X_j^2(x)\, dx}$$

$$= \dfrac{\dfrac{1}{\rho A} \sin\left(\omega_j \sqrt{\dfrac{\rho}{E}}\, L\right)}{\dfrac{m}{\rho A} \sin^2\left(\omega_j \sqrt{\dfrac{\rho}{E}}\, L\right) + \dfrac{L}{2} - \sqrt{\dfrac{E}{\rho}} \dfrac{1}{4\omega_j} \sin\left(2\omega_j \sqrt{\dfrac{\rho}{E}}\, L\right)}$$

The solution for $p_j(t)$ subject to $\dot{p}_j(0) = 0$ and $p_j(0) = 0$ is

$$p_j(t) = \dfrac{C_j F_0}{\omega_j^2 + \alpha^2}\left(e^{-\alpha t} - \cos \omega_j t + \dfrac{\alpha}{\omega_j}\sin \omega_j t\right)$$

7.21   Use Rayleigh's quotient with a trial function of

$$u(x) = B \sin \dfrac{\pi x}{L}$$

to approximate the lowest natural frequency of the system of Fig. 7-12.

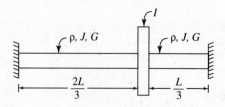

Fig. 7-12

The appropriate form of Rayleigh's quotient for the torsional system of Fig. 7-12 is

$$R(u) = \dfrac{\displaystyle\int_0^L JG\left(\dfrac{du}{dx}\right)^2 dx}{\displaystyle\int_0^L \rho J u^2(x)\, dx + I u^2\!\left(\dfrac{2}{3}L\right)}$$

$$= \dfrac{\displaystyle\int_0^L JG\left[B\dfrac{\pi}{L}\cos\left(\dfrac{\pi x}{L}\right)\right]^2 dx}{\displaystyle\int_0^L \rho J B^2 \sin^2\left(\dfrac{\pi x}{L}\right) dx + I B^2 \sin^2\left(\dfrac{2\pi}{3}\right)}$$

$$= \dfrac{\dfrac{\pi^2 GJB^2}{2L}}{\dfrac{\rho J B^2 L}{2} + \dfrac{3IB^2}{2}}$$

Hence an upper bound for the system's lowest natural frequency is

$$\omega_1 \leq \left(\dfrac{\pi^2 G}{\rho L^2 + \dfrac{3IL}{J}}\right)^{1/2}$$

7.22 Use Rayleigh's quotient to approximate the lowest natural frequency of longitudinal motion of the tapered bar of circular cross section shown in Fig. 7-13.

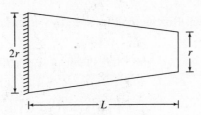

Fig. 7-13

The trial function

$$u(x) = B \sin \frac{\pi x}{2L}$$

satisfies the boundary conditions $u(0) = 0$ and $du/dx(L) = 0$. The geometric properties of the bar's cross section are

$$r(x) = r\left(1 - \frac{x}{2L}\right)$$

$$A(x) = \pi r^2\left(1 - \frac{x}{2L}\right)^2$$

Application of the appropriate form of Rayleigh's quotient to the trial function leads to

$$R(u) = \frac{\int_0^L EA(x)\left(\frac{du}{dx}\right)^2 dx}{\int_0^L \rho A(x) u^2(x) dx}$$

$$= \frac{\int_0^L E\pi r^2\left(1 - \frac{x}{2L}\right)^2\left[\frac{\pi}{2L}\cos\left(\frac{\pi x}{2L}\right)\right]^2 dx}{\int_0^L \rho \pi r^2\left(1 - \frac{x}{2L}\right)^2 \sin^2\left(\frac{\pi x}{2L}\right) dx}$$

$$= \frac{0.9017 EL\pi r^2}{0.2157 L\rho\pi r^2} = 4.205 \frac{E}{\rho L^2}$$

Thus, an upper bound for the lowest natural frequency of longitudinal oscillations is

$$\omega_1 < \frac{2.05}{L}\sqrt{\frac{E}{\rho}}$$

7.23 Use the Rayleigh-Ritz method to approximate the two lowest natural frequencies of the system of Fig. 7-13. Use the two lowest mode shapes for a uniform fixed-free bar as trial functions.

The two lowest mode shapes for a fixed-free uniform bar are obtained in Problem 7.3 as

$$\phi_1(x) = \sin\left(\frac{\pi x}{2L}\right), \qquad \phi_2(x) = \sin\left(\frac{3\pi x}{2L}\right)$$

Hence the Rayleigh-Ritz approximation to the mode shapes is

$$u(x) = C_1 \sin\left(\frac{\pi x}{2L}\right) + C_2 \sin\left(\frac{3\pi x}{2L}\right)$$

Application of the Rayleigh-Ritz method leads to two equations of the form of Eq. (7.20) with

$$\alpha_{ij} = \int_0^L EA(x)\left(\frac{d\phi_i}{dx}\right)\left(\frac{d\phi_j}{dx}\right) dx$$

$$\beta_{ij} = \int_0^L \rho A(x)\, \phi_i(x)\, \phi_j(x)\, dx$$

The coefficients for this problem are determined as

$$\alpha_{11} = \int_0^L E\pi r^2\left(1 - \frac{x}{2L}\right)^2 \left[\frac{\pi}{2L}\cos\left(\frac{\pi x}{2L}\right)\right]^2 dx = \frac{0.9017 E\pi r^2}{L}$$

$$\alpha_{12} = \alpha_{21} = \int_0^L E\pi r^2\left(1 - \frac{x}{2L}\right)^2 \left[\frac{\pi}{2L}\cos\left(\frac{\pi x}{2L}\right)\right]\left[\frac{3\pi}{2L}\cos\left(\frac{3\pi x}{2L}\right)\right] dx$$

$$= \frac{0.6094 E\pi r^2}{L}$$

$$\alpha_{22} = \int_0^L E\pi r^2\left(1 - \frac{x}{2L}\right)^2 \left[\frac{3\pi}{2L}\cos\left(\frac{3\pi x}{2L}\right)\right]^2 dx = \frac{6.664 E\pi r^2}{L}$$

$$\beta_{11} = \int_0^L \rho\pi r^2\left(1 - \frac{x}{2L}\right)^2 \sin^2\left(\frac{\pi x}{2L}\right) dx = 0.2157\rho\pi r^2 L$$

$$\beta_{12} = \beta_{21} = \int_0^L \rho\pi r^2\left(1 - \frac{x}{2L}\right)^2 \sin\left(\frac{\pi x}{2L}\right)\sin\left(\frac{3\pi x}{2L}\right) dx = 0.0697\rho\pi r^2 L$$

$$\beta_{22} = \int_0^L \rho\pi r^2\left(1 - \frac{x}{2L}\right)^2 \sin^2\left(\frac{3\pi x}{2L}\right) dx = 0.28322\rho\pi r^2 L$$

Upon substitution and simplification, Eqs. (7.20) become

$$(0.9017 - 0.2157\phi)C_1 + (0.6094 - 0.0697\phi)C_2 = 0$$

$$(0.6094 - 0.0697\phi)C_1 + (6.664 - 0.2833\phi)C_2 = 0$$

where

$$\phi = \left(\frac{\rho L^2}{E}\right)\omega^2$$

A nontrivial solution of the above equations exists if and only if the determinant of the coefficient matrix of the system is zero. To this end

$$\begin{vmatrix} 0.9017 - 0.2157\phi & 0.6094 - 0.0697\phi \\ 0.6094 - 0.0697\phi & 6.664 - 0.2833\phi \end{vmatrix} = 0$$

$$(0.9017 - 0.2157\phi)(6.664 - 0.2833\phi) - (0.6094 - 0.0697\phi)^2 = 0$$

$$0.0562\phi^2 - 1.608\phi + 5.637 = 0$$

The solutions of the above quadratic equation are $\phi = 4.093, 24.53$ leading to

$$\omega_1 = 2.0228\sqrt{\frac{E}{\rho L^2}}, \qquad \omega_2 = 4.949\sqrt{\frac{E}{\rho L^2}}$$

7.24 Use the Rayleigh-Ritz method to approximate the two lowest natural frequencies of the torsional system of Fig. 7-14. Use cubic polynomials as trial functions.

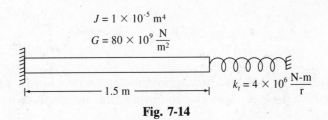

$$J = 1 \times 10^{-5} \text{ m}^4$$

$$G = 80 \times 10^9 \frac{\text{N}}{\text{m}^2}$$

$$k_t = 4 \times 10^6 \frac{\text{N-m}}{\text{r}}$$

1.5 m

Fig. 7-14

The Rayleigh-Ritz method can be applied using trial functions satisfying only the geometric boundary conditions (i.e., boundary conditions developed solely from geometric considerations). Polynomials satisfying the boundary conditions for a fixed-free shaft are

$$\phi_1(x) = x^3 - 3L^2 x, \qquad \phi_2(x) = x^2 - 2Lx$$

The coefficients used in Eq. (7.20) are

$$\alpha_{ij} = \int_0^L JG \left(\frac{d\phi_i}{dx} \right) \left(\frac{d\phi_j}{dx} \right) dx + k_t \phi_i(L) \, \phi_j(L)$$

$$\beta_{ij} = \int_0^L \rho J \phi_i(x) \, \phi_j(x) \, dx$$

Evaluation of these coefficients leads to

$$\alpha_{11} = \int_0^L JG(3x^2 - 3L^2)^2 \, dx + k_t(L^3 - 3L^3)^2 = 2.114 \times 10^8$$

$$\alpha_{12} = \int_0^L JG(3x^2 - 3L^2)(2x - 2L) \, dx + k_t(L^3 - 3L^3)(L^2 - 2L^2) = 6.379 \times 10^7$$

$$\alpha_{13} = \int_0^L JG(2x - 2L)^2 \, dx + k_t(L^2 - 2L^2) = 2.115 \times 10^7$$

$$\beta_{11} = \int_0^L \rho J(x^3 - 3L^3)^2 \, dx = 2.589$$

$$\beta_{12} = \int_0^L \rho J(x^3 - 3L^2 x)(x^2 - 2Lx) \, dx = 0.903$$

$$\beta_{22} = \int_0^L \rho J(x^2 - 2Lx)^2 \, dx = 0.3159$$

Equations (7.20) become

$$(2.114 \times 10^8 - 2.589\omega^2)C_1 + (6.379 \times 10^7 - 0.903\omega^2)C_2 = 0$$

$$(6.379 \times 10^7 - 0.903\omega^2)C_1 + (2.115 \times 10^7 - 0.3157\omega^2)C_2 = 0$$

A nontrivial solution is obtained if and only if the determinant of the coefficient matrix is set to zero, leading to

$$0.001369\omega^4 - 6.2533 \times 10^6\omega^2 + 4.0195 \times 10^{14} = 0$$

whose solutions are

$$\omega_1 = 8076\,\frac{\text{rad}}{\text{s}}, \qquad \omega_2 = 66430\,\frac{\text{rad}}{\text{s}}$$

7.25 Use the Rayleigh-Ritz method to approximate the two lowest natural frequencies of a uniform fixed-fixed beam. Use the following trial functions, which satisfy all boundary conditions:

$$\phi_1(x) = x^4 - 2Lx^3 + L^2x^2, \qquad \phi_2(x) = x^5 - 3L^2x^3 + 2L^3x^2$$

The Rayleigh-Ritz approximation to the mode shape is

$$w(x) = C_1\phi_1(x) + C_2\phi_2(x)$$

The appropriate form of the coefficients for Eqs. (7.20) is

$$\alpha_{ij} = \int_0^L EI\left(\frac{d^2\phi_i}{dx^2}\right)\left(\frac{d^2\phi_j}{dx^2}\right) dx$$

$$\beta_{ij} = \int_0^L \rho A \phi_i(x)\,\phi_j(x)\,dx$$

Using the suggested trial functions, the coefficients are calculated as

$$\alpha_{11} = \int_0^L EI(12x^2 - 12Lx + 2L^2)^2\,dx = 0.8EIL^5$$

$$\alpha_{12} = \int_0^L EI(12x^2 - 12Lx + 2L^2)(20x^3 - 18L^2x + 4L^3)\,dx = 2EIL^6$$

$$\alpha_{22} = \int_0^L EI(20x^3 - 18L^2x + 4L^3)^2\,dx = 5.1428EIL^7$$

$$\beta_{11} = \int_0^L \rho A(x^4 - 2Lx^3 + L^2x^2)^2\,dx = 0.001587\rho AL^9$$

$$\beta_{12} = \int_0^L \rho A(x^4 - 2Lx^3 + L^2x^2)(x^5 - 3L^2x^3 + 2L^3x^2)\,dx = 0.003968\rho AL^{10}$$

$$\beta_{22} = \int_0^L \rho A(x^5 - 3L^2x^3 + 2L^3x^2)^2\,dx = 0.0099567\rho AL^{11}$$

Substitution into Eqs. (7.20) and rearrangement yields

$$(0.8 - 0.001587\phi)C_1 + (2 - 0.003968\phi)C_2 = 0$$

$$(2 - 0.003968\phi)C_1 + (5.1428 - 0.0099567\phi)C_2 = 0$$

where
$$\phi = \omega^2 \frac{\rho A L^4}{EI}$$

A nontrivial solution of the above system exists if and only if the determinant of the coefficient matrix is set to zero. To this end

$$(0.8 - 0.001589\phi)(5.1428 - 0.0099567\phi) - (2 - 0.003968\phi)^2 = 0$$

$$5.63 \times 10^{-8}\phi^2 - 2.549 \times 10^{-4}\phi + 0.11424 = 0$$

The solutions of the above equations are 504.09 and 4028.23 leading to

$$\omega_1 = 22.4\sqrt{\frac{EI}{\rho A L^4}} \qquad \omega_2 = 63.47\sqrt{\frac{EI}{\rho A L^4}}$$

Supplementary Problems

7.26   How long does it take a wave to travel across a 30-m transmission line of tension 15,000 N and linear density 4.7 kg/m?

Ans.   0.531 s

7.27   Derive the partial differential equation governing the longitudinal vibrations of a uniform bar.

Ans.
$$E\frac{\partial^2 u}{\partial x^2} = \rho \frac{\partial^2 u}{\partial t^2}$$

7.28   Derive the partial differential equation governing the transverse vibrations of a taut string or cable.

Ans.
$$T\frac{\partial^2 u}{\partial x^2} = \mu \frac{\partial^2 u}{\partial t^2}$$

7.29   Determine the lowest torsional natural frequency of a 5-m-long steel $(G = 80 \times 10^9 \text{ N/m}^2,$ $E = 200 \times 10^9 \text{ N/m}^2,$ $\rho = 7800 \text{ kg/m}^3)$ annular shaft of inner diameter 20 mm and outer diameter 30 mm. The shaft is fixed at one end and free at its other end.

Ans.   1006 rad/s

7.30   Determine the lowest longitudinal natural frequency of the shaft of Problem 7.29.

Ans.   1590 rad/s

7.31   A pulley of moment of inertia 1.85 kg/m² is attached to the end of a 80-cm steel $(G = 80 \times 10^9 \text{ N/m}^2,$ $E = 210 \times 10^9 \text{ N/m}^2,$ $\rho = 7800 \text{ kg/m}^3)$ shaft of diameter 30 cm. What are the two lowest natural frequencies of torsional oscillation of the pulley?

Ans.   4655 rad/s, 15,000 rad/s

7.32   Determine the characteristic equation for the longitudinal oscillations of the bar of Fig. 7-15.

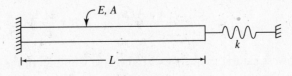

Fig. 7-15

Ans.

$$\frac{EA}{k}\,\omega\sqrt{\frac{\rho}{E}} = -\tan\left(\omega\sqrt{\frac{\rho}{E}}\,L\right)$$

7.33 Determine the two lowest natural frequencies for the system of Problem 7.32 if $E = 150 \times 10^9$ N/m^2, $\rho = 5000$ kg/m^3, $A = 1.5 \times 10^{-4}$ m, $L = 1.6$ m, and $k = 3.1 \times 10^6$ N/m.

Ans. 5.82×10^3 rad/s, 1.63×10^4 rad/s

7.34 Determine the characteristic equation for the system of Fig. 7-16.

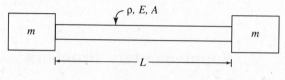

Fig. 7-16

Ans.

$$\omega \tan\left(\sqrt{\frac{\rho}{E}}\,\omega L\right) = \frac{2m\omega^2 A\sqrt{\rho E}}{m^2\omega^2 - \rho E A^2}$$

7.35 Show that the mode shapes for the system of Problem 7.32 satisfy the orthogonality condition, Eq. (7.6).

7.36 Develop the orthogonality condition satisfied by the mode shapes of the system of Problem 7.34.

Ans.

$$\rho A \int_0^L X_i(x)\, X_j(x)\, dx + m X_i(L)\, X_j(L) + m X_i(0)\, X_j(0) = 0$$

7.37 Develop the characteristic equation for a fixed-free beam.

Ans.

$$\cosh \phi \cos \phi = -1, \qquad \phi = \left(\frac{\omega^2 \rho A L^4}{EI}\right)^{1/4}$$

7.38 Develop the characteristic equation for a free-free beam.

Ans.

$$\cosh \phi \cos \phi = 1, \qquad \phi = \left(\frac{\omega^2 \rho A L^4}{EI}\right)^{1/4}$$

7.39 Develop the characteristic equation for a beam fixed at one end with a spring of stiffness k attached at its other end.

Ans.

$$\phi^3(\cosh \phi \cos \phi + 1) - \beta(\cos \phi \sinh \phi - \cosh \phi \sin \phi) = 0$$

$$\phi = \left(\frac{\omega^2 \rho A L^4}{EI}\right)^{1/4}, \qquad \beta = \frac{kL^3}{EI}$$

7.40 Determine the lowest natural frequency of transverse vibration of the system of Problem 7.29.

Ans. 6.41 rad/s

7.41   Determine the characteristic equation for a beam fixed at one end with a disk of negligible mass but a large moment of inertia $\tilde{I}$ attached at its other end.

Ans.

$$\cos \phi \cosh \phi + \beta(\sin \phi \cosh \phi + \cos \phi \sinh \phi) = -1$$

$$\phi = \left(\frac{\omega^2 \rho A L^4}{EI}\right)^{1/4}, \qquad \beta = \frac{\tilde{I}}{\rho A L^3}$$

7.42   Show that the mode shapes of the system of Problem 7.37 satisfy the orthogonality condition, Eq. (7.6).

7.43   Develop an orthogonality condition satisfied by the mode shapes of the system of Problem 7.41.

Ans.

$$\int_0^L X_i(x)X_j(x)\,dx + \beta X_i(L)X_j(L) = 0$$

7.44   Determine the steady-state amplitude of angular oscillation of the disk of Fig. 7-17.

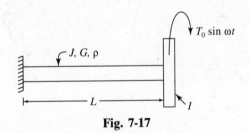

Fig. 7-17

Ans.

$$\frac{T_0 L \sin \phi}{JG(\phi \cos \phi - \beta \phi^2 \sin \phi)}, \qquad \beta = \frac{I}{\rho JL}, \qquad \phi = \omega L \sqrt{\frac{\rho}{G}}$$

7.45   Determine the steady-state amplitude of the end of the beam of Fig. 7-18.

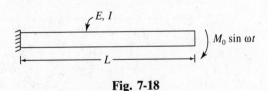

Fig. 7-18

Ans.

$$\frac{M_0(1 + \sinh^2 \beta L - \sin \beta L \sinh \beta L)}{EI\beta^2(1 + \cosh \beta L \cos \beta L)}$$

$$\beta = \left(\frac{\omega^2 \rho A}{EI}\right)^{1/4}$$

7.46   A torque T_0 is statically applied to the midspan of the shaft of Fig. 7-19. Determine the mathematical form of the time-dependent torsional oscillations when the torque is suddenly removed.

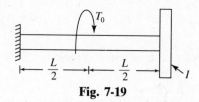

Fig. 7-19

Ans.

$$\theta(x,t) = \sum_{n=1}^{\infty} A_n \sin\left(\omega_n \sqrt{\frac{\rho}{G}}\, x\right) \sin \omega_n t$$

$$A_n = \frac{T_0}{JG} \cdot \frac{\dfrac{G}{\rho \omega_n^{\,2}} \sin\left(\omega_n \dfrac{L}{2}\sqrt{\dfrac{\rho}{G}}\right) - \dfrac{L}{2\omega_n}\sqrt{\dfrac{G}{\rho}}\cos\left(\omega_n \sqrt{\dfrac{\rho}{G}}\,L\right)}{\dfrac{1}{2}\left[L - \dfrac{1}{2\omega_n}\sqrt{\dfrac{G}{\rho}}\sin\left(\omega_n \sqrt{\dfrac{\rho}{G}}\,L\right)\right]}$$

$$\sqrt{\rho G}\,\frac{J}{I} = \omega_n \tan\left(\omega_n \sqrt{\frac{\rho}{G}}\,L\right)$$

7.47   Use modal superposition to determine the response of the pulley of Problem 7.31 when it is subject to an angular impulse H at $t = 0$.

Ans.

$$\frac{2H}{\rho J}\sum_{k=1}^{\infty} \frac{\sin\left(\omega_k \sqrt{\dfrac{\rho}{G}}\,L\right)}{L - \dfrac{1}{2\omega_k}\sqrt{\dfrac{G}{\rho}}\sin\left(2\omega_k \sqrt{\dfrac{\rho}{G}}\,L\right)}\, \sin\left(\omega_k \sqrt{\dfrac{\rho}{G}}\,x\right)\sin \omega_k t$$

7.48   Use modal superposition to determine the response of the system of Fig. 7-20.

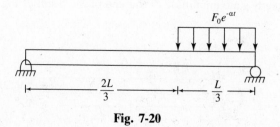

Fig. 7-20

Ans.

$$\frac{2}{\pi}F_0 \sum_{k=0}^{\infty} \frac{\cos(2/3)k\pi - \cos k\pi}{k(\omega_k^{\,2} + \alpha^2)}\left(e^{-\alpha t} - \cos \omega_k t + \frac{\alpha}{\omega_k}\sin \omega_k t\right)\sin\frac{k\pi x}{L}$$

$$\omega_k = k^2\pi^2\left(\frac{EI}{\rho A L^4}\right)^{1/4}$$

7.49   Determine the differential equations to be solved for p_k for a simply supported beam subject to a concentrated load of magnitude I that starts at the beam's left end at $t = 0$ and traverses the beam at a speed v.

Ans.

$$\ddot{p}_k + \left(\frac{k\pi}{L}\right)^2 p_k = I\sqrt{\frac{2}{L}}\left[u(t) - u\left(t - \frac{L}{v}\right)\right]\sin\left(\frac{k\pi vt}{L}\right)$$

7.50   Use Rayleigh's quotient to approximate the lowest natural frequency of longitudinal vibrations of the system of Fig. 7-21 using

$$\phi(x) = \sin\frac{\pi x}{2L}$$

as a trial function.

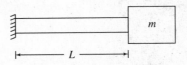

Fig. 7-21

Ans.

$$\omega < \sqrt{\frac{\pi^2 EA}{4\rho AL^2 + 8mL}}$$

7.51   Use Rayleigh's quotient with the trial function

$$\phi(x) = \sin\frac{\pi x}{L}$$

to approximate the lowest natural frequency of a simply supported beam with a concentrated mass m at its midspan.

Ans.

$$\omega < \sqrt{\frac{EI\pi^4}{\rho AL^4 + 2mL^3}}$$

7.52   Use the Rayleigh-Ritz method with trial functions

$$\phi_1(x) = L^3x - 2Lx^3 + x^4, \qquad \phi_2(x) = \tfrac{7}{3}L^4x - \tfrac{10}{3}L^2x^3 + x^5$$

to approximate the lowest natural frequency of a uniform simply supported beam.

Ans.

$$9.877\left(\frac{EI}{\rho AL}\right)^{1/4}$$

7.53   Use the Rayleigh-Ritz method with trial functions

$$\phi_1(x) = \sin\frac{\pi x}{L}, \qquad \phi_2(x) = \sin\frac{2\pi x}{L}$$

to approximate the lowest natural frequency of the system of Fig. 7-22.

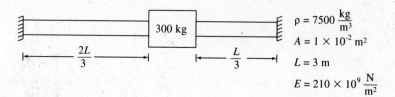

Fig. 7-22

Ans. 3.01×10^3 rad/s

Chapter 8

Vibration Control

Vibration control is the design or modification of a system to suppress unwanted vibrations or to reduce force or motion transmission. The design parameters include inertia properties, stiffness properties, damping properties, and even the system configuration including the number of degrees of freedom.

8.1 VIBRATION ISOLATION

Vibration isolators are used either to protect a foundation from large forces developed during operation of a machine or to protect a machine from large accelerations induced by motion of its base. The parallel elastic spring and viscous damper combination in the system of Fig. 8-1 serves as a vibration isolator. If the machine is subject to an excitation $F(t)$ which induces a displacement $x(t)$, the force transmitted to the foundation through the isolator is

$$F_T = kx + c\dot{x} \tag{8.1}$$

If the base of the system of Fig. 8-2 is subject to a displacement $y(t)$, then the acceleration transmitted to the machine of mass m is determined as

$$\ddot{x} = -\frac{c\dot{z} + kz}{m} \tag{8.2}$$

where $z(t)$ is the displacement of the machine relative to its base and is equal to the total displacement of the isolator.

Fig. 8-1

Fig. 8-2

The above shows that the two types of isolation problems are analogous, and the same theory is used to analyze both problems.

8.2 ISOLATION FROM HARMONIC EXCITATION

The steady-state response of the machine of Fig. 8-1 due to a harmonic excitation $F(t) = F_0 \sin \omega t$ is

$$x(t) = X \sin(\omega t - \phi) \tag{8.3}$$

where

$$X = \frac{F_0}{m\omega_n^2} M(r, \zeta) = \frac{F_0}{m\omega_n^2} \frac{1}{\sqrt{(1 - r^2)^2 + (2\zeta r)^2}} \tag{8.4}$$

235

where $r = \omega/\omega_n$ is the frequency ratio. Then the force transmitted to the foundation of Fig. 8-1 is determined using Eqs. (8.1) through (8.4) as

$$F_T = F \sin(\omega t - \lambda) \tag{8.5}$$

where

$$F_T = F_0 T(r, \zeta) = F_0 \sqrt{\frac{1 + (2\zeta r)^2}{(1 - r^2)^2 + (2\zeta r)^2}} \tag{8.6}$$

and

$$\lambda = \tan^{-1}\left[\frac{2\zeta r^3}{1 + (4\zeta^2 - 1)r^2}\right] \tag{8.7}$$

The function $T(r, \zeta)$ is called the *transmissibility ratio* and is plotted in Fig. 8-3. Note that vibration isolation occurs only when $T < 1$. Thus from Fig. 8-3, isolation from harmonic excitation occurs only when $r > \sqrt{2}$.

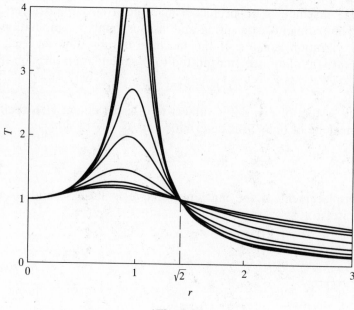

Fig. 8-3

8.3 SHOCK ISOLATION

Consider an excitation $F(t)$, characterized by parameters F_0 (perhaps its maximum value), and a characteristic time (perhaps the duration of the excitation) of t_0. The *displacement spectrum* for $F(t)$ is a nondimensional plot of $kx_{\max}/F_0$ on the vertical scale versus $\omega_n t_0/2\pi$ on the horizontal scale. The *force spectrum* is a plot of $F_{T,\max}/F_0$ on the vertical scale versus $\omega_n t_0/2\pi$ on the horizontal scale. The force spectrum is identical to the displacement spectrum for an undamped system. The force and displacement spectra are used in design and analysis applications for transient excitations.

8.4 IMPULSE ISOLATION

If a system is subject to a very short duration pulse, the shape of the pulse is insignificant in determining the maximum displacement and maximum transmitted force. An excitation applied to an elastic system can often be modeled as an impulsive excitation if the pulse duration t_0 is much less than the system's natural period T. In this circumstance, the important quantity is the total magnitude of the impulse applied to the system:

$$I = \int_0^{t_0} F(t)\, dt \tag{8.8}$$

Application of the impulse leads to a velocity change of

$$v = \frac{I}{m} \tag{8.9}$$

Isolators are designed to protect foundations from large impulsive forces. A nondimensional representation of the maximum force transmitted through an isolator to its foundation is

$$Q(\zeta) = \frac{F_{T_{max}}}{mv\omega_n} = \begin{cases} (e^{-((\zeta/\sqrt{1-\zeta^2})\,\tan^{-1}\{[\sqrt{1-\zeta^2}(1-4\zeta^2)/[\zeta(3-4\zeta^2)]]\})} & \zeta < 0.5 \\ 2\zeta & \zeta > 0.5 \end{cases} \tag{8.10}$$

Figure 8-4 shows that $Q(\zeta)$ is flat as its minimum and approximately equal to 0.81 for $\zeta = 0.24$.

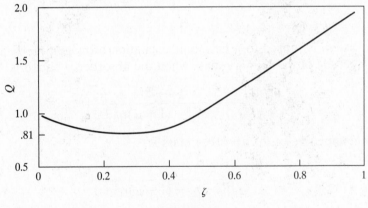

Fig. 8-4

Isolator efficiency for impulsive excitations is defined as

$$E(\zeta) = \frac{\frac{1}{2}mv^2}{F_{T_{max}} x_{max}} = \begin{cases} \frac{1}{2}e^{(\zeta/\sqrt{1-\zeta^2})\,\tan^{-1}\{[\zeta\sqrt{1-\zeta^2}(4-8\zeta^2)/(8\zeta^2-8\zeta^4-1)]\}} & \zeta < 0.5 \\ \frac{1}{4}e^{(\zeta/\sqrt{1-\zeta^2})\,\tan^{-1}[(\sqrt{1-\zeta^2}/\zeta)]} & \zeta > 0.5 \end{cases} \tag{8.11}$$

$E(\zeta)$ has a maximum of 0.96 for $\zeta = 0.4$. It is noted that only evaluations of the inverse tangent function between 0 and π are used in evaluating Eqs. (8.10) and (8.11).

8.5 VIBRATION ABSORBERS

Large amplitude steady-state vibrations exist when a system is subject to a harmonic excitation whose frequency of excitation is near the natural frequency of the system. The steady-state amplitude can be reduced by changing the system configuration by the addition of a *vibration absorber,* an auxiliary mass-spring system illustrated in Fig. 8-5. The addition of a vibration absorber adds 1 degree of freedom to the system and shifts the natural frequencies away from the excitation frequency. The lower of the new system's natural frequencies is less than the natural frequency of the primary system while the higher natural frequency is greater than the natural frequency of the primary system.

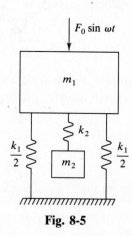

Fig. 8-5

If the primary system is subject to a harmonic excitation of magnitude F_0 and frequency ω, the steady-state amplitude of the primary mass when the absorber is added is

$$X_1 = \frac{F_0}{k_1} \left| \frac{1 - r_2^2}{r_1^2 r_2^2 - r_2^2 - (1 + \mu)r_1^2 + 1} \right| \qquad (8.12)$$

and the steady-state amplitude of the absorber mass is

$$X_2 = \frac{F_0}{k_1} \left| \frac{1}{r_1^2 r_2^2 - r_2^2 - (1 + \mu)r_1^2 + 1} \right| \qquad (8.13)$$

where

$$r_1 = \frac{\omega}{\omega_{11}}, \qquad \omega_{11} = \sqrt{\frac{k_1}{m_1}} \qquad (8.14)$$

$$r_2 = \frac{\omega}{\omega_{22}}, \qquad \omega_{22} = \sqrt{\frac{k_2}{m_2}} \qquad (8.15)$$

$$\mu = \frac{m_2}{m_1} \qquad (8.16)$$

As shown by Eq. (8.12) and illustrated in Fig. 8-6, if $r_2 = 1$, the steady-state amplitude of the primary mass is zero. In this situation

$$X_2 = \frac{F_0}{k_2} \qquad (8.17)$$

When a vibration absorber is added to a system with a harmonic excitation and the absorber is tuned to the excitation frequency, the point in the system where the absorber is added has zero steady-state amplitude.

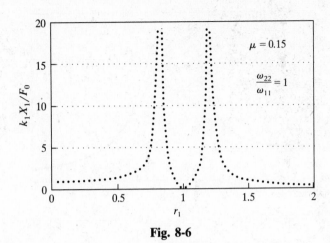

Fig. 8-6

8.6 DAMPED ABSORBERS

Damping may be added to a vibration absorber in order to alleviate two problems that exist for an undamped absorber:

(a) Since the lower natural frequency of the 2-degree-of-freedom system is less than the frequency to which the absorber is tuned, large amplitude transient vibrations occur during start-up.

(b) As illustrated in Fig. 8-6, the steady-state amplitude of the primary mass grows large for speeds slightly away from the tuned speed. Thus the absorber cannot be used when a machine operates at variable speeds.

When viscous damping is added to an absorber, as shown in Fig. 8-7, the steady-state amplitude of the primary mass is

$$X_1 = \frac{F_0}{k_1} \sqrt{\frac{(2\zeta r_1 q)^2 + (r_1^2 - q^2)^2}{\{r_1^4 - [1 + (1 + \mu)q^2]r_1^2 + q^2\} + (2\zeta r_1 q)^2[1 - r_1^2(1 + \mu)]^2}} \qquad (8.18)$$

where
$$q = \frac{\omega_{22}}{\omega_{11}}, \qquad \zeta = \frac{c}{2\sqrt{k_2 m_2}} \qquad (8.19)$$

Equation (8.18) is illustrated in Fig. 8-8 for several values of the parameters.

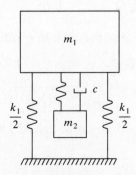

Fig. 8-7

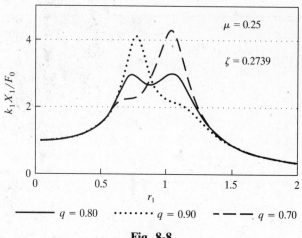

Fig. 8-8

The optimum damping ratio is defined as the damping ratio for which the peaks in the frequency response curve for the primary mass are approximately equal, leading to a wider operating range. This value is

$$\zeta_{\text{opt}} = \sqrt{\frac{3\mu}{8(1 + \mu)}} \tag{8.20}$$

8.7 HOUDAILLE DAMPERS

A *Houdaille damper,* illustrated in Fig. 8-9, is used in rotating devices such as engine crankshafts where absorption is needed over a wide range of speeds. The damper is inside a casing attached to the end of the shaft. The casing contains a viscous fluid and a mass that is free to rotate in the casing. If the shaft is subject to a harmonic moment of the form $M_0 \sin \omega t$, then when the Houdaille damper is added to the shaft, its steady-state amplitude of torsional oscillation is

$$\theta_1 = \frac{M_0}{k_t} \sqrt{\frac{4\zeta^2 + r^2}{4\zeta^2(r^2 + \mu r^2 - 1)^2 + (r^2 - 1)^2 r^2}} \tag{8.21}$$

where

$$r = \frac{\omega}{\sqrt{\frac{k_t}{J_1}}}, \qquad \zeta = \frac{c}{2J_2\sqrt{\frac{k_t}{J_1}}}, \qquad \mu = \frac{J_2}{J_1} \tag{8.22}$$

The optimum damping ratio is defined as the damping ratio for which the peak amplitude is smallest. It is determined as

$$\zeta_{\text{opt}} = \frac{1}{\sqrt{2(\mu + 1)(\mu + 2)}} \tag{8.23}$$

If the damping ratio of Eq. (8.23) is used in the design of a Houdaille damper,

$$\Theta_{\text{max}} = \frac{M_0}{k_t}\left(\frac{\mu + 2}{\mu}\right) \tag{8.24}$$

and occurs when

$$r = \sqrt{\frac{2}{2+\mu}} \qquad (8.25)$$

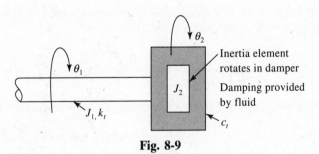

Inertia element
rotates in damper

Damping provided
by fluid

c_t

Fig. 8-9

8.8 WHIRLING

Whirling is a phenomenon that occurs when the center of mass of a rotor, attached to a rotating shaft, is not aligned with the axis of the shaft. The motion of the shaft and the eccentricity of the rotor cause an unbalanced inertia force in the rotor, pulling the shaft away from its centerline and causing it to bow. Whirling is illustrated in Fig. 8-10. For synchronous whirl, where the speed of the whirling is the same as the angular velocity of the shaft, the distance between the shaft's axis and its centerline, the amplitude of the whirl, is

$$\frac{X}{e} = \Lambda(r, \zeta) = \frac{r^2}{\sqrt{(1-r^2)^2 + (2\zeta r)^2}} \qquad (8.26)$$

where e is the eccentricity of the rotor, ζ is the damping ratio of the shaft, and $r = \omega/\omega_n$ where ω_n is the natural frequency of the rotor and shaft system.

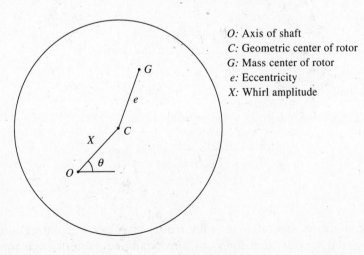

O: Axis of shaft
C: Geometric center of rotor
G: Mass center of rotor
e: Eccentricity
X: Whirl amplitude

Fig. 8-10

Solved Problems

8.1   What is the maximum stiffness of an undamped isolator to provide 81 percent isolation for a 200-kg fan operating at 1000 r/min?

For 81 percent isolation the maximum transmissibility ratio is 0.19. Using Eq. (8.6) with $\zeta = 0$ and noting that isolation only occurs when $r > \sqrt{2}$ lead to

$$0.19 \geq \frac{1}{r^2 - 1}$$

which is solved giving $r \geq 2.50$. The system's maximum allowable natural frequency is

$$\omega_n = \frac{\omega}{r_{min}} = \frac{1}{2.50}\left(1000\ \frac{\text{r}}{\text{min}}\right)\left(2\pi\ \frac{\text{rad}}{\text{r}}\right)\left(\frac{1\ \text{min}}{60\ \text{s}}\right) = 41.9\ \frac{\text{rad}}{\text{s}}$$

and thus the maximum isolator stiffness is

$$k = m\omega_n^2 = (200\ \text{kg})\left(41.9\ \frac{\text{rad}}{\text{s}}\right)^2 = 3.51 \times 10^5\ \frac{\text{N}}{\text{m}}$$

8.2   What is the minimum static deflection of an undamped isolator to provide 75 percent isolation to a pump that operates at speeds between 1500 and 2000 r/min?

For 75 percent isolation, the transmissibility ratio is 0.25. Then using Eq. (8.6) with $\zeta = 0$ and noting that isolation occurs only when $r > \sqrt{2}$ lead to

$$0.25 = \frac{1}{r^2 - 1}$$

whose solution is $r = 2.24$. From Fig. 8-3, is is noted that isolation is greater at higher speeds. Thus if 75 percent isolation is achieved at 1500 r/min, better than 75 percent isolation is achieved at higher speeds. Thus, the maximum natural frequency is

$$\omega_n = \frac{\omega}{r} = \frac{\left(1500\ \frac{\text{r}}{\text{min}}\right)\left(2\pi\ \frac{\text{rad}}{\text{r}}\right)\left(\frac{1\ \text{min}}{60\ \text{s}}\right)}{2.24} = 70.25\ \frac{\text{rad}}{\text{s}}$$

Then the minimum static deflection is

$$\Delta_{st} = \frac{g}{\omega_n^2} = \frac{9.81\ \frac{\text{m}}{\text{s}^2}}{\left(70.25\ \frac{\text{rad}}{\text{s}}\right)^2} = 1.99\ \text{mm}$$

8.3   A 150-kg sewing machine operates at 1200 r/min and has a rotating unbalance of 0.45 kg-m. What is the maximum stiffness of an undamped isolator such that the force transmitted to the machine's foundation is less than 2000 N?

The excitation frequency and magnitude are

$$\omega = \left(1200 \frac{r}{min}\right)\left(2\pi \frac{rad}{r}\right)\left(\frac{1 \ min}{60 \ s}\right) = 125.7 \frac{rad}{s}$$

$$F_0 = m_0 e \omega^2 = (0.45 \ \text{kg-m})\left(125.7 \frac{rad}{s}\right)^2 = 7.11 \times 10^3 \ \text{N}$$

The maximum transmissibility ratio such that the amplitude of the transmitted force is less than 2000 N is

$$T = \frac{F_T}{F_0} = \frac{2000 \ \text{N}}{7.11 \times 10^3 \ \text{N}} = 0.281$$

Application of Eq. (8.6) with $\zeta = 0$ and noting $T < 1$ only when $r > 1$ lead to

$$0.281 = \frac{1}{r^2 - 1} \quad \rightarrow \quad r = 2.134$$

The minimum natural frequency and maximum stiffness are calculated as

$$\omega_n = \frac{\omega}{r} = \frac{125.7 \frac{rad}{s}}{2.134} = 58.9 \frac{rad}{s}$$

$$k = m\omega_n^2 = (150 \ \text{kg})\left(58.9 \frac{rad}{s}\right)^2 = 5.20 \times 10^5 \frac{\text{N}}{\text{m}}$$

8.4 A 20-kg laboratory experiment is to be mounted to a table that is bolted to the floor in a laboratory. Measurements indicate that due to the operation of a nearby pump that operates at 2000 r/min, the table has a steady-state displacement of 0.25 mm. What is the maximum stiffness of an undamped isolator, placed between the experiment and the table such that the experiment's acceleration amplitude is less than 4 m/s^2?

The excitation frequency is 2000 r/min = 209.4 rad/s. The magnitude of the table's acceleration is

$$\omega^2 Y = \left(209.4 \frac{rad}{s}\right)^2 (0.00025 \ \text{m}) = 10.97 \frac{\text{m}}{\text{s}^2}$$

The required transmissibility ratio is

$$T = \frac{A_{max}}{\omega^2 Y} = \frac{4 \frac{\text{m}}{\text{s}^2}}{10.97 \frac{\text{m}}{\text{s}^2}} = 0.365$$

The minimum frequency ratio is calculated by

$$0.365 = \frac{1}{r^2 - 1} \quad \rightarrow \quad r = 1.93$$

The maximum natural frequency and isolator stiffness are

$$\omega_n = \frac{\omega}{r} = \frac{209.4 \frac{rad}{s}}{1.93} = 108.5 \frac{rad}{s}$$

$$k = m\omega_n^2 = (20 \ \text{kg})\left(108.5 \frac{rad}{s}\right)^2 = 2.35 \times 10^5 \frac{\text{N}}{\text{m}}$$

8.5   A 100-kg turbine operates at 2000 r/min. What percent isolation is achieved if the turbine

is mounted on four identical springs in parallel, each of stiffness 3×10^5 N/m?

The equivalent stiffness of the parallel spring combination is

$$k_{eq} = 4k = 4\left(3 \times 10^5 \frac{N}{m}\right) = 1.2 \times 10^6 \frac{N}{m}$$

When the turbine is placed on the springs, the system's natural frequency is

$$\omega_n = \sqrt{\frac{k}{m}} = \sqrt{\frac{1.2 \times 10^6 \frac{N}{m}}{100 \text{ kg}}} = 109.5 \frac{rad}{s}$$

Noting that 2000 r/min = 209.5 /rad/s, the frequency ratio is

$$r = \frac{\omega}{\omega_n} = \frac{209.4 \frac{rad}{s}}{109.5 \frac{rad}{s}} = 1.912$$

The transmissibility ratio is

$$T = \frac{1}{r^2 - 1} = \frac{1}{(1.912)^2 - 1} = 0.376$$

and thus the percentage isolation is

$$100(1 - T) = 62.4 \text{ percent}$$

8.6   What can be done to the turbine of Problem 8.5 to achieve 81 percent isolation if the
same mounting system is used?

In Problem 8.1 it is shown that 81 percent isolation requires a minimum frequency ratio of
2.50. Thus for the system of Problem 8.5, the maximum natural frequency is

$$\omega_n = \frac{\omega}{r} = \frac{209.4 \frac{rad}{s}}{2.50} = 83.8 \frac{rad}{s}$$

Since the same mountings as in Problem 8.5 are to be used, the natural frequency is decreased only
if the mass is increased. The required mass is

$$m = \frac{k}{\omega_n^2} = \frac{1.2 \times 10^6 \frac{N}{m}}{\left(83.8 \frac{rad}{s}\right)^2} = 170.9 \text{ kg}$$

Thus 81 percent isolation is achieved if 70.9 kg is added to the turbine.

8.7   List one negative and two beneficial effects of adding damping to an isolator.

From Fig. 8-3, it is seen that in the range of isolation $(r > \sqrt{2})$, the best isolation is achieved for an undamped isolator. Thus a negative effect of adding damping is to require a larger frequency ratio to achieve the same isolation.

Since the range of isolation occurs for $r > 1$, resonance is experienced during start-up. Adding damping decreases the maximum start-up amplitude.

The addition of viscous damping leads to smaller isolator displacements.

8.8 An isolator of damping ratio ζ is to be designed to achieve a transmissibility ratio $T < 1$. Derive an expression, in terms of ζ and T, for the smallest frequency ratio to achieve appropriate isolation.

The relation between T, r, and ζ is given by Eq. (8.6):

$$T = \sqrt{\frac{1 + 4\zeta^2 r^2}{r^4 + (4\zeta^2 - 2)r^2 + 1}}$$

Squaring and rearranging the previous equation leads to

$$r^4 + \left(4\zeta^2 \frac{T^2 - 1}{T^2} - 2\right)r^2 + \frac{T^2 - 1}{T^2} = 0$$

The preceding equation is quadratic in r^2. Use of the quadratic formula leads to

$$r^2 = 1 - 2\zeta^2\left(\frac{T^2 - 1}{T^2}\right) \pm \sqrt{\left[2\zeta^2\left(\frac{T^2 - 1}{T^2}\right) - 1\right]^2 - \left(\frac{T^2 - 1}{T^2}\right)}$$

Since $T < 1$, only the choice of the plus sign leads to a positive r^2. Hence the smallest allowable frequency ratio is

$$r = \sqrt{1 - 2\zeta^2\left(\frac{T^2 - 1}{T^2}\right) + \sqrt{\left[2\zeta^2\left(\frac{T^2 - 1}{T^2}\right) - 1\right]^2 - \left(\frac{T^2 - 1}{T^2}\right)}}$$

8.9 Solve Problem 8.1 as if the isolator had a damping ratio of 0.1.

Using the results of Problem 8.8 with $\zeta = 0.1$ and $T = 0.19$ leads to

$$r = \left(1 - 2(0.1)^2\left[\frac{(0.19)^2 - 1}{(0.19)^2}\right]\right.$$

$$\left. + \sqrt{\left\{1 - 2(0.1)^2\left[\frac{(0.19)^2 - 1}{(0.19)^2}\right]\right\}^2 - \left[\frac{(0.19)^2 - 1}{(0.19)^2}\right]}\right)^{1/2}$$

$$= 2.63$$

Thus the maximum natural frequency and maximum allowable stiffness are

$$\omega_n = \frac{\omega}{r} = \frac{\left(1000 \frac{r}{min}\right)\left(2\pi \frac{rad}{r}\right)\left(\frac{1\ min}{60\ s}\right)}{2.63} = 39.82 \frac{rad}{s}$$

$$k = m\omega_n^2 = (200\ kg)\left(39.82 \frac{rad}{s}\right)^2 = 3.17 \times 10^5 \frac{N}{m}$$

8.10   Solve Problem 8.3 as if the isolator had a damping ratio of 0.08.

From Problem 8.3 the required transmissibility ratio is 0.281. Thus using the results of Problem 8.8 with $\zeta = 0.08$ and $T = 0.281$ leads to

$$r = \left(1 - 2(0.08)^2\left[\frac{(0.281)^2 - 1}{(0.281)^2}\right]\right.$$

$$\left. + \sqrt{\left\{2(0.08)^2\left[\frac{(0.281)^2 - 1}{(0.281)^2}\right] - 1\right\}^2 - \left[\frac{(0.281)^2 - 1}{(0.281)^2}\right]}\right)^{1/2}$$

$$= 2.18$$

The maximum natural frequency and maximum isolator stiffness are calculated as

$$\omega_n = \frac{\omega}{r} = \frac{125.7\ \frac{\text{rad}}{\text{s}}}{2.18} = 57.7\ \frac{\text{rad}}{\text{s}}$$

$$k = m\omega_n^2 = (150\ \text{kg})\left(57.7\ \frac{\text{rad}}{\text{s}}\right)^2 = 4.99 \times 10^5\ \frac{\text{N}}{\text{m}}$$

8.11   What are the maximum start-up amplitude and the steady-state amplitude of the system of Problem 8.10?

Recall from Chap. 3 that the amplitude is related to the magnification factor by

$$X = \frac{F_0}{k}M(r, \zeta) = \frac{F_0}{k}\frac{1}{\sqrt{(1 - r^2)^2 + (2\zeta r)^2}}$$

Substituting values calculated in Problems 8.3 and 8.10 leads to a steady-state amplitude of

$$X = \frac{7.11 \times 10^3\ \text{N}}{4.99 \times 10^5\ \frac{\text{N}}{\text{m}}}\frac{1}{\sqrt{[1 - (2.18)^2]^2 + [2(0.08)(2.18)]^2}} = 3.78\ \text{mm}$$

The maximum value of the magnification factor is

$$M_{\text{max}} = \frac{1}{2\zeta\sqrt{1 - \zeta^2}}$$

Thus the maximum amplitude during start-up is

$$X_{\text{max}} = \frac{F_0}{k}M_{\text{max}} = \frac{7.11 \times 10^3\ \frac{\text{N}}{\text{m}}}{4.99 \times 10^5\ \frac{\text{N}}{\text{m}}}\frac{1}{2(0.08)\sqrt{1 - (0.08)^2}} = 8.93\ \text{cm}$$

8.12   Design an isolator by specifying k and ζ for the system of Problem 8.3 such that the maximum start-up amplitude is 30 mm and the maximum transmitted force is 3000 N.

The isolator is to be designed by specifying k and c. Two constraints must be satisfied: The maximum start-up amplitude is 30 mm, which leads to

$$0.03\ \text{m} > \frac{7.11 \times 10^3\ \frac{\text{N}}{\text{m}}}{2\zeta k\sqrt{1 - \zeta^2}} \qquad (8.27)$$

and the maximum transmitted force must be less than 3000 N, which leads to

$$3000\ \text{N} > 7.11 \times 10^3\ \text{N}\sqrt{\frac{1 + (2\zeta r)^2}{(1 - r^2)^2 + (2\zeta r)^2}} \qquad (8.28)$$

Equation (8.27) can be rewritten in terms of r by noting $k = m\omega^2/r$. The result is

$$0.03 > \frac{7.11 \times 10^3 \dfrac{\text{N}}{\text{m}}}{(150 \text{ kg})\left(125.7 \dfrac{\text{rad}}{\text{s}}\right)^2} \frac{r^2}{2\zeta\sqrt{1-\zeta^2}} \qquad (8.29)$$

Equations (8.28) and (8.29) must be simultaneously satisfied. There are many solutions to Eqs. (8.28) and (8.29) which can be obtained by trial and error. One solution is $r = 1.98$ and $\zeta = 0.20$, which leads to $X_{\max} = 0.03$ m and $F_T = 2998$ N. Thus

$$k = \frac{m\omega^2}{r} = 6.05 \times 10^5 \frac{\text{N}}{\text{m}}$$

8.13 Using the isolator designed in Problem 8.12, what, if any, mass should be added to the machine to limit its steady-state amplitude to 3 mm?

The steady-state amplitude is calculated by

$$X = \frac{F_0}{k} \frac{1}{\sqrt{(1-r^2)^2 + (2\zeta r)^2}}$$

When mass is added to the machine, for a fixed k, the natural frequency is decreased, and hence the frequency ratio is increased. Using the values calculated in Problem 8.12,

$$0.003 \text{ m} = \frac{7.11 \times 10^3 \text{ N}}{6.05 \times 10^5 \dfrac{\text{N}}{\text{m}}} \frac{1}{\sqrt{(1-r^2)^2 + [2(0.08)r]^2}}$$

which is solved for $r = 2.21$, leading to

$$m = \frac{k}{\omega_n^2} = \frac{kr^2}{\omega^2} = \frac{\left(6.05 \times 10^5 \dfrac{\text{N}}{\text{m}}\right)(2.21)^2}{\left(125.7 \dfrac{\text{rad}}{\text{s}}\right)^2} = 187.0 \text{ kg}$$

Hence 37.0 kg must be added to the machine.

8.14 Solve Problem 8.4 as if the isolator had a damping ratio of 0.13.

The required transmissibility ratio is determined in Problem 8.4 as $T = 0.365$. Using the equation developed in Problem 8.8 with $T = 0.365$ and $\zeta = 0.13$ leads to

$$r = \left(1 - 2(0.13)^2\left[\frac{(0.365)^2 - 1}{(0.365)^2}\right]\right.$$
$$\left. + \sqrt{\left\{1 - 2(0.13)^2\left[\frac{(0.365)^2 - 1}{(0.365)^2}\right]\right\}^2 - \left[\frac{(0.365)^2 - 1}{(0.365)^2}\right]}\right)^{1/2}$$
$$= 2.011$$

The maximum natural frequency and maximum isolator stiffness are determined as

$$\omega_n = \frac{\omega}{r} = \frac{209.4 \dfrac{\text{rad}}{\text{s}}}{2.011} = 104.1 \frac{\text{rad}}{\text{s}}$$

$$k = m\omega_n^2 = (20 \text{ kg})\left(104.1 \frac{\text{rad}}{\text{s}}\right)^2 = 2.17 \times 10^5 \frac{\text{N}}{\text{m}}$$

8.15 For the isolator designed in Problem 8.14, what is the steady-state amplitude of the experiment, and what is the maximum deformation in the isolator?

Using the theory of Chap. 3, the steady-state amplitude of the experiment is

$$X = YT(r, \zeta) = YT(2.011, 0.13) = (0.00025 \text{ m})(0.365)$$

$$= 9.13 \times 10^{-5} \text{ m}$$

The maximum deformation in the isolator is the same as the steady-state amplitude of the relative displacement between the experiment and the table:

$$Z = Y\Lambda(2.011, 0.13)$$

$$= (0.00025) \frac{(2.011)^2}{\sqrt{[1 - (2.011)^2]^2 + [2(0.13)(2.011)]^2}}$$

$$= 3.27 \times 10^{-4} \text{ m}$$

8.16 A 200-kg turbine operates at speeds between 1000 and 2000 r/min. The turbine has a rotating unbalance of 0.25 kg-m. What is the required stiffness of an undamped isolator such that the maximum force transmitted to the turbine's foundation is 1000 N?

The rotating unbalance provides a frequency squared excitation to the machine of the form

$$F_0 = m_0 e \omega^2$$

Thus the transmitted force is of the form

$$F_T = m_0 e \omega^2 T(r, \zeta)$$

As r increases above $\sqrt{2}$, $T(r, \zeta)$ decreases. However, since F_T is also proportional to ω^2, the transmitted force decreases with increasing ω until a minimum is reached.

In view of the above, if the isolator is designed such that sufficient isolation is achieved at the lowest operating speed, the transmitted force must be checked at the highest operating speed. To this end, at the lowest operating speed,

$$F_0 = m_0 e \omega^2 = (0.25 \text{ kg-m})\left(104.7 \frac{\text{rad}}{\text{s}}\right)^2 = 2740 \text{ N}$$

$$T = \frac{1000 \text{ N}}{2740 \text{ N}} = 0.365 = \frac{1}{r^2 - 1} \quad \rightarrow \quad r = 1.93$$

$$\omega_n = \frac{\omega}{r} = \frac{104.7 \frac{\text{rad}}{\text{s}}}{1.93} = 54.2 \frac{\text{rad}}{\text{s}}$$

Checking at the upper operating speed,

$$r = \frac{\omega}{\omega_n} = \frac{209.4 \frac{\text{rad}}{\text{s}}}{54.2 \frac{\text{rad}}{\text{s}}} = 3.86$$

$$F_T = m_0 e \omega^2 \frac{1}{r^2 - 1} = \frac{(0.25 \text{ kg-m})\left(209.4 \frac{\text{rad}}{\text{s}}\right)^2}{(3.86)^2 - 1} = 789 \text{ N}$$

Hence the isolator design is acceptable with

$$k = m\omega_n^2 = (200 \text{ kg})\left(54.2 \frac{\text{rad}}{\text{s}}\right)^2 = 5.88 \times 10^5 \frac{\text{N}}{\text{m}}$$

8.17 Repeat Problem 8.16 as if the isolator had a damping ratio of 0.1.

The solution procedure is as described in Problem 8.16. Setting $T(r, 0.1) = 0.365$, using the equation derived in Problem 8.8, leads to

$$r = \left(1 - 2(0.1)^2\left[\frac{(0.365)^2 - 1}{(0.365)^2}\right]\right.$$
$$\left. + \sqrt{\left\{2(0.1)^2\left[\frac{(0.365)^2 - 1}{(0.365)^2}\right] - 1\right\}^2 - \left[\frac{(0.365)^2 - 1}{(0.365)^2}\right]}\right)^{1/2}$$
$$= 1.98$$

$$\omega_n = \frac{\omega}{r} = \frac{104.7 \frac{\text{rad}}{\text{s}}}{1.98} = 52.9 \frac{\text{rad}}{\text{s}}$$

Checking the transmitted force at the upper operating speed,

$$r = \frac{\omega}{\omega_n} = \frac{209.4 \frac{\text{rad}}{\text{s}}}{52.9 \frac{\text{rad}}{\text{s}}} = 3.95$$

$$F_T = m_0 e \omega^2 T(3.95, 0.1)$$
$$= (0.25 \text{ kg-m})\left(209.4 \frac{\text{rad}}{\text{s}}\right)^2 \sqrt{\frac{1 + [2(0.1)(3.95)]^2}{[1 - (3.95)^2]^2 + [2(0.1)(3.95)]^2}}$$
$$= 955 \text{ N}$$

Hence the isolator design is acceptable with

$$k = m\omega_n^2 = (200 \text{ kg})\left(52.9 \frac{\text{rad}}{\text{s}}\right)^2 = 5.60 \times 10^5 \frac{\text{N}}{\text{m}}$$

8.18 Repeat Problem 8.17 as if the upper operating speed were 2500 r/min.

The transmitted force at the upper operating speed for the turbine with the isolator design of Problem 8.17 is calculated as

$$r = \frac{\omega}{\omega_n} = \frac{261.8 \frac{\text{rad}}{\text{s}}}{52.9 \frac{\text{rad}}{\text{s}}} = 4.95$$

$$F_T = m_0 e \omega^2 T(4.95, 0.1)$$
$$= (0.25 \text{ kg-m})\left(261.8 \frac{\text{rad}}{\text{s}}\right)^2 \sqrt{\frac{1 + [2(0.1)(4.95)]^2}{[1 - (4.95)^2]^2 + [2(0.1)(4.95)]^2}}$$
$$= 1025 \text{ N}$$

Thus this isolator is not acceptable. An isolator cannot be designed such that sufficient isolation is achieved over the entire operating range.

8.19 Measurement indicates that the peak components of the table vibration of Problem 8.4 are a 0.25-mm component at 100 rad/s and a 0.4-mm component at 150 rad/s. An available isolator has a stiffness of 8×10^4 N/m and a damping ratio of 0.1. Will the acceleration felt by the apparatus exceed 6 m/s² when this isolator is installed?

Let $r = 100/\omega_n$. Then $150/\omega_n = 1.5r$. An upper bound on the acceleration felt by the apparatus is

$$a \leq (0.00025 \text{ m})\left(100 \frac{\text{rad}}{\text{s}}\right)^2 T(r, 0.1)$$

$$+ (0.0004 \text{ m})\left(150 \frac{\text{rad}}{\text{s}}\right)^2 T(1.5r, 0.1)$$

$$= 2.5\sqrt{\frac{1 + [2(0.1)r]^2}{(1 - r^2)^2 + [2(0.1)r]^2}}$$

$$+ 9\sqrt{\frac{1 + [2(0.1)(1.5r)]^2}{[1 - (1.5r)^2]^2 + [2(0.1)(1.5r)]^2}}$$

$$= 2.5\sqrt{\frac{1 + 0.04r^2}{r^4 - 1.96r^2 + 1}} + 9\sqrt{\frac{1 + 0.09r^2}{5.0625r^4 - 4.41r^2 + 1}}$$

If the proposed isolator is used, then

$$\omega_n = \sqrt{\frac{k}{m}} = \sqrt{\frac{8 \times 10^4 \frac{\text{N}}{\text{m}}}{20 \text{ kg}}} = 63.2 \frac{\text{rad}}{\text{s}} \qquad r = \frac{100 \frac{\text{rad}}{\text{s}}}{63.2 \frac{\text{rad}}{\text{s}}} = 1.58$$

and the upper bound on the acceleration is calculated as 3.86 m/s². Thus the available isolator is sufficient.

8.20 Repeat Problem 8.1 as if the isolator damping is assumed to be hysteretic with a hysteretic damping coefficient of 0.2.

The appropriate form for the transmissibility ratio for a system with hysteretic damping is

$$T_h(r, h) = \sqrt{\frac{1 + h^2}{(1 - r^2)^2 + h^2}}$$

Thus for 81 percent isolation with an isolator of $h = 0.2$,

$$0.19 = \sqrt{\frac{1 + (0.2)^2}{(1 - r^2)^2 + (0.2)^2}} = \sqrt{\frac{1.04}{r^4 - 2r^2 + 1.04}}$$

$$(0.19)^2(r^4 - 2r^2 + 1.04) = 1.04$$

which is solved yielding $r = 2.52$. Hence the maximum allowable natural frequency is

$$\omega_n = \frac{\omega}{r} = \frac{104.7 \frac{\text{rad}}{\text{s}}}{2.52} = 41.5 \frac{\text{rad}}{\text{s}}$$

from which the maximum isolator stiffness is calculated as

$$k = m\omega_n{}^2 = (200 \text{ kg})\left(41.6 \frac{\text{rad}}{\text{s}}\right)^2 = 3.45 \times 10^5 \frac{\text{N}}{\text{m}}$$

Problems 8.12 through 8.25 refer to the following: During testing, a 150-kg model of an automobile is subject to a triangular pulse, whose force and displacement spectra are shown in Fig. 8-11.

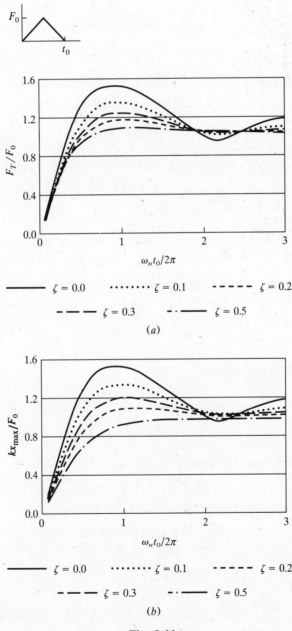

Fig. 8-11

8.21 If the model is mounted on an isolator of stiffness 5.4×10^5 N/m and damping ratio 0.1, what is the maximum transmitted force and maximum model displacement for a pulse of magnitude 2500 N and duration 0.12 s?

The system's natural frequency is

$$\omega_n = \sqrt{\frac{k}{m}} = \sqrt{\frac{5.4 \times 10^5\ \frac{N}{m}}{150\ kg}} = 60\ \frac{rad}{s}$$

The value of the parameter on the horizontal scale of the spectra is

$$\frac{\omega_n t_0}{2\pi} = \frac{\left(60\ \frac{rad}{s}\right)(0.12\ s)}{2\pi} = 1.15$$

From the force spectrum, $F_T/F_0 = 1.30$; hence

$$F_T = 1.30 F_0 = 1.30(2500\ N) = 3250\ N$$

From the displacement spectrum, $kx_{max}/F_0 = 1.35$; hence

$$x_{max} = 1.35 \frac{F_0}{k} = \frac{1.35(2500\ N)}{5.4 \times 10^5\ \frac{N}{m}} = 6.25\ mm$$

8.22 What is the maximum stiffness of an isolator such that the maximum transmitted force is less than 2000 N for a pulse of magnitude 2500 N and duration 0.12 s?

It is desired to set $F_T/F_0 = 2000/2500 = 0.8$. From the force spectra, this corresponds to a horizontal coordinate of 0.3. Hence

$$\frac{\omega_n t_0}{2\pi} = 0.3 \qquad \omega_n = \frac{0.3(2\pi)}{0.12\ s} = 15.71\ \frac{rad}{s}$$

The maximum allowable stiffness is

$$k = m\omega_n^2 = (150\ kg)\left(15.71\ \frac{rad}{s}\right)^2 = 3.70 \times 10^4\ \frac{N}{m}$$

8.23 What is the maximum displacement of the model with the isolator of Problem 8.22 installed?

For a horizontal coordinate of 0.3, $kx_{max}/F_0 = 0.8$; hence,

$$x_{max} = 0.8 \frac{F_0}{k} = \frac{0.8(2500\ N)}{3.70 \times 10^4\ \frac{N}{m}} = 0.054\ m$$

8.24 If the model is mounted on an isolator of stiffness 5.4×10^5 N/m and damping ratio 0.14, what is the maximum transmitted force if the pulse has a magnitude of 3000 N and a duration of 0.01 s?

The natural frequency for the model on this isolator is 60 rad/s; thus the natural period is

0.105 s. Since the pulse duration is much smaller than the natural period, a short duration pulse assumption is used. The magnitude of the impulse is the total area under the force time plot:

$$I = \int_0^{0.01\ s} F(t)\ dt = 2\tfrac{1}{2}(0.005\ \text{s})(3000\ \text{N}) = 15\ \text{N-s}$$

From Eq. (8.10), $Q(0.14) = 0.848$; thus,

$$F_T = 0.848 I \omega_n = 0.848(15\ \text{N-s})\left(60\ \frac{\text{rad}}{\text{s}}\right) = 7.63 \times 10^2\ \text{N}$$

8.25 What is the model's maximum displacement for the situation described in Problem 8.24?

The velocity induced by the application of the impulse is

$$v = \frac{I}{m} = \frac{15\ \text{N-s}}{150\ \text{kg}} = 0.1\ \frac{\text{m}}{\text{s}}$$

From Eq. (8.11), $E(0.14) = 1.39$. Note that the range of the inverse tangent function is taken from 0 to π. Then

$$x_{\max} = 1.39\ \frac{\frac{1}{2}mv^2}{F_T} = 1.39\ \frac{\frac{1}{2}(150\ \text{kg})\left(0.1\ \frac{\text{m}}{\text{s}}\right)^2}{763\ \text{N}} = 1.37\ \text{mm}$$

8.26 The 120-kg hammer of a 300-kg forge hammer is dropped from 1.3 m. Design an isolator for the hammer such that the maximum transmitted force is less than 15,000 N and the maximum displacement is a minimum.

The velocity of the hammer upon impact is

$$v_h = \sqrt{2gh} = \sqrt{2\left(9.81\ \frac{\text{m}}{\text{s}^2}\right)(1.3\ \text{m})} = 5.05\ \frac{\text{m}}{\text{s}}$$

The principle of impulse and momentum is used to determine the velocity of the machine induced by the impact:

$$v_m = \frac{m_h v_h}{m} = \frac{(120\ \text{kg})\left(5.05\ \frac{\text{m}}{\text{s}}\right)}{300\ \text{kg}} = 2.02\ \frac{\text{m}}{\text{s}}$$

For a specified transmitted force, the minimum maximum displacement is attained by choosing $\zeta = 0.4$. Then the required natural frequency is obtained by

$$\frac{F_T}{mv\omega_n} = Q(0.4)$$

$$\omega_n = \frac{F_T}{mvQ(0.4)} = \frac{15,000\ \text{N}}{(300\ \text{kg})\left(2.02\ \frac{\text{m}}{\text{s}}\right)(0.88)} = 28.1\ \frac{\text{rad}}{\text{s}}$$

Thus the maximum allowable stiffness is

$$k = m\omega_n^2 = (300\ \text{kg})\left(28.1\ \frac{\text{rad}}{\text{s}}\right)^2 = 2.37 \times 10^5\ \frac{\text{N}}{\text{m}}$$

8.27 A 200-kg machine is attached to a spring of stiffness 4×10^5 N/m. During operation the machine is subjected to a harmonic excitation of magnitude 500 N and frequency 50 rad/s. Design an undamped vibration absorber such that the steady-state amplitude of the primary mass is zero and the steady-state amplitude of the absorber mass is less than 2 mm.

The steady-state amplitude of the machine is zero when the absorber is tuned to the excitation frequency. Thus

$$r_2 = 1 \quad \rightarrow \quad \omega_{22} = \omega \quad \rightarrow \quad \sqrt{\frac{k_2}{m_2}} = \omega$$

When this occurs, the steady-state amplitude of the absorber mass is given by Eq. (8.17). Thus

$$0.002 \text{ m} \geq \frac{F_0}{k_2} \quad \rightarrow \quad k_2 \geq \frac{500 \text{ N}}{0.002 \text{ m}} = 2.5 \times 10^5 \frac{\text{N}}{\text{m}}$$

Using the minimum allowable stiffness, the required absorber mass is

$$m_2 = \frac{k_2}{\omega^2} = \frac{2.5 \times 10^5 \dfrac{\text{N}}{\text{m}}}{\left(50 \dfrac{\text{rad}}{\text{s}}\right)^2} = 100 \text{ kg}$$

Thus an absorber of stiffness 2.5×10^5 N/m and mass 100 kg can be used.

8.28 What are the natural frequencies of the system of Problem 8.27 with the absorber in place?

The natural frequencies of the 2-degree-of-freedom system with the absorber in place are the vales of ω such that the denominator of Eq. (8.12) is zero. Thus, noting that the mass ratio is $\mu = 100/200 = 0.5$,

$$r_1^2 r_2^2 - r_2^2 - 1.5 r_1^2 + 1 = 0$$

$$\frac{\omega^4}{\omega_{11}^2 \omega_{22}^2} - \omega^2 \left(\frac{1.5}{\omega_{11}^2} + \frac{1}{\omega_{22}^2} \right) + 1 = 0$$

$$\omega^4 - (1.5\omega_{22}^2 + \omega_{11}^2) + \omega_{11}^2 \omega_{22}^2 = 0$$

It is noted that $\omega_{22} = 50$ rad/s, and

$$\omega_{11} = \sqrt{\frac{k}{m_1}} = \sqrt{\frac{4 \times 10^5 \dfrac{\text{N}}{\text{m}}}{200 \text{ kg}}} = 44.72 \frac{\text{rad}}{\text{s}}$$

Substitution of these values leads to

$$\omega^4 - 5.75 \times 10^3 \omega^2 + 5 \times 10^6 = 0$$

which is solved for ω^2 using the quadratic equation. Taking the positive square roots of the roots leads to

$$\omega_1 = 32.698 \frac{\text{rad}}{\text{s}}, \qquad \omega_2 = 68.42 \frac{\text{rad}}{\text{s}}$$

8.29 A piping system experiences resonance when the pump supplying power to the system operates at 500 r/min. When a 5-kg absorber tuned to 500 r/min is added to the pipe, the system's new natural frequencies are measured as 380 and 624 r/min. What is the natural frequency of the piping system and its equivalent mass?

The system has natural frequencies corresponding to values of ω that makes the denominator of Eq. (8.12) zero. Using the definitions of Eqs. (8.14) and (8.15), this leads to

$$\omega^4 - [\omega_{11}^{\;2} + (1+\mu)\omega_{22}^{\;2}]\omega^2 + \omega_{11}^{\;2}\omega_{22}^{\;2} = 0 \qquad (8.30)$$

Noting that

$$\omega_{22} = 500 \text{ r/min} = 52.4 \; \frac{\text{rad}}{\text{s}}$$

and applying Eq. (8.30) with $\omega = 380$ r/min $= 39.8$ rad/s leads to

$$2.51 \times 10^6 + 1.17 \times 10^3 \omega_{11}^{\;2} - 4.35 \times 10^6 (1 + \mu) = 0$$

Application of Eq. (8.30) for $\omega = 624$ r/min $= 65.3$ rad/s leads to

$$1.82 \times 10^7 - 1.51 \times 10^3 \omega_{11}^{\;2} - 1.17 \times 10^7 (1 + \mu) = 0$$

Simultaneous solution of the previous two equations leads to

$$\omega_{11} = 49.2 \; \frac{\text{rad}}{\text{s}} \qquad \mu = 0.225$$

Hence the piping system's natural frequency is 49.2 rad/s, and its equivalent mass is

$$m_1 = \frac{m_2}{\mu} = \frac{5 \text{ kg}}{0.225} = 22.2 \text{ kg}$$

8.30 Redesign the absorber used in Problem 8.29 such that the system's natural frequencies are less than 350 r/min and greater than 650 r/min.

Applying Eq. (8.30) of Problem 8.29 with $\omega = 350$ r/min $= 36.7$ rad/s, $\omega_{11} = 49.2$ rad/s, and $\omega_{22} = 52.4$ rad/s leads to $\mu = 0.414$. Applying Eq. (8.30) of Problem 8.29 with $\omega = 650$ r/min $= 68.1$ rad/s, with the same values of ω_{11} and ω_{22} leads to $\mu = 0.330$. Thus in order for the natural frequencies of the system with the absorber added to be less than 350 r/min and greater than 650 r/min requires an absorber mass of at least

$$m_2 = \mu m_1 = (0.414)(22.2 \text{ kg}) = 9.19 \text{ kg}$$

Then the absorber stiffness is

$$k_2 = m_2 \omega_{22}^{\;2} = (9.19 \text{ kg})\left(52.4 \; \frac{\text{rad}}{\text{s}}\right)^2 = 2.52 \times 10^4 \; \frac{\text{N}}{\text{m}}$$

8.31 A 100-kg machine is placed at the midspan of a simply supported beam of length 3 m, elastic modulus 200×10^9 N/m^2, and moment of inertia 1.3×10^{-6} m^4. During operation the machine is subjected to a harmonic excitation of magnitude 5000 N at speeds between 600 and 700 r/min. Design an undamped vibration absorber such that the machine's steady-state amplitude is less than 3 mm at all operating speeds.

The beam's stiffness is

$$k = \frac{48EI}{L^3} = \frac{48\left(200 \times 10^9 \; \frac{\text{N}}{\text{m}^2}\right)(1.3 \times 10^{-6} \text{ m}^4)}{(3 \text{ m})^3} = 4.62 \times 10^5 \; \frac{\text{N}}{\text{m}}$$

The system's natural frequency is

$$\omega_n = \sqrt{\frac{k_1}{m_1}} = \sqrt{\frac{4.62 \times 10^5 \, \dfrac{\text{N}}{\text{m}}}{100 \text{ kg}}} = 68.0 \, \frac{\text{rad}}{\text{s}}$$

Assume that steady-state vibrations are to be eliminated at this speed; then,

$$\omega_{11} = \omega_{22} = 68.0 \, \frac{\text{rad}}{\text{s}}$$

Note that in Eq. (8.12), for $r < 1$, the numerator is positive and the denominator is negative; hence, in order to enforce $X_1 < 3$ mm for $\omega = 600$ r/min $= 62.8$ rad/s with $r_1 = r_2 = 62.8/68.0 = 0.923$,

$$-0.003 \text{ m} = \frac{5000 \text{ N}}{4.62 \times 10^5 \, \dfrac{\text{N}}{\text{m}}} \cdot \frac{1 - (0.923)^2}{(0.923)^2(0.923)^2 - (0.923)^2 - (1 + \mu)(0.923)^2 + 1}$$

which is solved for $\mu = 0.652$. For $r_2 > 1$, both the numerator and denominator of Eq. (8.12) are negative. Hence for $\omega = 700$ r/min $= 73.3$ rad/s and $r_1 = r_2 = 73.3/68.0 = 1.078$,

$$0.003 \text{ m} = \frac{5000 \text{ N}}{4.62 \times 10^5 \, \dfrac{\text{N}}{\text{m}}} \cdot \frac{1 - (1.078)^2}{(1.078)^2(1.078)^2 - (1.078)^2 - (1 + \mu)(1.078)^2 + 1}$$

which is solved for $\mu = 0.525$. Since the mass ratios calculated represent the minimum mass ratios for the amplitude to be less than 3 mm at the limits of the operating range, the larger mass ratio must be chosen. Hence

$$\mu = 0.652, \qquad m_2 = \mu m_1 = (0.652)(100 \text{ kg}) = 65.2 \text{ kg}$$

$$k_2 = m_2 \omega_{22}^2 = (65.2 \text{ kg})\left(68.0 \, \frac{\text{rad}}{\text{s}}\right)^2 = 3.014 \times 10^5 \, \frac{\text{N}}{\text{m}}$$

8.32 If an optimally designed damped vibration absorber is used on the system of Problem 8.31 with a mass ratio of 0.25, what is the machine's steady-state amplitude at 600 r/min?

Mathcad

The optimum absorber tuning is obtained from Eq. (8.19) as

$$q = \frac{1}{1 + \mu} = \frac{1}{1 + 0.25} = 0.8$$

The optimum damping ratio is calculated using Eq. (8.20):

$$\zeta = \sqrt{\frac{3\mu}{8(1 + \mu)}} = \sqrt{\frac{3(0.25)}{8(1 + 0.25)}} = 0.274$$

Using Eq. (8.18) with these values and $r_1 = 0.923$, $F_0 = 5000$ N, and $k_1 = 4.62 \times 10^5$ N/m leads to $X_1 = 2.9$ cm.

8.33 A 300-kg machine is placed at the end of a cantilever beam of length 1.8 m, elastic modulus 200×10^9 N/m^2, and moment of inertia 1.8×10^{-5} m^4. When the machine operates at 1000 r/min, it has a steady-state amplitude of 0.8 mm. What is the machine's steady-state amplitude when a 30-kg absorber of damping coefficient 650 N-s/m and stiffness 1.5×10^5 N/m is added to the end of beam?

The beam's stiffness is

$$k = \frac{3EI}{L^3} = \frac{3\left(200 \times 10^9 \frac{N}{m^2}\right)(1.8 \times 10^{-5} \text{ m}^4)}{(1.8 \text{ m})^3} = 1.85 \times 10^6 \frac{N}{m}$$

and the system's natural frequency is

$$\omega_{11} = \sqrt{\frac{k}{m}} = \sqrt{\frac{1.85 \times 10^6 \frac{N}{m}}{300 \text{ kg}}} = 78.5 \frac{rad}{s}$$

The frequency ratio for $\omega = 1000$ r/min = 104.7 rad/s is

$$r_1 = \frac{\omega}{\omega_{11}} = \frac{104.7 \frac{rad}{s}}{78.5 \frac{rad}{s}} = 1.33$$

The excitation amplitude is calculated from knowledge of the steady-state amplitude before the absorber is added:

$$F_0 = k_2 X_1(r_1^2 - 1) = \left(1.85 \times 10^6 \frac{N}{m}\right)(0.0008 \text{ m})[(1.33)^2 - 1]$$

$$= 1.14 \times 10^3 \text{ N}$$

The natural frequency of the absorber is

$$\omega_{22} = \sqrt{\frac{k_2}{m_2}} = \sqrt{\frac{1.5 \times 10^5 \frac{N}{m}}{30 \text{ kg}}} = 70.7 \frac{rad}{s}$$

Thus the parameters of the absorber design are

$$\mu = \frac{m_2}{m_1} = \frac{30 \text{ kg}}{300 \text{ kg}} = 0.1 \qquad q = \frac{\omega_{22}}{\omega_{11}} = \frac{70.7 \frac{rad}{s}}{78.5 \frac{rad}{s}} = 0.90$$

$$\zeta = \frac{c}{2\sqrt{m_2 k_2}} = \frac{650 \frac{N\text{-}s}{m}}{2\sqrt{\left(1.5 \times 10^5 \frac{N}{m}\right)(30 \text{ kg})}} = 0.153$$

Application of Eq. (8.18) with these values leads to $X_1 = 9.08 \times 10^{-4}$ m.

8.34 An engine has a moment of inertia of 3.5 kg-m² and a natural frequency of 100 Hz. Design a Houdaille damper such that the engine's maximum magnification factor is 4.8.

If the optimum damper design is used, then setting the maximum magnification factor to 4.8 and using Eq. (8.24) leads to

$$4.8 = \frac{\mu + 2}{\mu} \qquad \mu = 0.526$$

The optimum damping ratio is determined from Eq. (8.23) as

$$\zeta = \frac{1}{\sqrt{2(1.526)(2.526)}} = 0.360$$

The Houdaille damper parameters are determined from Eq. (8.22) as

$$J_2 = \mu J_1 = (0.526)(3.5 \text{ kg-m}^2) = 1.84 \text{ kg-m}^2$$

$$c = 2\zeta J_2 \omega_1 = 2(0.360)(1.84 \text{ kg-m}^2)\left(100 \frac{\text{cycle}}{\text{s}}\right)\left(\frac{2\pi \text{ rad}}{\text{cycle}}\right)$$

$$= 832 \frac{\text{N-s}}{\text{m}}$$

8.35 During operation the engine of Problem 8.34 is subjected to a harmonic torque of

magnitude 100 N-m at a frequency of 110 Hz. What is the engine's steady-state amplitude when the Houdaille damper designed in Problem 8.34 is used?

The frequency ratio is

$$r = \frac{\omega}{\omega_1} = \frac{110 \text{ Hz}}{100 \text{ Hz}} = 1.1$$

Thus from Eq. (8.21),

$$\Theta_1 = \frac{M_0}{J_1\omega_1^{2}} \sqrt{\frac{4\zeta^2 + r^2}{4\zeta^2(r^2 + \mu r^2 - 1)^2 + (r^2 - 1)^2 r^2}}$$

$$= \frac{100 \text{ N-m}}{(3.5 \text{ kg-m}^2)\left[100(2\pi) \frac{\text{rad}}{\text{s}}\right]^2} \sqrt{\frac{4(0.360)^2 + (1.1)^2}{4(0.360)^2[(1.1)^2 + 0.536(1.1)^2 - 1]^2 + [(1.1)^2 - 1]^2(1.1)^2}}$$

$$= 1.4 \times 10^{-4} \text{ rad}$$

8.36 A 40-kg rotor has an eccentricity of 1.2 cm. It is mounted on a shaft and bearing system whose stiffness is 3.2×10^5 N/m and has a damping ratio of 0.07. What is the amplitude of whirling when the rotor operates at 1000 r/min?

The shaft's natural frequency is

$$\omega_n = \sqrt{\frac{k}{m}} = \sqrt{\frac{3.2 \times 10^5 \frac{\text{N}}{\text{m}}}{40 \text{ kg}}} = 89.4 \frac{\text{rad}}{\text{s}}$$

The frequency ratio is

$$r = \frac{\omega}{\omega_n} = \frac{\left(1000 \frac{\text{r}}{\text{min}}\right)\left(2\pi \frac{\text{rad}}{\text{r}}\right)\left(\frac{1 \text{ min}}{60 \text{ s}}\right)}{89.4 \frac{\text{rad}}{\text{s}}} = 1.17$$

The amplitude of whirling is calculated using Eq. (8.27):

$$X = \frac{(0.012 \text{ m})(1.17)^2}{\sqrt{[1 - (1.17)^2]^2 + [2(0.07)(1.17)]^2}} = 4.07 \text{ cm}$$

8.37 An engine flywheel has an eccentricity of 1.2 cm and mass of 40 kg. Assuming a damping ratio of 0.05, what is the necessary stiffness of its bearings to limit its whirl amplitude to 1.2 mm at all speeds between 1000 and 2000 r/min?

The maximum allowable value of Λ is

$$\Lambda_{max} = \frac{X_{max}}{e} = \frac{1.2 \text{ mm}}{1.2 \text{ cm}} = 0.1$$

Then using Eq. (8.26),

$$0.1 < \frac{r^2}{\sqrt{r^4 - 1.99r^2 + 1}}$$
$$r < 0.302$$

Thus at all operating speeds, $r < 0.302$. Thus since the largest operating speed is 2000 r/min = 209.4 rad/s

$$\omega_n > \frac{209.4 \frac{\text{rad}}{\text{s}}}{0.302} = 693 \frac{\text{rad}}{\text{s}}$$

Hence the minimum bearing stiffness is

$$k = m\omega_n^2 = (40 \text{ kg})\left(693 \frac{\text{rad}}{\text{s}}\right)^2 = 1.93 \times 10^7 \frac{\text{N}}{\text{m}}$$

Supplementary Problems

8.38 What is the maximum stiffness of an undamped isolator to provide 81 percent isolation to a 350-kg sewing machine when it operates at 2100 r/min?

Ans. 2.70×10^6 N/m

8.39 What is the maximum stiffness of an undamped isolator to provide 70 percent isolation to a 200-kg pump that operates at speeds between 1000 and 1500 r/min?

Ans. 5.06×10^5 N/m

8.40 Repeat Problem 8.38 for an isolator with a damping ratio of 0.1.

Ans. 2.44×10^6 N/m

8.41 Repeat Problem 8.39 for an isolator with a damping ratio of 0.08.

Ans. 4.87×10^5 N/m

8.42 A 50-kg compressor operates at 200 rad/s. The only available isolator has a stiffness of 1.3×10^6 N/m and a damping ratio of 0.1. What is the minimum mass that must be added to the compressor to provide 68 percent isolation?

Ans. 91.7 kg

8.43 During operation at 1000 r/min, a 200-kg tumbler produces a harmonic force of magnitude

5000 N. What is the minimum static deflection of an isolator of damping ratio 0.12 such that the transmitted force is less than 2000 N?

Ans.   3.33 mm

8.44   What is the steady-state amplitude of the system of Problem 8.43 when the isolator with the minimum static deflection is installed?

Ans.   3.08 mm

8.45   A 15-kg flow meter is mounted on a table in a laboratory. Measurements indicate that the dominant frequency of surrounding vibrations is 250 rad/s. The amplitude at this frequency is 0.8 mm. What is the maximum stiffness of an isolator of damping ratio 0.1 such that the acceleration transmitted to the flow meter is 5 m/s²?

Ans. 7.01×10^4 N/m

8.46   A 60-kg engine operates at 2000 r/min and has a rotating unbalance of 0.2 kg-m. Can an isolator of damping ratio 0.1 be designed to limit the transmitted force to 1000 N and the steady-state amplitude to 3 mm?

Ans.   No, the maximum allowable isolator stiffness to limit the transmitted force to 1000 N is 2.27×10^5 N/m. If the stiffness is reduced below this value, the steady-state amplitude will always be greater than 3 mm.

8.47   What is the minimum mass that can be added to the engine of Problem 8.46 such that a steady-state amplitude of 3 mm can be attained when a transmitted force of 1000 N is attained using an isolator of damping ratio 0.1?

Ans.   12.8 kg

8.48   What is the maximum stiffness of an isolator of damping ratio 0.1 that limits the transmitted force to 1000 N when 12.8 is added to the engine of Problem 8.47?

Ans. 2.75×10^5 N/m

8.49   What is the maximum start-up amplitude of the system of Problem 8.48.

Ans.   0.16 m

8.50   Repeat Problem 8.4 as if the isolator had hysteretic damping with a damping coefficient of 0.15.

Ans. 2.33×10^5 N/m

8.51   What is the maximum stiffness of an isolator of damping ratio 0.1 such that the acceleration felt by the apparatus of Problem 8.19 is less than 6 m/s²?

Ans. 1.08×10^5 N/m

The systems of Problems 8.52 through 8.56 are subject to a pulse of the form of Fig. 8-12. The force and displacement spectra for this type of pulse are given in Fig. 8-13.

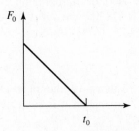

Fig. 8-12

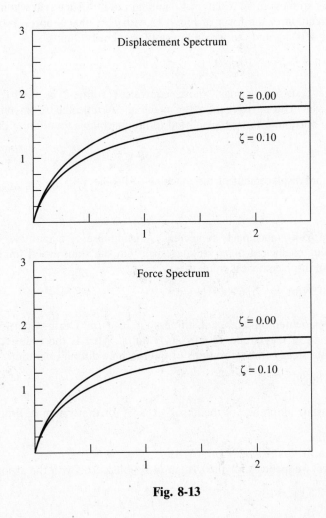

Fig. 8-13

8.52   A 50-kg machine is mounted on four parallel springs, each of stiffness 3×10^5 N/m. What is the maximum transmitted force when the machine is subject to an excitation of the form of Fig. 8-12 with $F_0 = 1200$ N and $t_0 = 0.05$ s?

Ans.   2040 N

8.53   What is the maximum displacement of the machine of Problem 8.52?

   Ans.   1.7 mm

8.54   What is the maximum stiffness of an isolator of damping ratio 0.1 such that the maximum transmitted force for a 100-kg machine is 1125 N when it is subjected to the excitation of Fig. 8-12 with $F_0 = 1500$ N and $t_0 = 0.04$ s?

   Ans. 2.22×10^5 N

8.55   What is the minimum stiffness of an isolator of damping ratio 0.1 such that the maximum displacement of a 150-kg machine is 2.2 mm when it is subjected to an excitation of the form of Fig. 8-12 with $F_0 = 2000$ N and $t_0 = 0.06$ s?

   Ans. 1×10^6 N/m

8.56   What is the range of stiffness of an isolator of damping ratio 0.1 such that when a 200-kg machine is subjected to an excitation of the form of Fig. 8-12 with $F_0 = 2000$ N and $t_0 = 0.05$ s, the maximum transmitted force is 1500 kg and the maximum displacement is 6 mm?

   Ans. 2×10^5 N/m $< k < 2.84 \times 10^5$ N/m

8.57   A 200-kg machine rests on springs whose equivalent stiffness is 1×10^5 N/m and damping coefficient 1500 N-s/m. During operation the machine is subjected to an impulse of magnitude 75 N-s. What is the maximum force transmitted to the machine's foundation due to the impulse?

   Ans. 1.59×10^3 N

8.58   What is the maximum displacement of the machine of Problem 8.57?

   Ans.   11.62 mm

8.59   During operation a 65-kg machine is subjected to an impulse of magnitude 100 N-s. Specify the stiffness and damping coefficient of an isolator such that the transmitted force is 4000 N and the machine's maximum displacement is minimized.

   Ans. $k = 1.34 \times 10^5$ N/m, $c = 2.36 \times 10^4$ N-s/m

8.60   A 50-kg machine is mounted on a table of stiffness 1×10^5 N/m. During operation it is subjected to a harmonic excitation of magnitude 1200 N at 45 rad/s. What is the required stiffness of a 5-kg absorber to eliminate steady-state vibrations of the machine during operation?

   Ans. 1.01×10^4 N/m

8.61   What is the steady-state amplitude of the absorber mass for the system of Problem 8.60?

   Ans.   11.9 cm

8.62   What are the natural frequencies for the system of Problem 8.60 with the absorber in place?

   Ans.   38.3 rad/s, 52.5 rad/s

8.63   For what range of frequencies near 45 rad/s is the steady-state amplitude of the machine of Problem 8.60 less than 5 mm when the absorber is in place?

   Ans.   44.1 rad/s $\leq \omega \leq 45.9$ rad/s

8.64   When a 10-kg undamped absorber tuned to 100 rad/s is added to a 1-degree-of-freedom structure of stiffness 5×10^6 N/m, the lowest natural frequency of the structure is 85.44 rad/s. What is the higher natural frequency of the structure?

Ans.   103 rad/s

8.65   A 15-kg undamped absorber tuned to 250 rad/s is added to a 150-kg machine mounted on a foundation of stiffness 1×10^7 N/m. At 250 rad/s, the amplitude of the absorber mass is 3.9 mm. What is the amplitude of the machine at 275 rad/s?

Ans. 9.01×10^{-4} m

8.66   A 50-kg machine is placed at the midspan of a 1.5-m simply supported beam of elastic modulus 210×10^9 N/m^2 and moment of inertia 1.5×10^{-6} m^4. When running at 3000 r/min, the machine's steady-state amplitude is measured as 1.2 cm. Design an undamped absorber such that the steady-state amplitude is less than 2 mm at all speeds between 2900 and 3100 r/min.

Ans.   A nonunique design is 2.77 kg, 2.77×10^9 N/m.

8.67   If an optimally designed 15-kg damped vibration absorber is used for the system of Problem 8.66, what is the steady-state amplitude of the machine when operating at 3000 r/min?

Ans.   3.39 mm

8.68   A 20-kg machine is mounted on a foundation of stiffness 1.3×10^6 N/m. What are the stiffness and damping coefficient of an optimally designed 4-kg damped vibration absorber?

Ans. 1.81×10^4 N/m, 134.3 N-s/m

8.69   With the absorber designed in Problem 8.68 in place, what is the steady-state amplitude of the machine when operating at 85 rad/s if the machine has a rotating unbalance of 0.5 kg-m?

Ans.   0.0925 m

8.70   A 110-kg machine is subjected to an excitation of magnitude 1500 N. The machine is mounted on a foundation of stiffness 3×10^6 N/m. What are the mass and damping coefficient of an optimally designed vibration damper such that the maximum amplitude is 3 mm?

Ans.   44 kg, 5624 N-s/m

8.71   What is the steady-state amplitude at 180 rad/s of the machine of Problem 8.70 when the optimally designed vibration damper is added?

Ans.   1.21 mm

8.72   An engine has a mass moment of inertia of 3.5 kg-m^2 and is mounted on a shaft of stiffness 1.45×10^5 N-m/rad. If the applied moment has a magnitude of 1000 N-m, what is the engine's steady-state amplitude at 2000 r/min when an optimally designed Houdaille damper of mass moment of inertia 1.1 kg-m^2 is added?

Ans.   1.60°

8.73   The center of gravity of a 12-kg rotor is 1.2 cm from its geometric center. The rotor is mounted on a shaft and spring-loaded bearings of stiffness 1.4×10^5 N/m. Assuming a damping ratio of 0.05, what is the amplitude of whirling when the rotor operates at 1500 r/min?

Ans.   2.26 cm

Chapter 9

Finite Element Method

The *finite element method* is used to provide discrete approximations to the vibrations of continuous systems. The finite element method is an application of the Rayleigh-Ritz method with the continuous system broken down into a finite number of discrete elements. The displacement function is assumed piecewise over each element. The displacement functions are chosen to satisfy *geometric boundary conditions* (i.e., displacements and slopes) and such that necessary continuity is attained between elements. It is sufficient to require displacement continuity for bars, while displacements and slopes must be continuous across element boundaries for beams.

9.1 GENERAL METHOD

Let ℓ be the length of an element. Define the *local coordinate* ξ: $0 \leq \xi \leq \ell$. Let $u(\xi, t)$ be the element displacement, chosen to satisfy appropriate continuity. If $u_1, u_2, \ldots, u_k$ represent the degrees of freedom for the element (end displacements, slopes, etc.), then

$$u(\xi, t) = \sum_{i=1}^{k} \phi_i(\xi) u_i(t) \tag{9.1}$$

where the $\phi_i(\xi)$ are called the *shape functions*. The potential energy for the element is calculated using Eq. (9.1) for the displacement and has the quadratic form

$$V = \tfrac{1}{2} \mathbf{u}^T \mathbf{k} \mathbf{u} \tag{9.2}$$

where $\mathbf{u} = [u_1 \quad u_2 \quad \cdots \quad u_k]^T$ and $\mathbf{k}$ is the derived *local stiffness matrix* or *element stiffness matrix*. The kinetic energy for the element is calculated using Eq. (9.1) and has the quadratic form

$$T = \tfrac{1}{2} \dot{\mathbf{u}}^T \mathbf{m} \dot{\mathbf{u}} \tag{9.3}$$

where $\mathbf{m}$ is the *local mass matrix* or *element mass matrix*.

The total number of degrees of freedom in the finite element model is $n =$ (number of elements)(number of degrees of freedom per element) − number of geometric boundary conditions. Define the *global displacement vector* $\mathbf{U} = [U_1 \quad U_2 \quad \cdots \quad U_n]^T$ where U_1, $U_2, \ldots, U_n$ represent the nonspecified model displacements. The total potential energy of the system has the quadratic form

$$V = \tfrac{1}{2} \mathbf{U}^T \mathbf{K} \mathbf{U} \tag{9.4}$$

where $\mathbf{K}$ is the *global stiffness matrix*, obtained by proper assembly of the local stiffness matrices. The total kinetic energy of the system has the quadratic form

$$T = \tfrac{1}{2} \dot{\mathbf{U}}^T \mathbf{M} \dot{\mathbf{U}} \tag{9.5}$$

where $\mathbf{M}$ is the *global mass matrix*, obtained by proper assembly of the local mass matrices.

The differential equations approximating the free vibrations of the continuous systems are written as

$$\mathbf{M} \ddot{\mathbf{U}} + \mathbf{K} \mathbf{U} = \mathbf{0} \tag{9.6}$$

The finite element approximations to the natural frequencies and mode shapes are obtained using the methods of Chap. 5. That is, the natural frequency approximations are the square roots of the eigenvalues of $\mathbf{M}^{-1}\mathbf{K}$, and the mode shapes are developed from their eigenvectors.

9.2 FORCED VIBRATIONS

If $F(x, t)$ represents the time-dependent external force applied to a continuous system, then the virtual work done by the external force due to variations in the global displacements is

$$\delta W = \int_0^\ell F(x, t)\, \delta u(x, t)\, dx$$

$$= \sum_{i=1}^{n} f_i(t)\, \delta U_i \tag{9.7}$$

Lagrange's equations are used to write the approximate differential equations in the form

$$\mathbf{M\ddot{U}} + \mathbf{KU} = \mathbf{F} \tag{9.8}$$

where $\mathbf{F} = [f_1 \quad f_2 \quad \cdots \quad f_n]^T$. The methods of Chap. 6 (modal analysis, etc.) can be used to approximate the system's forced response.

9.3 BAR ELEMENT

The local degrees of freedom for a bar element are the displacements of the ends of the element. Following Fig. 9-1, let u_1 be the displacement of the left end ($\xi = 0$) and u_2 be the displacement of the right end of the element ($\xi = \ell$). Then a finite element approximation for the bar element is

$$u(\xi, t) = \left(1 - \frac{\xi}{\ell}\right)u_1(t) + \frac{\xi}{\ell}u_2(t) \tag{9.9}$$

The potential energy of the element is

$$V = \frac{1}{2}\int_0^\ell EA\left(\frac{\partial u}{\partial \xi}\right)^2 d\xi = \frac{1}{2}\frac{EA}{\ell}\left(u_1^2 - 2u_1u_2 + u_2^2\right) \tag{9.10}$$

which for constant E and A leads to an element stiffness matrix of

$$\mathbf{k} = \frac{EA}{\ell}\begin{bmatrix} 1 & -1 \\ -1 & 1 \end{bmatrix} \tag{9.11}$$

The kinetic energy of the element is

$$T = \frac{1}{2}\int_0^\ell \rho A\left(\frac{\partial u}{\partial t}\right)^2 dx = \frac{1}{2}\frac{\rho A\ell}{3}\left(\dot{u}_1^2 + \dot{u}_1\dot{u}_2 + \dot{u}_2^2\right) \tag{9.12}$$

which for constant ρ and A leads to an element mass matrix of

$$\mathbf{m} = \frac{\rho A\ell}{6}\begin{bmatrix} 2 & 1 \\ 1 & 2 \end{bmatrix} \tag{9.13}$$

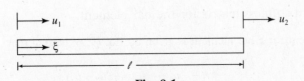

Fig. 9-1

9.4 BEAM ELEMENT

The beam element of a beam undergoing only transverse vibration has 4 degrees of freedom, the displacements and slopes at each end of the element. As illustrated in Fig. 9-2, let $u_1(t)$ be the displacement at $\xi = 0$, $u_2(t)$ the slope at $\xi = 0$, $u_3(t)$ the displacement at $\xi = \ell$, and $u_4(t)$ the slope at $\xi = \ell$. A finite element expression for the displacement across the beam element can be written as

$$u(\xi, t) = \left(1 - 3\frac{\xi^2}{\ell^2} + 2\frac{\xi^3}{\ell^3}\right)u_1 + \left(\frac{\xi}{\ell} - 2\frac{\xi^2}{\ell^2} + \frac{\xi^3}{\ell^3}\right)\ell u_2$$
$$+ \left(3\frac{\xi^2}{\ell^2} - 2\frac{\xi^3}{\ell^3}\right)u_3 + \left(-\frac{\xi^2}{\ell^2} + \frac{\xi^3}{\ell^3}\right)\ell u_4 \qquad (9.14)$$

The potential energy of the beam element is

$$V = \frac{1}{2}\int_0^\ell EI\left(\frac{\partial^2 u}{\partial \xi^2}\right)^2 d\xi \qquad (9.15)$$

which for constant E and I lead to an element stiffness matrix of

$$\mathbf{k} = \frac{EI}{\ell^3}\begin{bmatrix} 12 & 6\ell & -12 & 6\ell \\ 6\ell & 4\ell^2 & -6\ell & 2\ell^2 \\ -12 & -6\ell & 12 & -6\ell \\ 6\ell & 2\ell^2 & -6\ell & 4\ell^2 \end{bmatrix} \qquad (9.16)$$

The kinetic energy is given by Eq. (9.12), which for constant ρ and A lead to a mass matrix of

$$\mathbf{m} = \frac{\rho A \ell}{420}\begin{bmatrix} 156 & 22\ell & 54 & -13\ell \\ 22\ell & 4\ell^2 & 13\ell & -3\ell^2 \\ 54 & 13\ell & 156 & -22\ell \\ -13\ell & -3\ell^2 & -22\ell & 4\ell^2 \end{bmatrix} \qquad (9.17)$$

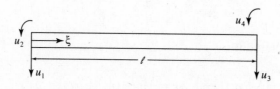

Fig. 9-2

Solved Problems

9.1   Derive the element stiffness matrix for the bar element.

The displacement for a bar element is given by Eq. (9.9). Noting that

$$\frac{\partial u}{\partial \xi} = -\frac{1}{\ell}u_1 + \frac{1}{\ell}u_2 = \frac{1}{\ell}(u_2 - u_1)$$

and substituting into the potential energy, Eq. (9.10),

$$V = \frac{1}{2} \int_0^{\ell} \frac{EA}{\ell^2} (u_2 - u_2)^2 \, d\xi$$

$$= \frac{1}{2} \frac{EA}{\ell^2} (u_2 - u_1)^2 \int_0^{\ell} d\xi$$

$$= \frac{1}{2} \frac{EA}{\ell} (u_1^2 - 2u_1 u_2 + u_2^2)$$

$$= \frac{1}{2} \frac{EA}{\ell} [u_1 \quad u_2] \begin{bmatrix} 1 & -1 \\ -1 & 1 \end{bmatrix} \begin{bmatrix} u_1 \\ u_2 \end{bmatrix}$$

Hence the element stiffness matrix is as given by Eq. (9.11).

9.2   Use a one-element finite element model to approximate the lowest natural frequency of a uniform fixed-free bar.

A one-element finite element model of a fixed-free bar has only 1 degree of freedom, the displacement of its free end. The potential and kinetic energies for this model bar are obtained using Eqs. (9.9) through (9.13) with $u_1 = 0$:

$$V = \frac{1}{2} \frac{EA}{\ell} u_2^2 \qquad T = \frac{1}{2} \frac{\rho A \ell}{3} \dot{u}_2^2$$

Energy methods are used to obtain the differential equation approximating the displacement of the bar's free end:

$$\frac{\rho A \ell}{3} \ddot{u}_2 + \frac{EA}{\ell} u_2 = 0$$

$$\ddot{u}_2 + \frac{3E}{\rho \ell^2} u_2 = 0$$

The approximation to the lowest natural frequency is

$$\omega_1 = \sqrt{\frac{3E}{\rho \ell^2}}$$

9.3   Determine the global stiffness matrix and global mass matrix for a four-element finite element model of a uniform fixed-free bar.

The four-element model of the fixed-free bar is illustrated in Fig. 9-3. In the global sense, the model uses 4 degrees of freedom. The global stiffness and mass matrices are 4×4 matrices. They are obtained by adding the potential and kinetic energies of the elements. When writing the differential equations, they will multiply the global displacement vector $\mathbf{U} = [U_1 \quad U_2 \quad U_3 \quad U_4]^T$ or its second time derivative. Their construction is illustrated below. (Recall that lowercase u's correspond to local coordinates while upper case U's refer to global coordinates.)

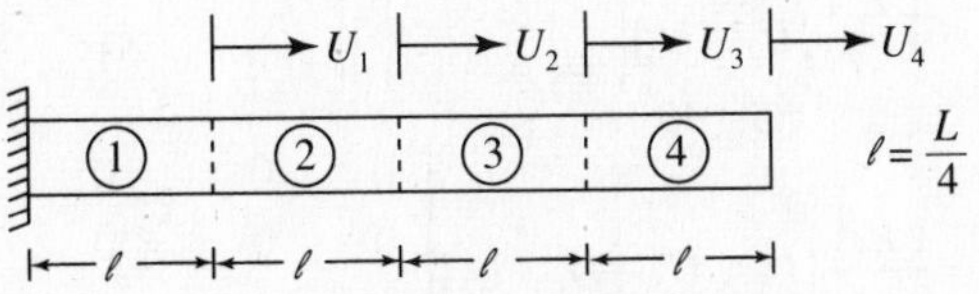

Fig. 9-3

Element 1: $u_1 = 0$, $u_2 = U_1$. Hence in terms of the global displacement vector,

$$V = \frac{1}{2}\frac{EA}{\ell}U_1^2 = \frac{1}{2}\frac{EA}{\ell}[U_1 \quad U_2 \quad U_3 \quad U_4]\begin{bmatrix} 1 & 0 & 0 & 0 \\ 0 & 0 & 0 & 0 \\ 0 & 0 & 0 & 0 \\ 0 & 0 & 0 & 0 \end{bmatrix}\begin{bmatrix} U_1 \\ U_2 \\ U_3 \\ U_4 \end{bmatrix}$$

$$T = \frac{1}{2}\frac{\rho A\ell}{6}2\dot{U}_1^2 = \frac{1}{2}\frac{\rho A\ell}{6}[\dot{U}_1 \quad \dot{U}_2 \quad \dot{U}_3 \quad \dot{U}_4]\begin{bmatrix} 2 & 0 & 0 & 0 \\ 0 & 0 & 0 & 0 \\ 0 & 0 & 0 & 0 \\ 0 & 0 & 0 & 0 \end{bmatrix}\begin{bmatrix} \dot{U}_1 \\ \dot{U}_2 \\ \dot{U}_3 \\ \dot{U}_4 \end{bmatrix}$$

Element 2: $u_1 = U_1$, $u_2 = U_2$

$$V = \frac{1}{2}\frac{EA}{\ell}[U_1 \quad U_2]\begin{bmatrix} 1 & -1 \\ -1 & 1 \end{bmatrix}\begin{bmatrix} U_1 \\ U_2 \end{bmatrix}$$

$$= \frac{1}{2}\frac{EA}{\ell}[U_1 \quad U_2 \quad U_3 \quad U_4]\begin{bmatrix} 1 & -1 & 0 & 0 \\ -1 & 1 & 0 & 0 \\ 0 & 0 & 0 & 0 \\ 0 & 0 & 0 & 0 \end{bmatrix}\begin{bmatrix} U_1 \\ U_2 \\ U_3 \\ U_4 \end{bmatrix}$$

$$T = \frac{1}{2}\frac{\rho A\ell}{6}[\dot{U}_1 \quad \dot{U}_2]\begin{bmatrix} 2 & 1 \\ 1 & 2 \end{bmatrix}\begin{bmatrix} \dot{U}_1 \\ \dot{U}_2 \end{bmatrix}$$

$$= \frac{1}{2}\frac{\rho A\ell}{6}[\dot{U}_1 \quad \dot{U}_2 \quad \dot{U}_3 \quad \dot{U}_4]\begin{bmatrix} 2 & 1 & 0 & 0 \\ 1 & 2 & 0 & 0 \\ 0 & 0 & 0 & 0 \\ 0 & 0 & 0 & 0 \end{bmatrix}\begin{bmatrix} \dot{U}_1 \\ \dot{U}_2 \\ \dot{U}_3 \\ \dot{U}_4 \end{bmatrix}$$

Element 3: $u_1 = U_2$, $u_2 = U_3$

$$V = \frac{1}{2}\frac{EA}{\ell}[U_2 \quad U_3]\begin{bmatrix} 1 & -1 \\ -1 & 1 \end{bmatrix}\begin{bmatrix} U_2 \\ U_3 \end{bmatrix}$$

$$= \frac{1}{2}\frac{EA}{\ell}[U_1 \quad U_2 \quad U_3 \quad U_4]\begin{bmatrix} 0 & 0 & 0 & 0 \\ 0 & 1 & -1 & 0 \\ 0 & -1 & 1 & 0 \\ 0 & 0 & 0 & 0 \end{bmatrix}\begin{bmatrix} U_1 \\ U_2 \\ U_3 \\ U_4 \end{bmatrix}$$

$$T = \frac{1}{2}\frac{\rho A\ell}{6}[\dot{U}_2 \quad \dot{U}_3]\begin{bmatrix} 2 & 1 \\ 1 & 2 \end{bmatrix}\begin{bmatrix} \dot{U}_2 \\ \dot{U}_3 \end{bmatrix}$$

$$= \frac{1}{2}\frac{\rho A\ell}{6}[\dot{U}_1 \quad \dot{U}_2 \quad \dot{U}_3 \quad \dot{U}_4]\begin{bmatrix} 0 & 0 & 0 & 0 \\ 0 & 2 & 1 & 0 \\ 0 & 1 & 2 & 0 \\ 0 & 0 & 0 & 0 \end{bmatrix}\begin{bmatrix} \dot{U}_1 \\ \dot{U}_2 \\ \dot{U}_3 \\ \dot{U}_4 \end{bmatrix}$$

Element 4: $u_1 = u_3$, $u_2 = U_4$

$$V = \frac{1}{2}\frac{EA}{\ell}[U_3 \quad U_4]\begin{bmatrix} 1 & -1 \\ -1 & 1 \end{bmatrix}\begin{bmatrix} U_3 \\ U_4 \end{bmatrix}$$

$$= \frac{1}{2}\frac{EA}{\ell}[U_1 \quad U_2 \quad U_3 \quad U_4]\begin{bmatrix} 0 & 0 & 0 & 0 \\ 0 & 0 & 0 & 0 \\ 0 & 0 & 1 & -1 \\ 0 & 0 & -1 & 1 \end{bmatrix}\begin{bmatrix} U_1 \\ U_2 \\ U_3 \\ U_4 \end{bmatrix}$$

$$T = \frac{1}{2} \frac{\rho A \ell}{6} [\dot{U}_3 \quad \dot{U}_4] \begin{bmatrix} 2 & 1 \\ 1 & 2 \end{bmatrix} \begin{bmatrix} \dot{U}_3 \\ \dot{U}_4 \end{bmatrix}$$

$$= \frac{1}{2} \frac{\rho A \ell}{6} [\dot{U}_1 \quad \dot{U}_2 \quad \dot{U}_3 \quad \dot{U}_4] \begin{bmatrix} 0 & 0 & 0 & 0 \\ 0 & 0 & 0 & 0 \\ 0 & 0 & 2 & 1 \\ 0 & 0 & 1 & 2 \end{bmatrix} \begin{bmatrix} \dot{U}_1 \\ \dot{U}_2 \\ \dot{U}_3 \\ \dot{U}_4 \end{bmatrix}$$

Hence the total potential energy is

$$V = \frac{1}{2} \frac{EA}{\ell} [U_1 \quad U_2 \quad U_3 \quad U_4] \begin{bmatrix} 2 & -1 & 0 & 0 \\ -1 & 2 & -1 & 0 \\ 0 & -1 & 2 & -1 \\ 0 & 0 & -1 & 1 \end{bmatrix} \begin{bmatrix} U_1 \\ U_2 \\ U_3 \\ U_4 \end{bmatrix}$$

The total kinetic energy is

$$T = \frac{1}{2} \frac{\rho A \ell}{6} [\dot{U}_1 \quad \dot{U}_2 \quad \dot{U}_3 \quad \dot{U}_4] \begin{bmatrix} 4 & 1 & 0 & 0 \\ 1 & 4 & 1 & 0 \\ 0 & 1 & 4 & 1 \\ 0 & 0 & 1 & 2 \end{bmatrix} \begin{bmatrix} \dot{U}_1 \\ \dot{U}_2 \\ \dot{U}_3 \\ \dot{U}_4 \end{bmatrix}$$

Hence the stiffness and mass matrices for the 4-degree-of-freedom model are

$$\mathbf{K} = \frac{EA}{\ell} \begin{bmatrix} 2 & -1 & 0 & 0 \\ -1 & 2 & -1 & 0 \\ 0 & -1 & 2 & -1 \\ 0 & 0 & -1 & 1 \end{bmatrix} \qquad \mathbf{M} = \frac{\rho A \ell}{6} \begin{bmatrix} 4 & 1 & 0 & 0 \\ 1 & 4 & 1 & 0 \\ 0 & 1 & 4 & 1 \\ 0 & 0 & 1 & 2 \end{bmatrix}$$

9.4 Use a 4-degree-of-freedom model to approximate the two lowest natural frequencies and mode shapes for a uniform fixed-free bar. Compare the finite element mode shapes to the exact mode shapes.

The natural frequency approximations are the square roots of the eigenvalues of $\mathbf{M}^{-1}\mathbf{K}$. Using the methods of Chap. 5 and the global mass and stiffness matrices derived in Problem 9.3, the two lowest natural frequencies are

$$\omega_1 = \frac{0.395}{\ell} \sqrt{\frac{E}{\rho}} \qquad \omega_2 = \frac{1.247}{\ell} \sqrt{\frac{E}{\rho}}$$

Note that ℓ is the element length and is equal to $L/4$ where L is the total length of the bar. Thus in terms of L,

$$\omega_1 = \frac{1.581}{L} \sqrt{\frac{E}{\rho}} \qquad \omega_2 = \frac{4.987}{L} \sqrt{\frac{E}{\rho}}$$

The corresponding eigenvectors are

$$\mathbf{X}_1 = \begin{bmatrix} 0.112 \\ 0.207 \\ 0.270 \\ 0.292 \end{bmatrix} \qquad \mathbf{X}_2 = \begin{bmatrix} 0.299 \\ 0.229 \\ -0.124 \\ -0.324 \end{bmatrix}$$

The eigenvectors represent the nodal displacements for the modes.

The two lowest exact natural frequencies and corresponding mode shapes are

$$\omega_2 = \frac{\pi}{2L} \sqrt{\frac{E}{\rho}} \qquad u_1(x) = \sin \frac{\pi x}{2L}$$

$$\omega_2 = \frac{3\pi}{2L} \sqrt{\frac{E}{\rho}} \qquad u_2(x) = \sin \frac{3\pi x}{2L}$$

The error in the first natural frequency approximation is 0.66 percent while the error in the second natural frequency approximation is 5.8 percent. The approximate mode shapes are plotted in Fig. 9-4 where the maximum displacement is set to 1.

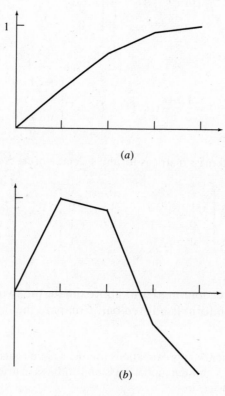

Fig. 9-4

9.5 Use a two-element finite element model to approximate the lowest natural frequency of the system of Fig. 9-5.

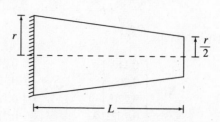

Fig. 9-5

The bar is divided into two elements of equal length, $\ell = L/2$, as shown in Fig. 9-6. In the global coordinate system the equation for the cross-sectional area is

$$A(x) = \pi r^2 \left(1 - \frac{x}{2L}\right)^2$$

Since the area varies over the length of the element, the element stiffness and mass matrices must be derived using the bar element of Eq. (9.9).

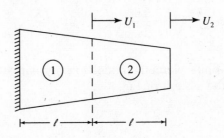

Fig. 9-6

Consider element 1: $\xi = x$, $u_1 = 0$, $u_2 = U_1$.

$$V = \frac{1}{2}\frac{1}{\ell^2}(u_2{}^2 - u_1{}^2)\int_0^\ell E\pi r^2\left(1 - \frac{\xi}{4\ell}\right)^2 d\xi$$

$$= \frac{1}{2}\frac{37}{48}\frac{E\pi r^2}{\ell}(u_1{}^2 - 2u_1 u_2 + u_2{}^2)$$

$$T = \frac{1}{2}\int_0^\ell \rho\pi r^2\left(1 - \frac{\xi}{4\ell}\right)^2\left[\left(1 - \frac{\xi}{\ell}\right)\dot{u}_1 + \frac{\xi}{\ell}\dot{u}_2\right]^2 d\xi$$

$$= \frac{1}{2}\frac{\rho\pi r^2\ell}{480}(141\dot{u}_1{}^2 + 123\dot{u}_1\dot{u}_2 + 106\dot{u}_2{}^2)$$

Hence the element mass and stiffness matrices for element 1 are

$$\mathbf{k}_1 = \frac{37}{48}\frac{E\pi r^2}{\ell}\begin{bmatrix} 1 & -1 \\ -1 & 1 \end{bmatrix} \qquad \mathbf{m}_1 = \frac{\rho\pi r^2\ell}{480}\begin{bmatrix} 141 & 61.5 \\ 61.5 & 106 \end{bmatrix}$$

Consider element 2: $\xi = x - \ell$, $u_1 = U_1$, $u_2 = U_2$.

$$V = \frac{1}{2}\frac{1}{\ell^2}(u_2 - u_1)^2\int_0^\ell E\pi r^2\left(\frac{3}{4} - \frac{\xi}{4\ell}\right)^2 d\xi$$

$$= \frac{1}{2}\frac{19}{48}\frac{E\pi r^2}{\ell}(u_1{}^2 - 2u_1 u_2 + u_2{}^2)$$

$$T = \frac{1}{2}\int_0^\ell \rho\pi r^2\left(\frac{3}{4} - \frac{\xi}{4\ell}\right)^2\left[\left(1 - \frac{\xi}{\ell}\right)\dot{u}_1 + \frac{\xi}{\ell}\dot{u}_2\right]^2 d\xi$$

$$= \frac{1}{2}\frac{\rho\pi r^2\ell}{480}(76\dot{u}_1{}^2 + 63\dot{u}_1\dot{u}_2 + 51\dot{u}_2{}^2)$$

Hence the element mass and stiffness matrices for element 2 are

$$\mathbf{k}_2 = \frac{19}{48}\frac{E\pi r^2}{\ell}\begin{bmatrix} 1 & -1 \\ -1 & 1 \end{bmatrix} \qquad \mathbf{m}_2 = \frac{\rho\pi r^2\ell}{480}\begin{bmatrix} 76 & 31.5 \\ 31.5 & 51 \end{bmatrix}$$

The global mass and stiffness matrices are formed as

$$\mathbf{K} = \frac{E\pi r^2}{48\ell} \left\{ \begin{bmatrix} 37 & 0 \\ 0 & 0 \end{bmatrix} + \begin{bmatrix} 19 & -19 \\ -19 & 19 \end{bmatrix} \right\} = \frac{E\pi r^2}{48\ell} \begin{bmatrix} 56 & -19 \\ -19 & 19 \end{bmatrix}$$

$$\mathbf{M} = \frac{\rho\pi r^2 \ell}{480} \left\{ \begin{bmatrix} 106 & 0 \\ 0 & 0 \end{bmatrix} + \begin{bmatrix} 76 & 31.5 \\ 31.5 & 51 \end{bmatrix} \right\} = \frac{\rho\pi r^2 \ell}{480} \begin{bmatrix} 182 & 31.5 \\ 31.5 & 51 \end{bmatrix}$$

The natural frequency approximations are the square roots of the eigenvalues of $\mathbf{M}^{-1}\mathbf{K}$ which lead to

$$\omega_1 = \frac{1.03}{\ell} \sqrt{\frac{E}{\rho}} = \frac{2.06}{L} \sqrt{\frac{E}{\rho}} \qquad \omega_2 = \frac{2.83}{\ell} \sqrt{\frac{E}{\rho}} = \frac{5.66}{L} \sqrt{\frac{E}{\rho}}$$

9.6 Use a three-element finite element model to approximate the lowest natural frequency and mode shape for the system of Fig. 9-7.

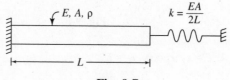

$$E, A, \rho \qquad k = \frac{EA}{2L}$$

$$|\longleftarrow L \longrightarrow|$$

Fig. 9-7

The bar is divided into three elements of equal length, $\ell = L/3$, as illustrated in Fig. 9-8. The element stiffness matrix for elements 1 and 2 is given by Eq. (9-11). The element mass matrix for all three elements is given by Eq. (9-13). The element stiffness matrix for element 3 must be revised to take into account the discrete spring. In terms of local coordinates for element 3,

$$V = \frac{1}{2}\frac{EA}{\ell}(u_1^2 - 2u_1u_2 + u_2^2) + \frac{1}{2}\frac{EA}{6\ell}u_2^2 = \frac{1}{2}\frac{EA}{\ell}\left(u_1^2 - 2u_1u_2 + \frac{7}{6}u_2^2\right)$$

Hence the element stiffness matrix is

$$\mathbf{k}_3 = \frac{EA}{\ell} \begin{bmatrix} 1 & -1 \\ -1 & \frac{7}{6} \end{bmatrix}$$

The global mass and stiffness matrices are assembled as in Problem 9.3, leading to

$$\mathbf{M} = \frac{\rho A\ell}{6} \begin{bmatrix} 4 & 1 & 0 \\ 1 & 4 & 1 \\ 0 & 1 & 2 \end{bmatrix} \qquad \mathbf{K} = \frac{EA}{\ell} \begin{bmatrix} 2 & -1 & 0 \\ -1 & 2 & -1 \\ 0 & -1 & \frac{7}{6} \end{bmatrix}$$

The natural frequency approximations are the square roots of the eigenvalues of $\mathbf{M}^{-1}\mathbf{K}$. The approximation to the lowest natural frequency is

$$\omega_1 = \frac{0.622}{\ell} \sqrt{\frac{E}{\rho}} = \frac{1.867}{L} \sqrt{\frac{E}{\rho}}$$

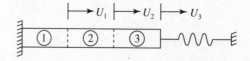

$$\longmapsto U_1 \longmapsto U_2 \longmapsto U_3$$

Fig. 9-8

9.7 Use a two-element finite element model to approximate the lowest natural frequency of the torsional system of Fig. 9-9.

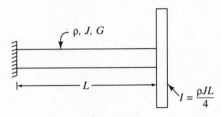

Fig. 9-9

The system is modeled using two elements of equal length, $\ell = L/2$. The forms of the element mass and stiffness matrices for the torsional system are analogous to those of an axial system. For a uniform element without a discrete stiffness element (elements 1 and 2),

$$\mathbf{k} = \frac{JG}{\ell}\begin{bmatrix} 1 & -1 \\ -1 & 1 \end{bmatrix}$$

and for a uniform element without a discrete inertia element (element 1),

$$\mathbf{m} = \frac{\rho J \ell}{6}\begin{bmatrix} 2 & 1 \\ 1 & 2 \end{bmatrix}$$

The kinetic energy for element 2 is

$$T = \frac{1}{2}\int_0^\ell \rho J \left[\left(1 - \frac{\xi}{\ell}\right)\dot{u}_1 + \frac{\xi}{\ell}\dot{u}_2\right]^2 d\xi + \frac{1}{2}I\dot{u}_2^2$$

$$= \frac{1}{2}\frac{\rho J \ell}{6}(2\dot{u}_1^2 + 2\dot{u}_1\dot{u}_2 + 2\dot{u}_2^2) + \frac{1}{2}\frac{\rho J \ell}{2}\dot{u}_2^2$$

$$= \frac{1}{2}\frac{\rho J \ell}{6}(2\dot{u}_1^2 + 2\dot{u}_1\dot{u}_2 + 5\dot{u}_2^2)$$

Hence the mass matrix for element 2 is

$$\mathbf{m}_2 = \frac{\rho J \ell}{6}\begin{bmatrix} 2 & 1 \\ 1 & 5 \end{bmatrix}$$

The global mass and stiffness matrices are assembled using the procedure of Problem 9.3 leading to

$$\mathbf{M} = \frac{\rho J \ell}{6}\begin{bmatrix} 4 & 1 \\ 1 & 5 \end{bmatrix} \qquad \mathbf{K} = \frac{JG}{\ell}\begin{bmatrix} 2 & -1 \\ -1 & 1 \end{bmatrix}$$

The natural frequency approximations are the square roots of the eigenvalues of $\mathbf{M}^{-1}\mathbf{K}$ leading to

$$\omega_1 = \frac{0.639}{\ell}\sqrt{\frac{G}{\rho}} = \frac{1.278}{L}\sqrt{\frac{G}{\rho}}$$

$$\omega_2 = \frac{2.155}{\ell}\sqrt{\frac{G}{\rho}} = \frac{4.31}{L}\sqrt{\frac{G}{\rho}}$$

9.8 Set up the differential equations governing the forced motion of the system of Fig. 9-10 using a two-element finite element model.

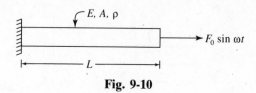

Fig. 9-10

The global mass and stiffness matrices for a two-element element model of the system of Fig. 9-10 are

$$\mathbf{M} = \frac{\rho A \ell}{6} \begin{bmatrix} 4 & 1 \\ 1 & 2 \end{bmatrix} \qquad \mathbf{K} = \frac{EA}{\ell} \begin{bmatrix} 2 & -1 \\ -1 & 1 \end{bmatrix}$$

The work done by the external force is

$$\delta W = \int_0^{\ell} \left[\left(1 - \frac{\xi}{\ell} \right) \delta U_1 + \frac{\xi}{\ell} \delta U_2 \right] F_0 \sin \omega t \, \delta(\xi - \ell) \, d\xi = F_0 \sin \omega t \, \delta U_2$$

Hence the governing differential equations are

$$\frac{\rho A \ell}{6} \begin{bmatrix} 4 & 1 \\ 1 & 2 \end{bmatrix} \begin{bmatrix} \ddot{U}_1 \\ \ddot{U}_2 \end{bmatrix} + \frac{EA}{\ell} \begin{bmatrix} 2 & -1 \\ -1 & 1 \end{bmatrix} = \begin{bmatrix} 0 \\ F_0 \sin \omega t \end{bmatrix}$$

9.9 Derive Eq. (9.14).

The static transverse deflection of an element of a beam not subject to transverse loads satisfies

$$\frac{d^4 u}{d\xi^4} = 0$$

which is integrated to yield

$$u(\xi) = C_1 \xi^3 + C_2 \xi^2 + C_3 \xi + C_4 \qquad (9.18)$$

The constants C_1, C_2, C_3, and C_4 are obtained by requiring the slope and transverse deflection specified at $\xi = 0$ and $\xi = \ell$. To this end, using the notation of Fig. 9-2,

$$u(0) = u_1 \qquad u(\ell) = u_3 \qquad \frac{du}{d\xi}(0) = u_2 \qquad \frac{du}{d\xi}(\ell) = u_4 \qquad (9.19)$$

Substituting the conditions of Eq. (9.19) into Eq. (9.18) leads to

$$u(0) = 0 \; \rightarrow \; C_4 = u_1$$

$$\frac{du}{d\xi}(0) = u_2 \; \rightarrow \; C_3 = u_2$$

$$u(\ell) = u_3 \; \rightarrow \; C_1 \ell^3 + C_2 \ell^2 + C_3 \ell + C_4 = u_3$$

$$\frac{du}{d\xi}(\ell) = u_4 \; \rightarrow \; 3C_1 \ell^2 + 2C_2 \ell + C_3 = u_4$$

Solving the last two of the previous equations simultaneously leads to

$$C_1 = \frac{1}{\ell^3}(2u_1 + \ell u_2 - 2u_3 + \ell u_4)$$

$$C_2 = \frac{1}{\ell^2}(-3u_1 - 2\ell u_2 + 3u_3 - \ell u_4)$$

Substituting for the determined constants in Eq. (9.18) and rearranging leads to Eq. (9.14).

9.10 Derive the m_{34} element of the local mass matrix for a uniform beam element.

Note from Eq. (9.14),

$$\frac{\partial u}{\partial t} = \left(1 - 3\frac{\xi^2}{\ell^2} + 2\frac{\xi^3}{\ell^3}\right)\dot{u}_1 + \left(\frac{\xi}{\ell} - 2\frac{\xi^2}{\ell^2} + \frac{\xi^3}{\ell^3}\right)\ell\dot{u}_2$$

$$+ \left(3\frac{\xi^2}{\ell^2} - 2\frac{\xi^3}{\ell^3}\right)\dot{u}_3 + \left(-\frac{\xi^2}{\ell^2} + \frac{\xi^3}{\ell^3}\right)\ell\dot{u}_4$$

The above expression is substituted into Eq. (9.12). The term that leads to m_{34} in the element mass matrix is

$$2\int_0^\ell \rho A\left(3\frac{\xi^2}{\ell^2} - 2\frac{\xi^3}{\ell^3}\right)\dot{u}_3\left(-\frac{\xi^2}{\ell^2} + \frac{\xi^3}{\ell^3}\right)\ell\dot{u}_4 \, d\xi$$

$$= 2\rho A\, \ell\dot{u}_3\dot{u}_4 \int_0^\ell \left(-3\frac{\xi^4}{\ell^4} + 5\frac{\xi^5}{\ell^5} - 2\frac{\xi^6}{\ell^6}\right) d\xi$$

$$= 2\rho A\, \ell\dot{u}_3\dot{u}_4\left(-\frac{3}{5} + \frac{5}{6} - \frac{2}{7}\right)$$

$$= -\frac{11}{105}\rho A\ell^2\dot{u}_3\dot{u}_4$$

When the quadratic form of the kinetic energy is expanded, it includes a term $2m_{34}\dot{u}_3\dot{u}_4$. Thus

$$m_{34} = -\tfrac{11}{210}\ell\rho A\ell$$

9.11 Use a one-element finite element model to approximate the lowest natural frequency and mode shape for a uniform fixed-free beam.

Since the slope and deflection at the fixed end is zero, $u_1 = u_2 = 0$. Thus, a one-element model of a fixed-free beam has 2 degrees of freedom. The global mass and stiffness matrices are obtained by simply setting $u_1 = u_2 = 0$. This is accomplished by eliminating the first and second rows and columns of the element mass and stiffness matrices of Eqs. (9.16) and (9.17). Thus

$$\mathbf{M} = \frac{\rho AL}{420}\begin{bmatrix} 156 & -22L \\ -22L & 4L^2 \end{bmatrix} \qquad \mathbf{K} = \frac{EI}{L^3}\begin{bmatrix} 12 & -6L \\ -6L & 4L^2 \end{bmatrix}$$

The natural frequency approximations are the square roots of the eigenvalues of $\mathbf{M}^{-1}\mathbf{K}$. The lowest natural frequency is approximated as

$$\omega_1 = 3.53\sqrt{\frac{EI}{\rho AL^4}}$$

The eigenvector corresponding to this first mode is $\mathbf{u} = [1 \quad 1.378/L]^T$. Thus from Eq. (9.14) the mode shape approximation is

$$u(x) = 3\frac{x^2}{L^2} - 2\frac{x^3}{L^3} + \left(-\frac{x^2}{L^2} + \frac{x^3}{L^3}\right)L\left(\frac{1.378}{L}\right)$$

$$= 1.622\frac{x^2}{L^2} - 0.622\frac{x^3}{L^3}$$

which is illustrated in Fig. 9-11.

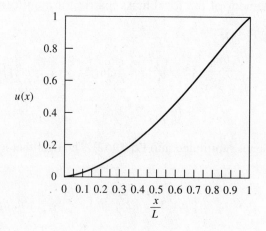

Fig. 9-11

9.12 Determine the global stiffness matrix for a two-element finite element approximation to the system of Fig. 9-12.

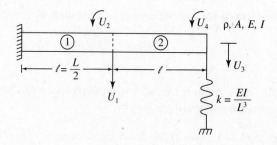

Fig. 9-12

The two-element model of the beam of Fig. 9-12 has 4 degrees of freedom with the global definitions of nodal displacements illustrated. Then for element 1: $u_1 = 0$, $u_2 = 0$, $u_3 = U_1$, $u_4 = U_2$. The contribution to the global stiffness matrix from element 1 is

$$\mathbf{K}_1 = \frac{EI}{\ell^3}\begin{bmatrix} 12 & -6\ell & 0 & 0 \\ -6\ell & 4\ell^2 & 0 & 0 \\ 0 & 0 & 0 & 0 \\ 0 & 0 & 0 & 0 \end{bmatrix}$$

Now consider element 2: $u_1 = U_1$, $u_2 = U_2$, $u_3 = U_3$, $u_4 = U_4$. The stiffness matrix for element 2 must be modified to account for the potential energy developed in the spring:

$$V_s = \frac{1}{2} k u_3^2 = \frac{1}{2} \frac{EI}{L^3} u_3^2 = \frac{1}{2} \frac{EI}{8\ell^3} u_3^2$$

Thus the element stiffness matrix for element 2 is

$$\mathbf{K}_2 = \frac{EI}{\ell^3} \begin{bmatrix} 12 & 6\ell & -12 & 6\ell \\ 6\ell & 4\ell^2 & -6\ell & 2\ell^2 \\ -12 & -6\ell & 12.125 & -6\ell \\ 6\ell & 2\ell^2 & -6\ell & 4\ell^2 \end{bmatrix}$$

Hence the global stiffness matrix is

$$\mathbf{K} = \frac{EI}{L^3} \begin{bmatrix} 24 & 0 & -12 & 6\ell \\ 0 & 4\ell^2 & -6\ell & 2\ell^2 \\ -12 & -6\ell & 12.125 & -6\ell \\ 6\ell & 2\ell^2 & -6\ell & 4\ell^2 \end{bmatrix}$$

9.13 Determine the mass matrix for a one-element finite element approximation for the system of Fig. 9-13.

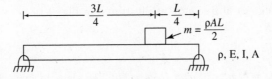

Fig. 9-13

Using a one-element finite element model for the simply supported beam, it is noted that $u_1 = u_3 = 0$. Then from, Eq. (9.14), at $\xi = 3L/4$,

$$u\left(\frac{3}{4} L\right) = \left[\frac{3}{4} - 2\left(\frac{3}{4}\right)^2 + \left(\frac{3}{4}\right)^3\right] L u_2 + \left[-\left(\frac{3}{4}\right)^2 + \left(\frac{3}{4}\right)^3\right] L u_4$$

$$= \frac{L}{64}(3u_2 - 9u_4)$$

Hence the kinetic energy of the block is

$$T_b = \frac{1}{2} m \left[\frac{L}{64}(3\dot{u}_2 - 9\dot{u}_4)\right]^2$$

$$= \frac{1}{2} \frac{mL^2}{4096}(9\dot{u}_2^2 - 54\dot{u}_2\dot{u}_4 + 81\dot{u}_4^2)$$

Noting that in the global system $U_1 = u_2$ and $U_2 = u_4$, the global mass matrix becomes

$$\mathbf{M} = \frac{\rho AL}{420} \begin{bmatrix} 4L^2 & -3L^2 \\ -3L^2 & 4L^2 \end{bmatrix} + \frac{\rho AL^3}{2(4096)} \begin{bmatrix} 9 & -27 \\ -27 & 81 \end{bmatrix}$$

$$= \rho AL^3 \begin{bmatrix} 0.0106 & -0.0104 \\ -0.0104 & 0.0194 \end{bmatrix}$$

9.14 Use a two-element finite element model to approximate the two lowest natural frequencies for the system of Fig. 9-14.

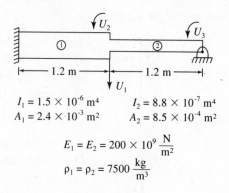

$$I_1 = 1.5 \times 10^{-6} \text{ m}^4 \qquad I_2 = 8.8 \times 10^{-7} \text{ m}^4$$
$$A_1 = 2.4 \times 10^{-3} \text{ m}^2 \qquad A_2 = 8.5 \times 10^{-4} \text{ m}^2$$

$$E_1 = E_2 = 200 \times 10^9 \, \frac{\text{N}}{\text{m}^2}$$

$$\rho_1 = \rho_2 = 7500 \, \frac{\text{kg}}{\text{m}^3}$$

Fig. 9-14

When a two-element finite element model is used with one element covering each segment, the system has 3 degrees of freedom, as illustrated in Fig. 9-14.

Element 1: $u_1 = 0$, $u_2 = 0$, $u_3 = U$, $u_4 = U_2$. The element mass and stiffness matrices are

$$\mathbf{k}_1 = \frac{\left(200 \times 10^9 \, \frac{\text{N}}{\text{m}^2}\right)(1.5 \times 10^{-6} \text{ m}^4)}{(1.2 \text{ m})^3} \begin{bmatrix} 12 & 6(1.2) & -12 & 6(1.2) \\ 6(1.2) & 4(1.2)^2 & -6(1.2) & 2(1.2)^2 \\ -12 & -6(1.2) & 12 & -6(1.2) \\ 6(1.2) & 2(1.2)^2 & -6(1.2) & 4(1.2)^2 \end{bmatrix}$$

$$= 10^6 \begin{bmatrix} 2.08 & 1.25 & -2.08 & 1.25 \\ 1.25 & 1.0 & -1.25 & 0.500 \\ -2.08 & -1.25 & 2.08 & -1.25 \\ 1.25 & 0.500 & -1.25 & 1.0 \end{bmatrix}$$

$$\mathbf{m}_1 = \frac{\left(7500 \, \frac{\text{kg}}{\text{m}^3}\right)(2.4 \times 10^{-3} \text{ m}^2)(1.2 \text{ m})}{420} \begin{bmatrix} 156 & 22(1.2) & 54 & -13(1.2) \\ 22(1.2) & 4(1.2)^2 & 13(1.2) & -3(1.2)^2 \\ 54 & 13(1.2) & 156 & -22(1.2) \\ -13(1.2) & -3(1.2)^2 & -22(1.2) & 4(1.2)^2 \end{bmatrix}$$

$$= \begin{bmatrix} 8.02 & 1.36 & 2.77 & -0.802 \\ 1.36 & 0.296 & -0.802 & -0.222 \\ 2.77 & 0.802 & 8.02 & -1.36 \\ -0.802 & -0.222 & -1.36 & 0.296 \end{bmatrix}$$

Element 2: $u_1 = U_1$, $u_2 = U_2$, $u_3 = 0$, $u_4 = U_3$. The element mass and stiffness matrices are

$$\mathbf{k}_2 = \frac{\left(200 \times 10^9 \frac{\text{N}}{\text{m}^2}\right)(8.8 \times 10^{-7} \text{ m}^4)}{(1.2 \text{ m})^3}$$

$$\begin{bmatrix} 12 & 6(1.2) & -12 & 6(1.2) \\ 6(1.2) & 4(1.2)^2 & -6(1.2) & 2(1.2)^2 \\ -12 & -6(1.2) & 12 & -6(1.2) \\ 6(1.2) & 2(1.2)^2 & -6(1.2) & 4(1.2)^2 \end{bmatrix}$$

$$= 10^6 \begin{bmatrix} 1.22 & 0.733 & -1.22 & 0.733 \\ 0.733 & 0.586 & -0.733 & 0.293 \\ -1.22 & -0.733 & 1.22 & -0.733 \\ 0.733 & 0.293 & -0.733 & 0.586 \end{bmatrix}$$

$$\mathbf{m}_2 = \frac{\left(7500 \frac{\text{kg}}{\text{m}^3}\right)(8.5 \times 10^{-4} \text{ m}^2)(1.2 \text{ m})}{420}$$

$$\begin{bmatrix} 156 & 22(1.2) & 54 & -13(1.2) \\ 22(1.2) & 4(1.2)^2 & 13(1.2) & -3(1.2)^2 \\ 54 & 13(1.2) & 156 & -22(1.2) \\ -13(1.2) & -3(1.2)^2 & -22(1.2) & 4(1.2)^2 \end{bmatrix}$$

$$= \begin{bmatrix} 2.84 & 0.480 & 0.983 & -0.284 \\ 0.480 & 0.105 & 0.284 & -0.0786 \\ 0.983 & 0.284 & 2.84 & -0.480 \\ -0.284 & -0.0786 & -0.480 & 0.105 \end{bmatrix}$$

The global matrices are constructed as

$$\mathbf{K} = 10^6 \begin{bmatrix} 2.09 & -1.25 & 0 \\ -1.25 & 1.0 & 0 \\ 0 & 0 & 0 \end{bmatrix} + 10^6 \begin{bmatrix} 1.22 & 0.733 & 0.733 \\ 0.733 & 0.586 & 0.293 \\ 0.733 & 0.293 & 0.586 \end{bmatrix}$$

$$= 10^6 \begin{bmatrix} 3.31 & -0.517 & 0.733 \\ -0.517 & 1.586 & 0.293 \\ 0.733 & 0.293 & 0.586 \end{bmatrix}$$

$$\mathbf{M} = \begin{bmatrix} 8.02 & -1.36 & 0 \\ -1.36 & 0.296 & 0 \\ 0 & 0 & 0 \end{bmatrix} + \begin{bmatrix} 2.84 & 0.480 & -0.284 \\ 0.480 & 0.105 & -0.0786 \\ -0.284 & -0.0786 & 0.105 \end{bmatrix}$$

$$= \begin{bmatrix} 10.86 & -0.880 & -0.284 \\ -0.880 & 0.401 & -0.0786 \\ -0.284 & -0.0786 & 0.105 \end{bmatrix}$$

The natural frequency approximations are the square roots of the eigenvalues of $\mathbf{M}^{-1}\mathbf{K}$. The two lowest natural frequencies are calculated as $\omega_1 = 404.7$ rad/s, $\omega_2 = 1524$ rad/s.

9.15 Use a three-element finite element model to set up the differential equations governing the forced vibration of the system of Fig. 9-15.

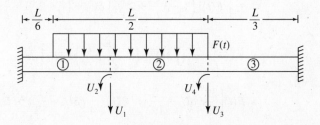

Fig. 9-15

A three-element model of the fixed-free beam of Fig. 9-15 leads to a 4-degree-of-freedom system. The global coordinates are illustrated in Fig. 9-15. The relations between the local and global coordinates for each element are:

Element 1: $u_1 = 0,\ u_2 = 0,\ u_3 = U_1,\ u_4 = U_2$

Element 2: $u_1 = U_1,\ u_2 = U_2,\ u_3 = U_3,\ u_4 = U_4$

Element 3: $u_1 = U_3,\ u_2 = U_4,\ u_3 = 0,\ u_4 = 0$

The global mass and stiffness matrices are constructed using the element mass and stiffness matrices by

$$
\mathbf{K} = \frac{EI}{\ell^3}\left\{
\begin{bmatrix}
12 & -6\ell & 0 & 0 \\
-6\ell & 4\ell^2 & 0 & 0 \\
0 & 0 & 0 & 0 \\
0 & 0 & 0 & 0
\end{bmatrix}
+
\begin{bmatrix}
12 & 6\ell & -12 & 6\ell \\
6\ell & 4\ell^2 & -6\ell & 2\ell^2 \\
-12 & -6\ell & 12 & -6\ell \\
6\ell & 2\ell^2 & -6\ell & 4\ell^2
\end{bmatrix}
+
\begin{bmatrix}
0 & 0 & 0 & 0 \\
0 & 0 & 0 & 0 \\
0 & 0 & 12 & 6\ell \\
0 & 0 & 6\ell & 4\ell^2
\end{bmatrix}
\right\}
$$

$$
= \frac{EI}{\ell^3}
\begin{bmatrix}
24 & 0 & -12 & 6\ell \\
0 & 8\ell^2 & -6\ell & 2\ell^2 \\
-12 & -6\ell & 24 & 0 \\
6\ell & 2\ell^2 & 0 & 8\ell^2
\end{bmatrix}
$$

$$
\mathbf{M} = \frac{\rho A \ell}{420}\left\{
\begin{bmatrix}
156 & -22\ell & 0 & 0 \\
-22\ell & 4\ell^2 & 0 & 0 \\
0 & 0 & 0 & 0 \\
0 & 0 & 0 & 0
\end{bmatrix}
\right.
$$

$$
\left.
+
\begin{bmatrix}
156 & 22\ell & 54 & -13\ell \\
22\ell & 4\ell^2 & 13\ell & -3\ell^2 \\
54 & 13\ell & 156 & -22\ell \\
-13\ell & -3\ell^2 & -22\ell & 4\ell^2
\end{bmatrix}
+
\begin{bmatrix}
0 & 0 & 0 & 0 \\
0 & 0 & 0 & 0 \\
0 & 0 & 156 & 22\ell \\
0 & 0 & 22\ell & 4\ell^2
\end{bmatrix}
\right\}
$$

$$
= \frac{\rho A \ell}{420}
\begin{bmatrix}
312 & 0 & 54 & -13\ell \\
0 & 8\ell^2 & 13\ell & -3\ell^2 \\
54 & 13\ell & 312 & 0 \\
-13\ell & -3\ell^2 & 0 & 8\ell^2
\end{bmatrix}
$$

where $\ell = L/3$.

The virtual work done by the distributed loading is

$$
\delta W = \int_{1/2}^{\ell} F(t)\left[\left(3\frac{\xi^2}{\ell^2} - 2\frac{\xi^3}{\ell^3}\right)\delta U_1 + \left(-\frac{\xi^2}{\ell^2} + \frac{\xi^3}{\ell^3}\right)\ell\,\delta U_2\right]d\xi
$$

$$
+ \int_0^{\ell} F(t)\left[\left(1 - 3\frac{\xi^2}{\ell^2} + 2\frac{\xi^3}{\ell^3}\right)\delta U_1 + \left(\frac{\xi}{\ell} - 2\frac{\xi^2}{\ell^2} + \frac{\xi^3}{\ell^3}\right)\ell\,\delta U_3\right.
$$

$$
\left. + \left(3\frac{\xi^2}{\ell^2} - \frac{\xi^3}{\ell^3}\right)\delta U_3 + \left(-\frac{\xi^2}{\ell^2} + \frac{\xi^3}{\ell^3}\right)\ell\,\delta U_4\right]d\xi
$$

$$
= F(t)\left[\frac{29}{32}\ell\,\delta U_1 + \frac{5}{192}\ell^2\,\delta U_2 + \frac{3}{4}\ell\delta U_3 - \frac{1}{12}\ell^2\,\delta U_4\right]
$$

Hence the differential equations for a three-element finite element approximation for the system of Fig. 9-15 are

$$
\frac{\rho A \ell}{420}
\begin{bmatrix}
312 & 0 & 54 & -13\ell \\
0 & 8\ell^2 & 13\ell & -3\ell^2 \\
54 & 13\ell & 312 & 0 \\
-13\ell & -3\ell^2 & 0 & 8\ell^2
\end{bmatrix}
\begin{bmatrix}
\ddot{U}_1 \\
\ddot{U}_2 \\
\ddot{U}_3 \\
\ddot{U}_4
\end{bmatrix}
+
\frac{EI}{\ell^3}
\begin{bmatrix}
24 & 0 & -12 & 6\ell \\
0 & 8\ell^2 & -6\ell & 2\ell^2 \\
-12 & -6\ell & 24 & 0 \\
6\ell & 2\ell^2 & 0 & 8\ell^2
\end{bmatrix}
\begin{bmatrix}
U_1 \\
U_2 \\
U_3 \\
U_4
\end{bmatrix}
$$

$$
=
\begin{bmatrix}
\dfrac{29}{32}\ell \\[2mm]
\dfrac{5}{192}\ell^2 \\[2mm]
\dfrac{3}{4}\ell \\[2mm]
-\dfrac{1}{12}\ell^2
\end{bmatrix}
F(t)
$$

Supplementary Problems

9.16   Derive the element mass matrix for a uniform bar element, Eq. (9.13).

9.17   Use a one-element finite element model to approximate the lowest natural frequency of the system of Fig. 9-7.

Ans.

$$
\frac{3}{L}\sqrt{\frac{E}{2\rho}}
$$

9.18   Use a one-element finite element model to approximate the lowest nonzero torsional natural frequency of a shaft free at both ends.

Ans.

$$
\frac{2}{L}\sqrt{\frac{3G}{\rho}}
$$

9.19   Use a two-element finite element model to approximate the lowest natural frequency of the system of Fig. 9-16.

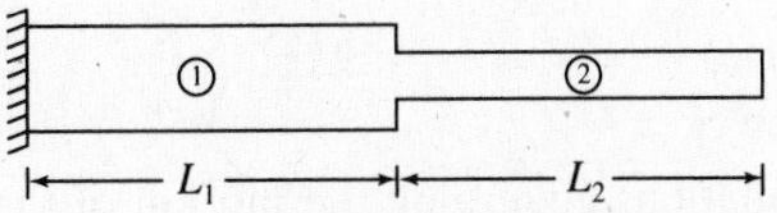

$$E_1 = E_2 = 200 \times 10^9 \,\frac{\text{N}}{\text{m}^2}$$

$$\rho_1 = \rho_2 = 7500 \,\frac{\text{kg}}{\text{m}^3}$$

$A_1 = 1.4 \times 10^{-3} \text{ m}^2$ $A_2 = 8.7 \times 10^{-4} \text{ m}^2$
$L_1 = 65 \text{ cm}$ $L_2 = 80 \text{ cm}$

Fig. 9-16

Ans. 6.68×10^3 rad/s

9.20 Derive the global stiffness matrix for the system of Fig. 9-17 using three elements to model the uniform bar and an additional degree of freedom for the discrete mass.

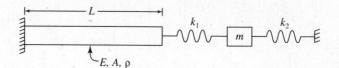

Fig. 9-17

Ans.

$$
\begin{bmatrix}
\dfrac{6EA}{L} & -\dfrac{3EA}{L} & 0 & 0 \\[2mm]
-\dfrac{3EA}{L} & \dfrac{6EA}{L} & -\dfrac{3EA}{L} & 0 \\[2mm]
0 & -\dfrac{3EA}{L} & \dfrac{6EA}{L} + k_1 & -k_1 \\[2mm]
0 & 0 & -k_1 & k_1 + k_2
\end{bmatrix}
$$

9.21 Derive the global mass matrix for the system of Problem 9.20.

Ans.

$$
\begin{bmatrix}
\dfrac{2\rho AL}{9} & \dfrac{\rho AL}{18} & 0 & 0 \\[2mm]
\dfrac{\rho AL}{18} & \dfrac{2\rho AL}{9} & \dfrac{\rho AL}{18} & 0 \\[2mm]
0 & \dfrac{\rho AL}{18} & \dfrac{\rho AL}{9} & 0 \\[2mm]
0 & 0 & 0 & m
\end{bmatrix}
$$

9.22 Approximate the lowest natural frequency of the system of Fig. 9-18 using one element to model each bar.

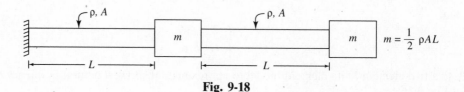

Fig. 9-18

Ans.

$$
\frac{0.755}{L}\sqrt{\frac{E}{\rho}}
$$

9.23 Derive the differential equations governing the motion of the shaft of Fig. 9-19 as it is subject to a time-dependent uniform torque loading. Use two elements to model the shaft.

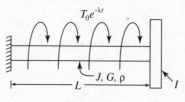

Fig. 9-19

Ans.

$$
\begin{bmatrix} \dfrac{\rho JL}{3} & \dfrac{\rho JL}{12} \\[3mm] \dfrac{\rho JL}{12} & \dfrac{\rho JL}{6}+I \end{bmatrix} \begin{bmatrix} \ddot{\theta}_1 \\[2mm] \ddot{\theta}_2 \end{bmatrix} + \begin{bmatrix} \dfrac{4JG}{L} & -\dfrac{2JG}{L} \\[3mm] -\dfrac{2JG}{L} & \dfrac{2JG}{L} \end{bmatrix} \begin{bmatrix} \theta_1 \\[2mm] \theta_2 \end{bmatrix} = \begin{bmatrix} 1 \\ 1 \\ \tfrac{1}{2} \end{bmatrix} T_0 L e^{-\lambda t}
$$

9.24   Derive the element k_{13} of the element stiffness matrix for a uniform beam element.
  Ans.

$$
-\frac{12EI}{\ell^3}
$$

9.25   Derive the element m_{12} of the element mass matrix for a uniform beam element.
  Ans.

$$
\frac{11\rho A\ell^2}{210}
$$

9.26   Approximate the lowest natural frequency of a simply supported beam using one element to model the beam.
  Ans.

$$
10.95\sqrt{\frac{EI}{\rho AL^4}}
$$

9.27   Derive the global mass matrix for the system of Fig. 9-20.

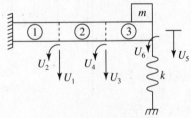

Fig. 9-20

Ans.

$$
\frac{\rho A\ell}{420}
\begin{bmatrix}
312 & 0 & 54 & -13\ell & 0 & 0 \\
0 & 8\ell^2 & 13\ell & -3\ell^2 & 0 & 0 \\
54 & 13\ell & 312 & 0 & 54 & -13\ell \\
-13\ell & -3\ell^2 & 0 & 8\ell^2 & 13\ell & -3\ell^2 \\
0 & 0 & 54 & 13\ell & 156+\dfrac{420m}{\rho A\ell} & -22\ell \\
0 & 0 & -13\ell & -3\ell^2 & -22\ell & 4\ell^2
\end{bmatrix}
$$

9.28   Derive the global stiffness matrix for the system of Fig. 9-20.
  Ans.

$$
\frac{EI}{\ell^3}
\begin{bmatrix}
24 & 0 & -12 & 6\ell & 0 & 0 \\
0 & 8\ell^2 & -6\ell & 2\ell^2 & 0 & 0 \\
-12 & -6\ell & 24 & 0 & -12 & 6\ell \\
6\ell & 2\ell^2 & 0 & 8\ell^2 & -6\ell & 2\ell^2 \\
0 & 0 & -12 & -6\ell & 12+\dfrac{k\ell^2}{EI} & -6\ell \\
0 & 0 & 6\ell & 2\ell^2 & -6\ell & 4\ell^2
\end{bmatrix}
$$

9.29 Approximate the two lowest natural frequencies of a uniform fixed-pinned beam using two elements of equal length to model the beam.

Ans.

$$15.56\sqrt{\frac{EI}{\rho AL^4}}, \qquad 58.41\sqrt{\frac{EI}{\rho AL^4}}$$

9.30 Write the differential equations governing the motion of the system of Fig. 9-21 when two elements are used to model the beam.

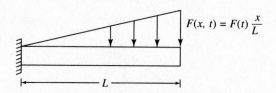

$$F(x, t) = F(t)\frac{x}{L}$$

Fig. 9-21

Ans.

$$\frac{\rho AL}{840}\begin{bmatrix} 312 & 0 & 54 & -\dfrac{13}{2}L \\ 0 & 2L^2 & \dfrac{13}{2}L & -\dfrac{3}{4}L^2 \\ 54 & \dfrac{13}{2}L & 156 & -11L \\ -\dfrac{13}{2}L & -\dfrac{3}{4}L^2 & -11L & L^2 \end{bmatrix}\begin{bmatrix} \ddot{U}_1 \\ \ddot{U}_2 \\ \ddot{U}_3 \\ \ddot{U}_4 \end{bmatrix} + \frac{8EI}{L^3}\begin{bmatrix} 24 & 0 & -12 & 3L \\ 0 & 2L^2 & -3L & \dfrac{L^2}{2} \\ -12 & -3L & 12 & -3L \\ 3L & \dfrac{L^2}{2} & -3L & L^2 \end{bmatrix}\begin{bmatrix} U_1 \\ U_2 \\ U_3 \\ U_4 \end{bmatrix}$$

$$= \begin{bmatrix} \dfrac{L}{4} \\ \dfrac{L^2}{120} \\ \dfrac{17}{80}L \\ -\dfrac{L^2}{60} \end{bmatrix} F(t)$$

Chapter 10

Nonlinear Systems

10.1 DIFFERENCES FROM LINEAR SYSTEMS

Some of the differences between a linear system and a nonlinear system are:

1. The behavior of a nonlinear system is governed by a nonlinear differential equation. Exact solutions do not exist for many nonlinear differential equations.

2. A nonlinear system may have more than one equilibrium point. An equilibrium point may be stable or unstable.

3. Steady-state behavior, if it exists for a nonlinear system, is dependent upon initial conditions.

4. The period of free vibration of a nonlinear system is dependent upon initial conditions. This implies that the frequency of free vibration is dependent upon the free vibration amplitude.

5. A nonlinear system exhibits resonance at excitation frequencies different from the system's linear natural frequency. A superharmonic resonance exists in a system with a cubic nonlinearity when the excitation frequency is one-third of the system's linear natural frequency. A subharmonic resonance exists when the excitation frequency is nearly three times the system's linear natural frequency.

6. The principle of linear superposition cannot be used to analyze a nonlinear system subject to a multifrequency excitation. A combination resonance can exist for appropriate combinations of excitation frequencies.

7. Internal resonances can exist in multi-degree-of-freedom and continuous systems for appropriate combinations of natural frequencies.

8. A periodic excitation may lead to a nonperiodic response in a nonlinear system. Such chaotic motion occurs in many nonlinear systems for certain parameter values.

10.2 QUALITATIVE ANALYSIS

The *state plane* or *phase plane* is a plot of velocity versus displacement during the history of motion. The nature and stability of equilibrium points can be examined from linearizing the governing differential equation in the vicinity of the equilibrium point (see Problem 10.2). Types of equilibrium points are shown in Fig. 10-1.

10.3 DUFFING'S EQUATION

Duffing's equation

$$\ddot{x} + 2\mu\dot{x} + x + \varepsilon x^3 = F \sin rt \qquad (10.1)$$

is a nondimensional equation that serves as a model for systems with cubic nonlinearities. If ε is positive, it models the response of a system with a *hardening spring* whereas if ε is negative, Duffing's equation models the response of a system with a *softening spring*. For free vibrations

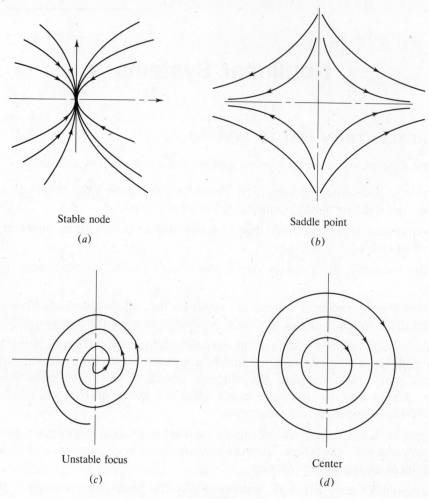

Stable node
(a)

Saddle point
(b)

Unstable focus
(c)

Center
(d)

Fig. 10-1

the frequency amplitude relation for a system governed by Duffing's equation is approximated using a *perturbation method* as

$$\omega = 1 + \tfrac{3}{8}\varepsilon A^2 + O(\varepsilon^2) \tag{10.2}$$

where ω is the nondimensional natural frequency ($\omega = 1$ for a linear system) and A is the amplitude. The forced response of Duffing's equation is analyzed near resonance by assuming

$$r = 1 + \varepsilon\sigma \tag{10.3}$$

Then the equation defining the steady-state amplitude is approximated as

$$4A^2[\mu^2 + (\sigma - \tfrac{3}{8}A^2)^2] = F^2 \tag{10.4}$$

The plot of Eq. (10.4) in Fig. 10-2 for $\varepsilon > 0$ illustrates the *backbone curve* and the *jump phenomenon*. For certain values of σ, Eq. (10.4) has three real and positive solutions for A^2

leading to three possible steady-state solutions. The intermediate solution is unstable, leading to the jump phenomenon.

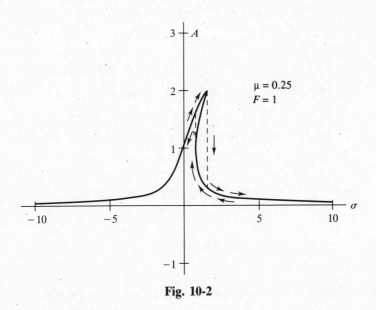

Fig. 10-2

10.4 SELF-EXCITED VIBRATIONS

Self-excited oscillations are oscillations that are excited by the motion of the system. Self-excited oscillations are induced by nonlinear forms of damping where the damping term is negative over a certain range of motion. A mechanical system that exhibits *negative damping*, where the free oscillations amplitude grows, is shown in Fig. 10-3. A model for some self-excited systems is the *van der Pol equation*:

$$\ddot{x} + \mu(x^2 - 1)\dot{x} + x = 0 \qquad (10.5)$$

The phase plane, Fig. 10-4, for the free oscillations of the van der Pol oscillator illustrates a *limit cycle*.

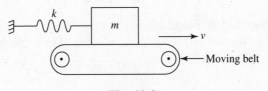

Fig. 10-3

Solved Problems

10.1 The nondimensional form of the nonlinear equation governing the motion of a pendulum is

$$\ddot{\theta} + \sin \theta = 0$$

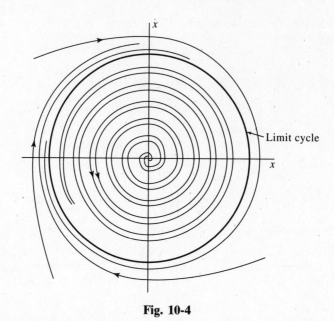

Fig. 10-4

(*i*)   Derive the general equation defining the phase plane for this motion.

(*ii*)   Determine the trajectory for the condition that $\dot{\theta} = 1$ when $\theta = 0$.

(*iii*) What is the maximum angle through which the pendulum will swing?

(*i*)   Define $v = \dot{\theta}$. Then

$$\ddot{\theta} = \frac{d\dot{\theta}}{dt} = \frac{dv}{dt} = \frac{dv}{d\theta}\frac{d\theta}{dt} = v\frac{dv}{d\theta}$$

Thus the differential equation can be written as

$$v\frac{dv}{d\theta} + \sin\theta = 0$$

Integrating with respect to θ leads to

$$\tfrac{1}{2}v^2 - \cos\theta = C$$

where C is a constant of integration.

(*ii*)   Requiring $v = 1$ when $\theta = 0$ leads to $C = -1/2$. Then solving for v,

$$v = \sqrt{2\cos\theta - 1}$$

(*iii*) $v = 0$ when $\theta = 60°$.

10.2 Let $x = x_0$ represent the equilibrium position for a nonlinear system. The motion of the system in the vicinity of the equilibrium point is analyzed by letting $x = x_0 + \Delta x$. Show how the type of the equilibrium point and its stability can be established by linearizing the differential equation about the equilibrium point.

Assume the governing differential equation has the form

$$\ddot{x} + f(x, \dot{x}) = 0$$

If $x = x_0$ represents an equilibrium point, then $f(x_0, 0) = 0$. Substituting $x = x_0 + \Delta x$ into the differential equation leads to

$$\Delta \ddot{x} + f(x_0 + \Delta x, \Delta \dot{x}) = 0$$

Using a Taylor series expansion,

$$\Delta \ddot{x} + f(x_0, 0) + \frac{\partial f}{\partial x}(x_0, 0)\, \Delta x + \frac{\partial f}{\partial \dot{x}}(x_0, 0)\, \Delta \dot{x} + \cdots + = 0$$

Imposing the equilibrium condition and linearizing by ignoring higher-order terms leads to

$$\Delta \ddot{x} + \alpha\, \Delta \dot{x} + \beta\, \Delta x = 0$$

$$\alpha = \frac{\partial f}{\partial \dot{x}}(x_0, 0) \qquad \beta = \frac{\partial f}{\partial x}(x_0, 0)$$

The solution of the previous equation can be written as

$$\Delta x = C_1 e^{\lambda_1 t} + C_2 e^{\lambda_2 t}$$

where λ_1 and λ_2 are the roots of $\lambda^2 + \alpha \lambda + \beta = 0$. The stability and type of equilibrium point are interpreted as follows:

(1) If either λ_1 or λ_2 have a positive real part, then the perturbation from equilibrium grows without bound and the solution is unstable.

(2) If λ_1 and λ_2 are real and have the same sign, the equilibrium point is a node (stable or unstable).

(3) If λ_1 and λ_2 are real and have opposite signs, the equilibrium point is a saddle point (unstable).

(4) If λ_1 and λ_2 are complex conjugates, the equilibrium point is a focus (stable or unstable).

(5) If λ_1 and λ_2 are purely imaginary, the equilibrium point is a center.

10.3 Determine the type and stability of all equilibrium points of the pendulum equation.

The nonlinear differential equation governing the motion of the pendulum is

$$\ddot{\theta} + \sin \theta = 0$$

Using the notation of Problem 10.2,

$$f(\theta, \dot{\theta}) = \sin \theta$$

and $$f(\theta_0, 0) = 0 \;\rightarrow\; \sin \theta_0 = 0 \;\rightarrow\; \theta_0 = n\pi, \qquad n = 0, \pm 1, \pm 2, \ldots$$

Now let

$$\theta = n\pi + \Delta\theta$$

Substitution into the governing equation leads to

$$\Delta \ddot{\theta} + \sin(n\pi + \Delta\theta) = 0$$

Using a Taylor series expansion, keeping only through the linear terms leads to

$$\Delta \ddot{\theta} + \cos(n\pi)\, \Delta\theta = 0$$

$$\Delta \ddot{\theta} + (-1)^n\, \Delta\theta = 0$$

Using the notation of Problem 10.2, the general solution of the above equation is

$$\Delta\theta = C_1 e^{(-1)^{(n-1)/2} t} + C_2 e^{-(-1)^{(n-1)/2} t}$$

and $$\lambda_1 = (-1)^{(n-1)/2}, \qquad \lambda_2 = -(-1)^{(n-1)/2}$$

Hence for odd n, λ_1, and λ_2 are real and of opposite signs. These equilibrium points are saddle points. For even n, λ_1, and λ_2 are purely imaginary. These equilibrium points are centers.

10.4   Sketch the phase plane for the pendulum motion.

The sketch of the phase plane using the results of Problem 10.3 is shown in Fig. 10-5.

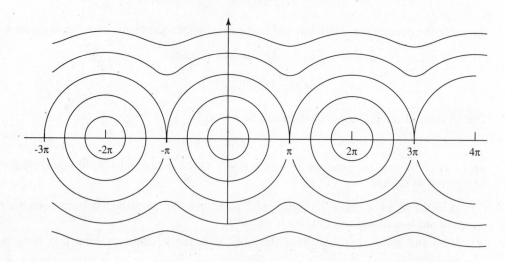

Fig. 10-5

10.5   The differential equation governing the motion of a particle on a rotating parabola, Fig. 10-6, is

$$(1 + 4p^2x^2)\ddot{x} + (2gp - \omega^2)x + 4p^2x\dot{x}^2 = 0$$

If $\omega = 10$ rad/s, for what values of p is the equilibrium point $x = 0$, a saddle point?

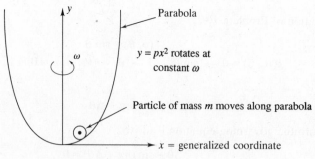

Fig. 10-6

Using the notation of Problem 10.2,

$$f(x, \dot{x}) = \frac{2gp - \omega^2}{1 + 4p^2x^2}x + \frac{4p^2}{1 + 4p^2x^2}x\dot{x}^2$$

Note that $x = 0$ is indeed an equilibrium point. To examine the behavior of phase plane trajectories in its vicinity, let

$$x = \Delta x$$

Using the notation of Problem 10.2,

$$\alpha = \frac{\partial f}{\partial \dot{x}}(0, 0) = 0$$

$$\beta = \frac{\partial f}{\partial x}(0, 0) = 2gp - \omega^2$$

Hence the differential equation governing phase plane trajectories near $x = 0$ is

$$\Delta\ddot{x} + (2gp - \omega^2)\,\Delta x = 0$$

The equilibrium point is a saddle point when

$$2gp < \omega^2$$

Hence for $\omega = 10$ rad/s,

$$p < \frac{\left(10\,\dfrac{\text{rad}}{\text{s}}\right)^2}{2\left(9.81\,\dfrac{\text{m}}{\text{s}^2}\right)} = 5.10\ \text{m}^{-1}$$

Problems 10.6 through 10.8 and Problems 10.11 and 10.12 refer to the system of Fig. 10-7. The force displacement relation for the spring is

$$F = k_1 y - k_3 y^3 \qquad k_1 = 1 \times 10^6\,\frac{\text{N}}{\text{m}} \qquad k_3 = 1 \times 10^{12}\,\frac{\text{N}}{\text{m}^3}$$

where y is measured from the spring's unstretched length.

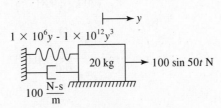

$$1 \times 10^6 y - 1 \times 10^{12} y^3$$

20 kg → 100 sin 50t N

$$100\,\frac{\text{N-s}}{\text{m}}$$

Fig. 10-7

10.6 Let $x = y/\Delta$ where y is the dimensional displacement from the spring's unstretched length and $\Delta = mg/k_1$. Write the differential equation governing the motion of the system of Fig. 10-7 in the form of Eq. (10.1), identifying ε, μ, F, and r.

The dimensional form of the differential equation is obtained by applying Newton's law to the block, resulting in

$$m\ddot{y} + c\dot{y} + k_1 y - k_3 y^3 = F_0 \sin \omega t \qquad\qquad (10.6)$$

The natural frequency of the linearized system is

$$\omega_n = \sqrt{\frac{k_1}{m}} = \sqrt{\frac{1 \times 10^6\,\dfrac{\text{N}}{\text{m}}}{20\ \text{kg}}} = 223.6\,\frac{\text{rad}}{\text{s}}$$

Also,
$$\Delta = \frac{mg}{k_1} = \frac{(20 \text{ kg})\left(9.81 \frac{\text{m}}{\text{s}^2}\right)}{1 \times 10^6 \frac{\text{N}}{\text{m}}} = 1.96 \times 10^{-4} \text{ m}$$

Define
$$\tau = \omega_n t$$

The chain rule yields
$$\frac{d}{dt} = \frac{d}{d\tau}\frac{d\tau}{dt} = \omega_n \frac{d}{dt}$$

Equation (10.6) is rewritten using nondimensional variables as

$$m\omega_n^2 \Delta \frac{d^2x}{d\tau^2} + c\omega_n \Delta \frac{dx}{d\tau} + k_1 \Delta x - k_3 \Delta^3 x^3 = F_0 \sin \frac{\omega}{\omega_n}\tau$$

$$\frac{d^2x}{d\tau^2} + \frac{c}{m\omega_n}\frac{dx}{d\tau} + x - \frac{k_3}{k_1}\Delta^2 x^3 = \frac{F_0}{k_1 \Delta}\sin\frac{\omega}{\omega_n}\tau$$

Which is of the form of Eq. (10.1) with

$$\varepsilon = -\frac{k_3}{k_1}\Delta^2 = -\frac{1 \times 10^{12} \frac{\text{N}}{\text{m}^3}}{1 \times 10^6 \frac{\text{N}}{\text{m}}}(1.96 \times 10^{-4} \text{ m})^2 = -0.0384$$

$$\mu = \frac{c}{2m\omega_n} = \frac{100 \frac{\text{N-s}}{\text{m}}}{2(20 \text{ kg})\left(223.6 \frac{\text{rad}}{\text{s}}\right)} = 0.0112$$

$$F = \frac{F_0}{k_1 \Delta} = \frac{F_0}{mg} = \frac{100 \text{ N}}{(20 \text{ kg})\left(9.81 \frac{\text{m}}{\text{s}^2}\right)} = 0.510$$

$$r = \frac{\omega}{\omega_n} = \frac{150 \frac{\text{rad}}{\text{s}}}{223.6 \frac{\text{rad}}{\text{s}}} = 0.671$$

10.7   Determine the nature and stability of the equilibrium positions of the system of Fig. 10-7.

Using the notation of Problem 10.2,
$$f(x, \dot{x}) = 2\mu\dot{x} + x + \varepsilon x^3$$

Then
$$f(x, 0) = 0 = x + \varepsilon x^3 \rightarrow x = 0, \pm\sqrt{-\frac{1}{\varepsilon}}$$

Note that
$$\frac{\partial f}{\partial \dot{x}} = 2\mu \qquad \frac{\partial f}{\partial x} = 1 + 3\varepsilon x^2$$

First consider the equilibrium point $x = 0$:

$$\alpha = 2\mu \qquad \beta = 1 \rightarrow \Delta\ddot{x} + 2\mu \Delta\dot{x} + \Delta x = 0$$
$$\lambda_{1,2} = -\mu \pm \sqrt{\mu^2 - 1} = -0.0112 \pm 0.999i$$

Since the values of λ are complex conjugates with negative real parts, the equilibrium point $x = 0$ is a stable focus.

Since ε is negative, the system also has equilibrium points corresponding to $x = \pm\sqrt{-1/\varepsilon}$. In either case,

$$a = 2\mu \qquad \beta = 1 + 3\varepsilon\left(-\frac{1}{\varepsilon}\right) = -2$$

$$\Delta\ddot{x} + 2\mu\,\Delta\dot{x} - 2\,\Delta x = 0$$
$$\lambda_{1,2} = -\mu \pm \sqrt{2 + \mu^2} = 1.413,\ -1.415$$

Thus these equilibrium points are saddle points and thus unstable.

10.8 Let $\mu = 0$, and determine an integral expression for the natural period assuming $x = x_0$ and $\dot{x} = 0$ when $t = 0$.

Mathcad

Duffing's equation for free vibrations with $\mu = 0$ is

$$\ddot{x} + x + \varepsilon x^3 = 0$$

Define $v = \dot{x}$. Then

$$v\frac{dv}{dx} + x + \varepsilon x^3 = 0$$

Integrating with respect to x leads to

$$\frac{1}{2}v^2 + \frac{1}{2}x^2 + \frac{\varepsilon}{4}x^4 = C$$

Application of initial conditions and solving for v leads to

$$v = \pm\sqrt{x_0^2 + \frac{\varepsilon}{2}x_0^4 - x^2 - \frac{\varepsilon}{2}x^4}$$

Noting that $v = dx/dt$,

$$dt = \pm\frac{dx}{\sqrt{x_0^2 + \frac{\varepsilon}{2}x_0^4 - x^2 - \frac{\varepsilon}{2}x^4}}$$

One-quarter of the period is the time the block returns to $x = 0$ from its initial position. During this time the velocity is negative. Hence integrating between x_0 and 0 leads to

$$T = 4\int_{x_0}^{0} \frac{dx}{-\sqrt{x_0^2 + \frac{\varepsilon}{2}x_0^4 - x^2 - \frac{\varepsilon}{2}x^4}}$$

10.9 Use a straightforward perturbation expansion to develop a two-term approximation to the solution of Duffing's equation with $F = 0$.

Assume

$$x = x_0(t) + \varepsilon x_1(t) + O(\epsilon^2) \tag{10.7}$$

Substituting Eq. (10.7) into the unforced Duffing's equation leads to

$$\ddot{x}_0 + \varepsilon\ddot{x}_1 + \cdots + x_0 + \varepsilon x_1 + \cdots + \varepsilon(x_0 + \varepsilon x_1 + \cdots +)^3 = 0$$
$$\ddot{x}_0 + x_0 + \varepsilon(\ddot{x}_1 + x_1 + x_0^3) + O(\varepsilon^2) = 0$$

Setting coefficients of like powers of ε to zero independently leads to

$$\ddot{x}_0 + x_0 = 0 \rightarrow x_0 = A\sin(t + \phi)$$
$$\ddot{x}_1 + x_1 = -x_0^3 \rightarrow \ddot{x}_1 + x_1 = -A^3\sin^3(t + \phi)$$

Use of trigonometric identities leads to

$$\ddot{x}_1 + x_1 = -\frac{A^3}{4}[3\sin(t+\phi) - \sin 3(t+\phi)]$$

$$x_1 = \frac{3}{8}A^3 t\cos(t+\phi) - \frac{A^3}{32}\sin 3(t+\phi)$$

Hence, $$x(t) = A\sin(t+\phi) + \varepsilon\left[\frac{3}{8}A^3 t\cos(t+\phi) - \frac{A^3}{32}\sin 3(t+\phi)\right] + O(\varepsilon^2)$$

10.10 The solution developed in Problem 10.9 is not periodic. Why, and what can be done to correct the situation?

In Problem 10.8 it is shown that the natural period of the nonlinear system is dependent on initial conditions. The perturbation solution of Problem 10.9 has no mechanism to allow for this dependence. Indeed, the response is developed at the same period as a linearized system. The situation can be corrected by the introduction of a time scale that is dependent on the amplitude:

$$t = w(1 + \varepsilon\lambda_1 + \varepsilon^2\lambda_2 + \cdots +)$$

The above expression can be introduced before making the straightforward expansion, in which case the method is called the *Linstedt-Poincaré method*. It can also be introduced after the aperiodic straightforward expansion is obtained in an effort to render it periodic. This latter method is called the *method of renormalization*. In either case the results are

$$t = w(1 - \tfrac{3}{8}\varepsilon A^2 + \cdots +)$$

$$x = A\sin(t+\phi) - \varepsilon\frac{A^3}{32}\sin 3(t+\phi) + \cdots +$$

10.11 The block of the system of Fig. 10-7 is displaced 1.0 mm from equilibrium and released. Determine the period of the resulting motion.

Using the nondimensionalization of Problem 10.6, the nondimensional initial conditions are

$$x(0) = \frac{0.001\ \text{m}}{1.96\times 10^{-4}\ \text{m}} = 5.10 \qquad \dot{x}(0) = 0$$

Application of the initial conditions to the two-term uniform expansion developed in Problem 10.10 and noting from Problem 10.5 that $\varepsilon = -0.0384$ lead to

$$\phi = \frac{\pi}{2}$$

$$x(0) = A\sin\frac{\pi}{2} - \frac{\varepsilon A^3}{32}\sin\frac{3\pi}{2}$$

$$5.10 = A - 0.0012A^3 \rightarrow A = 5.28$$

The nondimensional frequency is

$$w = 1 + \tfrac{3}{8}\varepsilon A^2 = 1 + \tfrac{3}{8}(-0.0384)(5.28)^2 = 0.599$$

The dimensional frequency and period are

$$\omega_n = 0.599\left(223.6\ \frac{\text{rad}}{\text{s}}\right) = 133.9\ \frac{\text{rad}}{\text{s}} \rightarrow T = 0.074\ \text{s}$$

10.12 Use a perturbation method to approximate the forced response of the system of Fig. 10-7.

To avoid sign confusion, define $\delta = -\varepsilon$. Since the damping is small, it is ordered with the nonlinearity. To this end

$$\mu = \delta\zeta \rightarrow \zeta = \frac{\mu}{\delta} = \frac{0.0112}{0.0384} = 0.292$$

Then the equation becomes

$$\ddot{x} + 0.584\,\delta\dot{x} + x - \delta x^3 = 0.510\sin 0.671t$$

A straightforward perturbation solution is assumed as

$$x(t) = x_0(t) + \delta x_1(t)$$

Substitution into the governing equation and setting coefficients of like powers of δ to zero lead to

$$\ddot{x}_0 + x_0 = 0.510\sin 0.671t$$

$$x_0 = \frac{0.510}{1 - (0.671)^2}\sin 0.671t = 0.928\sin 0.671t$$

$$\ddot{x}_1 + x_1 = -0.584\dot{x}_0 + x_0^3$$

$$= -0.364\cos 0.671t + 0.799\sin^3 0.671t$$

$$= -0.364\cos 0.671t + 0.599\sin 0.671t - 0.200\sin 2.103t$$

$$x_1(t) = -\frac{0.364}{1 - (0.671)^2}\cos 0.671t + \frac{0.599}{1 - (0.671)^2}\sin 0.671t$$

$$\qquad -\frac{0.200}{1 - (2.013)^2}\sin 2.103t$$

$$= -0.662\cos 0.671t + 1.09\sin 0.671t + 0.0655\sin 2.103t$$

10.13 Discuss quantitative tools that can be used to determine if the motion of a nonlinear system is chaotic.

(a) The trajectory in the phase plane will not repeat itself for chaotic motion.

(b) If a spectral analysis of the time history of motion yields a continuous spectrum, the motion is chaotic.

(c) If the response is sampled at regular intervals, the sampled response of a chaotic motion will appear to be random.

10.14 The Runge-Kutta method has been used to develop the phase planes for Duffing's equation for various values of the parameters, as illustrated in Fig. 10-8. Which of these motions appear to be chaotic?

The motion in Fig. 10-8a appears to be chaotic as there is no discernible pattern to the motion. The motion in Fig. 10-8b is not chaotic as it settles down into a steady state after an initial transient period.

10.15 The Runge-Kutta method has been used to develop *Poincaré sections* for the solution of Duffing's equation, shown in Fig. 10-9. Poincaré sections are samples of the phase plane at regular intervals. Comment on the motion for each Poincaré section.

(a) Since the Poincaré section is a collection of apparently random points, the motion is probably chaotic.

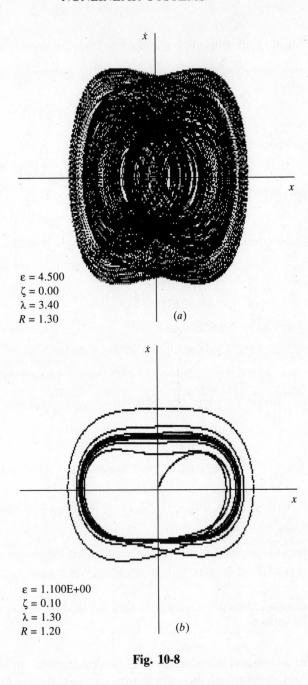

$\varepsilon = 4.500$
$\zeta = 0.00$
$\lambda = 3.40$
$R = 1.30$ (a)

$\varepsilon = 1.100E+00$
$\zeta = 0.10$
$\lambda = 1.30$
$R = 1.20$ (b)

Fig. 10-8

(b) Since the Poincaré section is a closed curve, the motion is periodic, but the sampling frequency is incommensurate with the frequency of motion.

(c) Since the Poincaré section only consists of three points, the motion is periodic, and the period of motion is three times the sampling period.

10.16 Use the van der Pol equation to qualitatively explain the phenomenon of *limit cycles*.

When x is small, the coefficient multiplying $\dot{x}$ in van der Pol's equation is negative. Thus energy is being added to the system through self-excitation. This causes the response to grow. However, when x grows above 1, the damping coefficient becomes positive, and energy is dissipated causing the motion to decay. This continual buildup and decay of amplitude through self-excitation lead to the limit cycle. This limit cycle is independent of initial conditions.

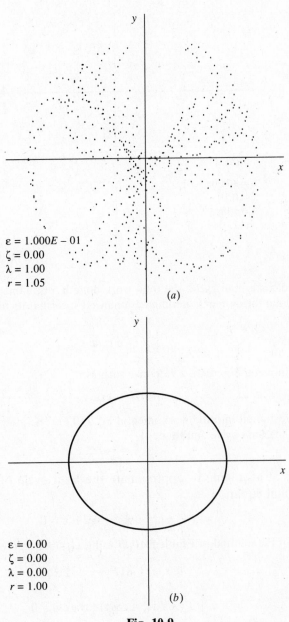

$\varepsilon = 1.000E - 01$
$\zeta = 0.00$
$\lambda = 1.00$
$r = 1.05$

(a)

$\varepsilon = 0.00$
$\zeta = 0.00$
$\lambda = 0.00$
$r = 1.00$

(b)

Fig. 10-9

10.17 Show how the *method of averaging*, or the *Galerkin method*, can be used to approximate the amplitude of a limit cycle.

Let $F(x, \dot{x})$ represent the nonconservative forces in the system. The work done by these forces over 1 cycle of motion is

$$W = \int F(x, \dot{x})\, dx = \int F(x, \dot{x})\dot{x}\, dt$$

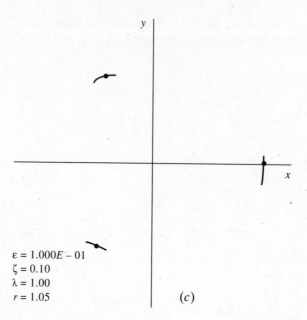

$$\varepsilon = 1.000E - 01$$
$$\zeta = 0.10$$
$$\lambda = 1.00$$
$$r = 1.05 \qquad\qquad (c)$$

Fig. 10-9 (*Continued*)

If the system develops a limit cycle, the total work done by the nonconservative forces over each cycle is zero. Assume the system is nondimensionalized such that its linear period is 2π then

$$\int_0^{2\pi} F(x, \dot{x})\dot{x}\, dt = 0 \qquad\qquad (10.7)$$

When the Galerkin method is used, a response such as

$$x(t) = A \sin t$$

is assumed and substituted into the work integral Eq. (10.7). The integral is evaluated, yielding an approximation to the limit cycle amplitude A.

10.18 Use the method of averaging to approximate the limit cycle of the system governed by the nondimensional equation

$$\ddot{x} + \alpha(\dot{x}^2 + x^2 - 1)\dot{x} + x = 0$$

Application of the method of Problem 10.17 using $x(t) = A \sin t$ leads to

$$F(x, \dot{x}) = \alpha(\dot{x}^2 + x^2 - 1)\dot{x}$$

$$\int_0^{2\pi} F(A \sin t, A \cos t)A \cos t\, dt = 0$$

$$\int_0^{2\pi} \alpha[A^2 \cos^2 t + A^2 \sin^2 t - 1](A \cos t)^2\, dt = 0$$

$$\alpha(A^2 - 1)A^2 \int_0^{2\pi} \cos^2 t\, dt = 0$$

$$\pi\alpha(A^2 - 1)A^2 = 0$$

$$A = 1$$

Supplementary Problems

10.19 Develop the general equation for the trajectory in the phase plane for a system governed by

$$\ddot{x} + x + \varepsilon x \cos x = 0$$

Ans.

$$\dot{x} = \sqrt{C - x^2 - 2x \sin x - 2 \cos x}$$

10.20 Develop the general equation for a trajectory in the phase plane for a system governed by the equation

$$\ddot{x} + x - \alpha x^2 = 0$$

Ans.

$$\dot{x} = \sqrt{C - x^2 + \tfrac{2}{3}\alpha x^3}$$

10.21 Determine the equilibrium points and their type for the system of Problem 10.20.

Ans. $x = 0$ is a center; $x = \alpha$ is a saddle point.

10.22 Sketch the phase plane for the system of Problem 10.20.

10.23 Determine the equilibrium points and their type for a system governed by

$$\ddot{x} + 2\zeta\dot{x} + x + \varepsilon x^2 = 0$$

Ans. $x = 0$ is a stable focus for $\zeta < 1$ and a stable node for $\zeta > 1$; $x = -1/\varepsilon$ is a saddle point.

10.24 Determine the equilibrium points and their type for a system governed by

$$\ddot{x} + 2\zeta\dot{x} - x + \varepsilon x^3 = 0$$

Ans. $x = 0$ is a saddle point; $x = \pm\sqrt{1/\varepsilon}$ are stable foci for $\zeta < \sqrt{2}$ and stable nodes for $\zeta > \sqrt{2}$.

10.25 Derive an integral expression for the period of motion of the nonlinear system governed by

$$\ddot{\theta} + \sin\theta(1 - \cos\theta) = 0$$

subject to $\theta = \theta_0$ when $\dot{\theta} = 0$.

Ans.

$$T = 4 \int_{\theta_0}^{0} \frac{d\theta}{-\sqrt{-\tfrac{1}{2}\cos 2\theta_0 + 2\cos\theta_0 + \tfrac{1}{2}\cos 2\theta - 2\cos\theta}}$$

10.26 A 50-kg block is attached to a spring whose force displacement relation is

$$F = 2000x + 6000x^3$$

for x in meters and F in newtons. The block is displaced 25 cm and released. What is the period of the ensuing oscillations?

Ans. 0.907 s

10.27 Use the perturbation method to obtain a two-term approximation to the response of a system governed by

$$\ddot{x} + \mu\varepsilon\dot{x} + x + \varepsilon x^2 = F \sin \omega t$$

Ans.

$$\frac{F}{1-\omega^2}\sin\omega t + \varepsilon\left[-\frac{1}{2}\left(\frac{F}{1-\omega^2}\right)^2\left(1-\frac{1}{1-4\omega^2}\cos 2\omega t\right)\right.$$

$$\left.-\frac{\mu\omega F}{(1-\omega^2)^2}\cos\omega t\right]$$

10.28 Use Galkerkin's method to approximate the amplitude of the limit cycle of van der Pol's equation.

Ans.   2

10.29 Explain the jump phenomenon from Fig. 10-2.

10.30 Discuss how the Fourier transform of a response can be used to determine if the response is chaotic.

Chapter 11

Computer Applications

Vibration analysis often requires much mathematical analysis and computation. Digital computation can be used in lieu of manual computation for many of the tedious tasks performed in vibration analysis. Computer algebra can be used to perform tedious mathematical analysis. However, the user must understand the sequence of the steps and how the results are used.

The focus of this chapter is the use of applications software for vibration analysis. It is worthwhile to know how to program using a higher-order programming language such as C, PASCAL, or FORTRAN, and programs can be written in these languages to solve many vibrations problems. However, much of the analysis used in the preceding chapters can be performed on personal computers using applications software.

The finite element method, a powerful method for approximating the solution of continuous vibrations problems when an exact solution is difficult to attain, is illustrated in Chap. 9. However, for the sake of illustration and for brevity, the examples presented here use at most four elements. When more elements are used, digital computation is essential in obtaining a solution. Many difficulties are encountered in the development of a large-scale finite element model. These range from efficient methods of assembly of the global mass and stiffness matrices to solution of the resulting differential equations using modal analysis. Thus large-scale finite element programs have been developed. Some are available for use on the personal computer. However, they often require pre- and postprocessor programs and are beyond the scope of this book.

11.1 SOFTWARE SPECIFIC TO VIBRATIONS APPLICATIONS

Software written specifically for vibrations applications is available. The programs in the software package *VIBES*, which accompanies the McGraw-Hill text *Fundamentals of Mechanical Vibrations* by Kelly, include programs that simulate the free and forced response of 1- and multi-degree-of-freedom systems. *VIBES* also has programs that numerically integrate the convolution integral, develop force and displacement spectra, perform modal analysis for continuous systems, and aid in the design of vibration isolators and vibration absorbers. Many of the files are executable programs while several require user-provided BASIC subprograms to allow for any type of excitation.

11.2 SPREADSHEET PROGRAMS

Spreadsheets allow the development of relationships between variables and parameters in tabular form. Spreadsheets also have graphical capabilities for presentation of results. The columns and graphs in a spreadsheet are automatically updated when the value of a parameter is changed. Thus the spreadsheet is a useful tool in "what-if" situations such as design applications. Examples of popular spreadsheets are Lotus Development Corporation's *Lotus 1–2–3*, Microsoft's *Excel*, Borland's *Paradox*, and WordPerfect's *Quatro Pro*.

11.3 ELECTRONIC NOTEPADS

When using an electronic notepad, the user develops the solution on the computer screen as if she or he were using pen, paper, and calculator. Electronic notepads such as MathSoft's *Mathcad* and The Math Works, Inc.'s, *MATLAB* provide mechanisms for performing complex sets of calculations. Electronic notepads have built-in algorithms that allow the user to quickly perform complicated calculations. These include numerical integrations and matrix eigenvalue algorithms. Electronic notepads also have automatic update, so that when the value of a parameter is changed, all subsequent calculations involving the parameter are recalculated. Electronic notepads also have graphical capabilities and allow for limited symbolic processing.

11.4 SYMBOLIC PROCESSORS

Symbolic processors such as *MAPLE V, MACSYMA,* and *Mathematica* perform symbolic manipulations. Examples of symbolic manipulations include differentiation with respect to a variable, indefinite integration, partial fraction decompositions, and solving equations for solutions in terms of parameters. Computer algebra software can also be used for linear algebra and solutions of differential equations.

Solved Problems

11.1   Use *VIBES* to plot the response of a 1-degree-of-freedom system of mass 100 kg, natural frequency 100 rad/s, and damping ratio 0.3 subject to the excitation

$$F(t) = 1000 \sin 125t \text{ N}$$

The *VIBES* program *FORCED* is used to develop the response as shown in Fig. 11-1. The excitation is plotted simultaneously with the response for comparison. The plot illustrates the transient response giving way to a steady-state response. The plots also illustrate the difference in period between the excitation and response and the phase difference.

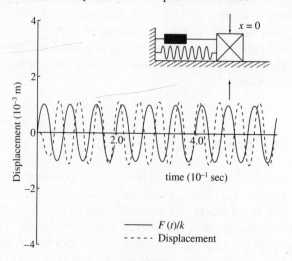

Fig. 11-1

11.2   Use *VIBES* to determine approximations to the natural frequencies and mode shapes of a uniform fixed-pinned beam when 4 degrees of freedom are used to model the beam.

The 4-degree-of-freedom model is illustrated in Fig. 11-2. The flexibility matrix is obtained using the *VIBES* program *BEAM*. Unit values of beam properties are used as input in *BEAM*. Thus the numerical values obtained in this example must be multiplied by L^3/EI to obtain the elements of the flexibility matrix. The output from *BEAM* is shown in Fig. 11-3. The *VIBES* program *MITER* uses matrix iteration to determine natural frequencies and normalized mode shapes of a multi-degree-of-freedom system. The flexibility matrix obtained from *BEAM* is used as input as well as the mass matrix

$$M = \begin{bmatrix} \frac{1}{5} & 0 & 0 & 0 \\ 0 & \frac{1}{5} & 0 & 0 \\ 0 & 0 & \frac{1}{5} & 0 \\ 0 & 0 & 0 & \frac{1}{5} \end{bmatrix} \rho A L$$

Again unit values of the properties are used. Hence the numerical values shown in Fig. 11-4 obtained using *MITER* in this example are nondimensional. The dimensional natural frequency approximations are obtained by multiplying these values by $EI/\rho AL^4$. The mode shape vectors determined using *MITER* have been normalized with respect to the mass matrix.

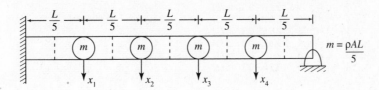

Fig. 11-2

11.3   Use *VIBES* to determine the three lowest natural frequencies and mode shape plots for the beam of Fig. 11-5.

The *VIBES* program *CFREQ* is used to determine the natural frequencies and mode shapes of the continuous system. Note that $\beta = m/\rho AL$. The natural frequency and mode shapes generated by *CFREQ* are shown in Figs 11.6 and 11.7.

11.4   A 100-kg reciprocating machine, which operates at 250 r/min, has a rotating unbalance of magnitude 0.5 kg-m. What is the maximum stiffness of a vibration isolator of damping ratio 0.1 to limit the transmitted force to 5000 N? What is the required static deflection of the isolator? What is the maximum deflection of the isolator during operation? Use *Mathcad* for the calculations.

The electronic notepad developed using *Mathcad* follows (Fig. 11-8). The methods used are those developed in Chap. 8. Note that m could not be used as the variable name for mass since *Mathcad* reserves its use to represent the units of meters. In addition, e could not be used as the variable name for eccentricity since *Mathcad* reserves its use for the base of the natural logarithm. When finding the root of a single equation, *Mathcad* requires an initial guess for the root.

11.5   A 100-kg structure of natural frequency 100 rad/s and damping ratio 0.05 is at rest in equilibrium when it is subject to an excitation of the form

$$F(t) = 12{,}500e^{-1.5t^{1.5}} \, \text{N}$$

Use *Mathcad* to develop the response of the system using the convolution integral.

FLEXIBILITY MATRIX FOR A   4 -DEGREE-OF-FREEDOM MODEL OF A BEAM THAT IS
FIXED-PINNED

THE BEAM'S PROPERTIES ARE:
LENGTH = 1   m
ELASTIC MODULUS= 1   N/m^2
MOMENT OF INERTIA= 1   m^4

THE LOCATIONS OF THE NODAL POINTS ARE:
X(1)= .2   m
X(2)= .4   m
X(3)= .6   m
X(4)= .8   m

```
    A( 1   , 1   )=   1.621E-03 m/N
    A( 1   , 2   )=   2.784E-03 m/N
    A( 1   , 3   )=   2.603E-03 m/N
    A( 1   , 4   )=   1.525E-03 m/N
    A( 2   , 1   )=   2.784E-03 m/N
    A( 2   , 2   )=   6.912E-03 m/N
    A( 2   , 3   )=   7.381E-03 m/N
    A( 2   , 4   )=   4.523E-03 m/N
    A( 3   , 1   )=   2.603E-03 m/N
    A( 3   , 2   )=   7.381E-03 m/N
    A( 3   , 3   )=   9.792E-03 m/N
    A( 3   , 4   )=   6.624E-03 m/N
    A( 4   , 1   )=   1.525E-03 m/N
    A( 4   , 2   )=   4.523E-03 m/N
    A( 4   , 3   )=   6.624E-03 m/N
    A( 4   , 4   )=   5.461E-03 m/N
```

Fig. 11-3


```
THE CONTROL MESSAGE IS ERR = 0
The natural frequency for mode 1 is  1.541E+01
The corresponding normalized mode shape is
  1  .4558122
  2  1.206697
  3  1.505346
  4  1.034433

The natural frequency for mode 2 is  4.969E+01
The corresponding normalized mode shape is
  1 -1.076292
  2 -1.314738
  3  .4244277
  4  1.390295

The natural frequency for mode 3 is  1.005E+02
The corresponding normalized mode shape is
  1  1.465718
  2 -.2379109
  3 -1.111364
  4  1.248975

The natural frequency for mode 4 is  1.518E+02
The corresponding normalized mode shape is
  1  1.218811
  2 -1.326177
  3  1.148331
  4 -.6611273
```

Fig. 11-4

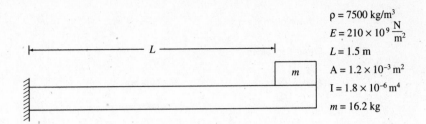

$$\rho = 7500 \ \text{kg/m}^3$$
$$E = 210 \times 10^9 \ \frac{\text{N}}{\text{m}^2}$$
$$L = 1.5 \ \text{m}$$
$$A = 1.2 \times 10^{-3} \ \text{m}^2$$
$$I = 1.8 \times 10^{-6} \ \text{m}^4$$
$$m = 16.2 \ \text{kg}$$

Fig. 11-5

```
Natural frequencies and mode shapes for a fixed-attached mass beam
                    with beta= 1.200


BEAM PROPERTIES
     mass density=  7.500E+03 kg/m^3
     elastic modulus= 2.100E+11 N/m^2
     length=  1.500E+00 m
     area= 1.200E-03 m^2
     moment of inertia= 1.800E-06 m^4
```

Mode Number	Dimensionless frequency	Natural frequency (rad/sec)	Normalization constant
1	1.83	167.03	0.712E+00
2	20.11	1831.36	0.520E+00
3	59.61	5429.45	0.472E+00

Fig. 11-6

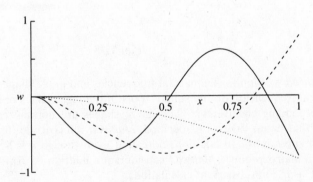

Fig. 11-7

Mathcad uses a Romberg integration scheme to numerically evaluate definite integrals. *Mathcad* uses a default tolerance for numerical integration of 0.001. The tolerance can be changed by the user.

Two methods of solution are presented (Figs 11-9 and 11-10). The first is a direct method where the integration is carried out over the entire time interval from 0 to t for each value of t. The alternate method uses the results of Problem 4.27 where the convolution integral is rewritten as the sum of two integrals. Using this formulation, the results of the previous integrations can be used and the new integration is carried out over only the new interval.

11.6 Use *Mathcad* to determine the natural frequencies and normalized mode shapes for the system of Fig. 11-11.

Solution of Problem 11.4

Parameter values

$\text{mass} := 100 \cdot \text{kg}$ Machine mass

$\omega := 250 \cdot \dfrac{\text{rad}}{\text{sec}}$ Operating speed

$\zeta := 0.1$ Damping ratio

$m_0 := 10 \cdot \text{kg}$ Unbalanced mass

$\text{ecc} := 0.05 \cdot \text{m}$ Eccentricity

$F_{max} := 5000 \cdot \text{newton}$ Maximum allowable force

$g := 9.81 \cdot \dfrac{\text{m}}{\text{sec}^2}$ Acceleration due to gravity

Function definitions

$$M(r, \zeta) := \frac{1}{\left[\left(1 - r^2\right)^2 + \left(2 \cdot \zeta \cdot r\right)^2 \right]^{\frac{1}{2}}}$$ Magnification factor

$$T(r, \zeta) := \left[\frac{1 + \left(2 \cdot \zeta \cdot r\right)^2}{\left(1 - r^2\right)^2 + \left(2 \cdot \zeta \cdot r\right)^2} \right]^{\frac{1}{2}}$$ Transmissibility ratio

Fig. 11-8

The *Mathcad* solution for the natural frequencies and mode shapes follows in Fig. 11-12. Note that the natural frequencies are the square roots of the eigenvalues of $\mathbf{M}^{-1}\mathbf{K}$ and the mode shapes are the corresponding eigenvectors. The mode shapes are normalized with respect to the mass matrix. Note that unless otherwise specified, *Mathcad* refers to the first row or first column of a matrix with a subscript 0. In addition, note that even though $q = \mathbf{X}^T\mathbf{M}\mathbf{X}$ is a scalar, since it is calculated as a matrix product, *Mathcad* considers it a matrix of 1 row and 1 column. Thus it must be referred to as $q_{0,0}$ in subsequent calculations.

11.7 Use *Mathcad* to help perform modal analysis to determine the steady-state response of the system of Fig. 11-12,

The modal analysis procedure of Chap. 6 is followed in developing the notepad presented in Fig. 11-13. The modal matrix is formed by augmenting the normalized mode shapes. The vector $\mathbf{G} = \mathbf{P}^T\mathbf{F}$ is formed, and the differential equations for the principal coordinates are

$$\ddot{p}_i + \omega_i^2 p_i = G_i \sin \omega t$$

The steady-state response for the principal coordinates is

$$p_i = \frac{G_i}{\omega_i^2 - \omega^2} \sin \omega t$$

The original generalized coordinates are then calculated from $\mathbf{x} = \mathbf{P}\mathbf{p}$.

Function graphs

$r := 0, 0.02 .. 3.0$

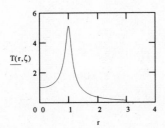

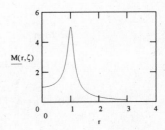

Problem Solution

$F_0 := m_0 \cdot ecc \cdot \omega^2$ $F_0 = 3.125 \cdot 10^4$ ·newton   Excitation amplitude

$T_{max} := \dfrac{F_{max}}{F_0}$ $T_{max} = 0.16$ Maximum transmissibility

$r_g := 3.2$ Initial guess for minimum r

$r_1 := root\left(T\left(r_g, \zeta\right) - T_{max}, r_g\right)$ Solution for minimum r

$r_1 = 2.8603$ Minimum allowable frequency ratio

$\omega_n := \dfrac{\omega}{r_1}$ $\omega_n = 8.7404 \cdot 10^1 \cdot \dfrac{rad}{sec}$ Maximum allowable natural frequency

$k := mass \cdot \omega_n^2$ $k = 7.6394 \cdot 10^5 \cdot \dfrac{newton}{m}$ Maximum allowable isolator stiffness

$\Delta := \dfrac{g}{\omega_n^2}$ $\Delta = 1.2841 \cdot 10^{-3}$ ·m   Minimum isolator static deflection

$x_{max} := \dfrac{F_0}{k} \cdot M\left(r_1, \zeta\right)$ $x_{max} = 5.6783 \cdot 10^{-3}$ ·m   Maximum isolator deflection

Fig. 11-8 (*Continued.*)

11.8 Use *Mathcad* to determine the finite element approximations to the longitudinal natural frequencies of a 2.9-m, fixed-free bar with $E = 210 \times 10^9$ N/m^2 and $\rho = 7100$ kg/m^3 when four elements are used to model the bar.

Mathcad

Solution of Problem 11.5 - Numerical evaluation of convolution integral

System parameters

$mass := 100 \cdot kg$ System mass

$\omega_n := 100 \cdot \dfrac{rad}{sec}$ Natural frequency

$\zeta := 0.05$ Damping ratio

Excitation

$F_0 := 12500 \cdot newton$

$F(t) := F_0 \cdot exp\left(-1.5 \cdot \dfrac{1}{sec^{1.5}} \cdot t^{1.5}\right)$

Impulsive response

$\omega_d := \omega_n \cdot \left(1 - \zeta^2\right)^{\frac{1}{2}}$ Damped natrual frequency

$h(t) := \dfrac{1}{mass \cdot \omega_d} \cdot exp\left(-\zeta \cdot \omega_n \cdot t\right) \cdot sin\left(\omega_d \cdot t\right)$ System response due to a unit impulse applied at t= 0

Convolution integral formula

$x(t) := \displaystyle\int_{0 \cdot sec}^{t} F(\tau) \cdot h(t - \tau) \, d\tau$

$t := 0 \cdot sec, 0.001 \cdot sec .. 0.3 \cdot sec$

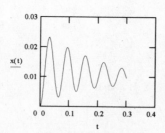

Fig. 11-9

The global mass and stiffness matrices for a four-element finite element model of a fixed-free bar are determined in Problem 9.3. Note that for convenience, nondimensional forms of these matrices are used in *Mathcad* calculations, as shown in Fig. 11-14.

Alternate solution of Problem 11.5 - Use of the method of Problem 4.27

System parameters

$mass := 100 \cdot kg$

$\omega_n := 100 \cdot \dfrac{rad}{sec}$

$\zeta := 0.05$

$\omega_d := \omega_n \cdot \left(1 - \zeta^2\right)^{\frac{1}{2}}$ $\omega_d = 99.875 \cdot \dfrac{rad}{sec}$

$F_0 := 12500 \cdot newton$

Excitation

$F(t) := F_0 \cdot \exp\left(-1.5 \cdot \dfrac{1}{sec^{1.5}} \cdot t^{1.5}\right)$

$i := 1, 2 \ldots 300$

$t_i := i \cdot 0.001 \cdot sec$

$h_1(t) := \dfrac{1}{mass \cdot \omega_d} \cdot \exp\left(\zeta \cdot \omega_n \cdot t\right) \cdot \cos\left(\omega_d \cdot t\right)$

$h_2(t) := \dfrac{1}{mass \cdot \omega_d} \cdot \exp\left(\zeta \cdot \omega_n \cdot t\right) \cdot \sin\left(\omega_d \cdot t\right)$

$G1_0 := 0 \cdot m$ $G2_0 := 0 \cdot m$

$G1_i := G1_{i-1} + \displaystyle\int_{t_{i-1}}^{t_i} F(\tau) \cdot h_1(\tau)\, d\tau$ $G2_i := G2_{i-1} + \displaystyle\int_{t_{i-1}}^{t_i} F(\tau) \cdot h_2(\tau)\, d\tau$

$x_i := \exp\left(-\zeta \cdot \omega_n \cdot t_i\right) \cdot \left(\sin\left(\omega_d \cdot t_i\right) \cdot G1_i - \cos\left(\omega_d \cdot t_i\right) \cdot G2_i\right)$

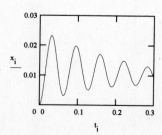

Fig. 11-10

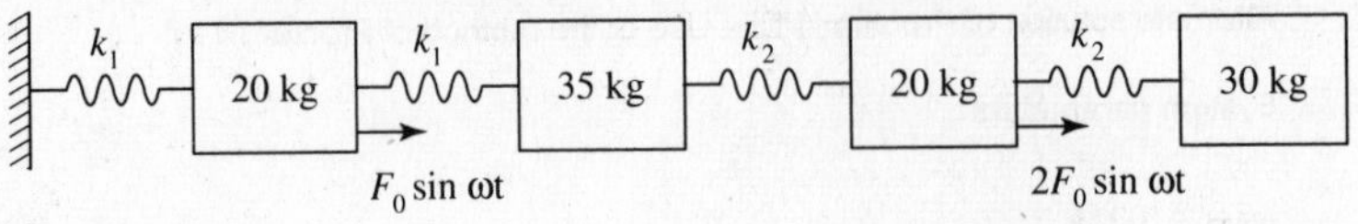

$k_1 = 10\,000$ N/m

$k_2 = 20\,000$ N/m

$F_0 = 1000$ N/m

$\omega = 35$ r/s

Fig. 11-11

Solution to Problem 11.6 - Natural frequencies and mode shapes for a 4 DOF system

Mass matrix

$$M := \begin{bmatrix} 20 & 0 & 0 & 0 \\ 0 & 35 & 0 & 0 \\ 0 & 0 & 20 & 0 \\ 0 & 0 & 0 & 30 \end{bmatrix}$$

Stiffness matrix

$$K := \begin{bmatrix} 20000 & -10000 & 0 & 0 \\ -10000 & 30000 & -20000 & 0 \\ 0 & -20000 & 40000 & -20000 \\ 0 & 0 & -20000 & 20000 \end{bmatrix}$$

Inverse of mass matrix

$$M^{-1} = \begin{bmatrix} 0.05 & 0 & 0 & 0 \\ 0 & 0.029 & 0 & 0 \\ 0 & 0 & 0.05 & 0 \\ 0 & 0 & 0 & 0.033 \end{bmatrix}$$

$$D := M^{-1} \cdot K$$

$$D = \begin{bmatrix} 1 \cdot 10^3 & -500 & 0 & 0 \\ -285.714 & 857.143 & -571.429 & 0 \\ 0 & -1 \cdot 10^3 & 2 \cdot 10^3 & -1 \cdot 10^3 \\ 0 & 0 & -666.667 & 666.667 \end{bmatrix}$$

Eigenvalues of D

$$\lambda := \text{eigenvals}(D)$$

$$\lambda = \begin{bmatrix} 2.665 \cdot 10^3 \\ 1.22 \cdot 10^3 \\ 589.073 \\ 49.733 \end{bmatrix}$$

Natural frequencies

$$\omega_1 := (\lambda_3)^{\frac{1}{2}} \qquad\qquad \omega_1 = 7.052$$

$$\omega_2 := (\lambda_2)^{\frac{1}{2}} \qquad\qquad \omega_2 = 24.271$$

$$\omega_3 := (\lambda_1)^{\frac{1}{2}} \qquad\qquad \omega_3 = 34.924$$

Fig. 11-12

$$\omega_4 := (\lambda_0)^{\frac{1}{2}} \qquad\qquad \omega_4 = 51.627$$

Normalized mode shapes

$$X_1 := \text{eigenvec}(D, \lambda_3) \qquad\qquad X_1 = \begin{bmatrix} 0.259 \\ 0.491 \\ 0.565 \\ 0.61 \end{bmatrix}$$

$$q := X_1{}^T \cdot M \cdot X_1 \qquad\qquad q = 27.347$$

$$X_1 := \dfrac{X_1}{\sqrt{q_{0,0}}} \qquad\qquad X_1 = \begin{bmatrix} 0.049 \\ 0.094 \\ 0.108 \\ 0.117 \end{bmatrix}$$

$$X_2 := \text{eigenvec}(D, \lambda_2) \qquad\qquad X_2 = \begin{bmatrix} -0.614 \\ -0.504 \\ 0.07 \\ 0.603 \end{bmatrix}$$

$$q := X_2{}^T \cdot M \cdot X_2 \qquad\qquad q = 27.457$$

$$X_2 := \dfrac{X_2}{\sqrt{q_{0,0}}} \qquad\qquad X_2 = \begin{bmatrix} -0.117 \\ -0.096 \\ 0.013 \\ 0.115 \end{bmatrix}$$

Fig. 11-12 (*Continued.*)

11.9 A 1500-kg machine is mounted on a foundation of stiffness 2×10^7 N/m. The machine has a vibration amplitude of 7.3 mm when it operates at 1000 r/min. It is desired to design a vibration absorber for the machine to eliminate steady-state vibrations at 1000 r/min. Use *Mathcad* to determine the following:

(*i*) The stiffness and mass of an undamped absorber to eliminate steady-state vibrations of the machine at 1000 r/min and to limit the steady-state amplitude of the absorber mass to 1.5 mm.

(*ii*) The natural frequencies of the system with the absorber in place.

(*iii*) The range of speeds near 1000 r/min such that the machine's steady-state amplitude is less than 2 mm.

The *Mathcad* notepad using the equations presented in Chap. 8 follows in Fig. 11-15. Please note the following regarding the solution: (1) Radians is a *Mathcad* defined unit, whereas revolutions is not. Thus a statement defining rev must be made before using it in an equation. (2) When the absorber is added, one natural frequency of the resulting 2-degree-of-freedom system is

$$X_3 := \text{eigenvec}(D, \lambda_1) \qquad X_3 = \begin{bmatrix} -0.873 \\ 0.383 \\ 0.193 \\ -0.233 \end{bmatrix}$$

$$q := X_3^T \cdot M \cdot X_3 \qquad q = 22.747$$

$$X_3 := \frac{X_3}{\sqrt{q_{0,0}}} \qquad X_3 = \begin{bmatrix} -0.183 \\ 0.08 \\ 0.04 \\ -0.049 \end{bmatrix}$$

$$X_4 := \text{eigenvec}(D, \lambda_0) \qquad X_4 = \begin{bmatrix} 0.09 \\ -0.299 \\ 0.901 \\ -0.301 \end{bmatrix}$$

$$q := X_4^T \cdot M \cdot X_4 \qquad q = 22.245$$

$$X_4 := \frac{X_4}{\sqrt{q_{0,0}}} \qquad X_4 = \begin{bmatrix} 0.019 \\ -0.063 \\ 0.191 \\ -0.064 \end{bmatrix}$$

Fig. 11-12 (*Continued.*)

less than ω_{22} while one is greater than ω_{22}. The root function uses an iteration to find the root. For a function with multiple roots, it will generally converge to the root nearest the initial guess. (3) A poor initial guess for the frequency where $X_1 = 2$ mm may lead to the iteration process used by the root function not to converge. This is due to the large derivatives of the function near the natural frequencies.

11.10 Use a Laplace transform solution using *MAPLE V* to find the response of a 1-degree-of-freedom mass-spring system initially at rest in equilibrium at $t = 0$ when subject to the excitation of Fig. 11-16.

The excitation of Fig. 11-16 is represented mathematically as

$$F(t) = F_0 e^{-\alpha t}\left[u(t) - u\left(t - \frac{1}{\alpha}\right)\right]$$

The unit step function used throughout this text is referred to as the *Heaviside* function in *MAPLE V*. The dsolve command with the laplace option solves the differential equation using the Laplace transform method. In this case (Fig. 11-17) the solution is returned in terms of inverse transforms

Solution to Problem 11.7 - Natural frequencies and normalized mode shapes calculated in solution of Problem 11.6.

Modal matrix

$P_1 := \text{augment}(X_1, X_2)$ $P_2 := \text{augment}(X_3, X_4)$

$P := \text{augment}(P_1, P_2)$ $P = \begin{bmatrix} 0.049 & -0.117 & -0.183 & 0.019 \\ 0.094 & -0.096 & 0.08 & -0.063 \\ 0.108 & 0.013 & 0.04 & 0.191 \\ 0.117 & 0.115 & -0.049 & -0.064 \end{bmatrix}$

Excitation vector   Right hand side vector

$F := \begin{bmatrix} 1000 \\ 0 \\ 2000 \\ 0 \end{bmatrix}$ $\omega := 35$ $G := P^T \cdot F$ $G = \begin{bmatrix} 265.487 \\ -90.305 \\ -102.007 \\ 401.193 \end{bmatrix}$

Steady-state amplitudes of principal coordinates

$p_0 := \dfrac{G_0}{\omega_1^2 - \omega^2}$ $p_0 = -0.226$

$p_1 := \dfrac{G_1}{\omega_2^2 - \omega^2}$ $p_1 = 0.142$

$p_2 := \dfrac{G_2}{\omega_2^2 - \omega^2}$ $p_2 = 0.16$

$p_3 := \dfrac{G_3}{\omega_4^2 - \omega^2}$ $p_3 = 0.279$

Steady-state amplitudes of generalized coordinates

$x := P \cdot p$ $x = \begin{bmatrix} -0.052 \\ -0.04 \\ 0.037 \\ -0.036 \end{bmatrix}$

Fig. 11-13

Solution to Problem 11.8 - Finite element model of bar

Global mass matrix   **Global stiffness matrix**

$$M := \begin{bmatrix} 4 & 1 & 0 & 0 \\ 1 & 4 & 1 & 0 \\ 0 & 1 & 4 & 1 \\ 0 & 0 & 1 & 2 \end{bmatrix}$$

$$K := \begin{bmatrix} 2 & -1 & 0 & 0 \\ -1 & 2 & -1 & 0 \\ 0 & -1 & 2 & -1 \\ 0 & 0 & -1 & 1 \end{bmatrix}$$

$$D := M^{-1} \cdot K$$

$$D = \begin{bmatrix} 0.608 & -0.433 & 0.124 & -0.031 \\ -0.433 & 0.732 & -0.495 & 0.124 \\ 0.124 & -0.495 & 0.856 & -0.464 \\ -0.062 & 0.247 & -0.928 & 0.732 \end{bmatrix}$$

Eigenvalues of D

$$\lambda := \text{eigenvals}(D)$$

$$\lambda = \begin{bmatrix} 1.788 \\ 0.855 \\ 0.259 \\ 0.026 \end{bmatrix}$$

Natural frequency calculations

$$E := 210 \cdot 10^9 \cdot \frac{newton}{m^2}$$ Elastic modulus

$$\rho := 7100 \cdot \frac{kg}{m^3}$$ Mass density

$$L := 2.9 \cdot m$$ Length

$$C := \sqrt{\frac{6 \cdot E}{\rho \cdot \left(\frac{L}{4}\right)^2}}$$ $$C = 1.837 \cdot 10^4 \cdot \frac{rad}{sec}$$

$$\omega = C \lambda^{1/2}$$

$$\omega_1 := C \cdot (\lambda_3)^{\frac{1}{2}}$$ $$\omega_1 = 2.965 \cdot 10^3 \cdot \frac{rad}{sec}$$

$$\omega_2 := C \cdot (\lambda_2)^{\frac{1}{2}}$$ $$\omega_2 = 9.353 \cdot 10^3 \cdot \frac{rad}{sec}$$

$$\omega_3 := C \cdot (\lambda_1)^{\frac{1}{2}}$$ $$\omega_3 = 1.699 \cdot 10^4 \cdot \frac{rad}{sec}$$

$$\omega_4 := C \cdot (\lambda_0)^{\frac{1}{2}}$$ $$\omega_4 = 2.457 \cdot 10^4 \cdot \frac{rad}{sec}$$

Fig. 11-14

Solution of Problem 11.9 - vibration absorber design

$m_1 := 1500 \cdot kg$ Mass of primary system

$k_1 := 2 \cdot 10^7 \cdot \dfrac{newton}{m}$ Stiffness of primary system

$rev := 2 \cdot \pi \cdot rad$ Unit conversion

$\omega := 1000 \cdot \dfrac{rev}{min} \cdot \dfrac{1 \cdot min}{60 \cdot sec}$ $\omega = 104.72 \cdot \dfrac{rad}{sec}$ Operating speed

$X := 0.0073 \cdot m$ Steady-state amplitude of primary system without absorber

$X_{2max} := 0.0015 \cdot m$ Maximum absorber amplitude

$X_{1max} := 0.002 \cdot m$ Maximum amplitude of primary system with absorber

Calculations

$\omega_{11} := \sqrt{\dfrac{k_1}{m_1}}$ $\omega_{11} = 115.47 \cdot \dfrac{rad}{sec}$ Natural frequency of primary system

$\omega_{22} := \omega$ $\omega_{22} = 104.72 \cdot \dfrac{rad}{sec}$ Natural frequency of absorber

$F_0 := k_1 \cdot X \cdot \left[1 - \left(\dfrac{\omega}{\omega_{11}} \right)^{\frac{1}{2}} \right]$ $F_0 = 6.962 \cdot 10^3 \cdot newton$ Excitation amplitude

Absorber design

$k_2 := \dfrac{F_0}{X_{2max}}$ $k_2 = 4.642 \cdot 10^6 \cdot \dfrac{newton}{m}$ Absorber stiffness

$m_2 := \dfrac{k_2}{\omega_{22}^2}$ $m_2 = 423.259 \cdot kg$ Absorber mass

Fig. 11-15

that *MAPLE V* was not able to invert. Note that *MAPLE V* applied the first shifting theorem. The Laplace transform of the Heaviside function is known, and the shifting theorem applied. Thus using the results of *MAPLE V*, the system response is

$$y(t) = \frac{F_0}{\alpha^2 + \omega^2} \left(\frac{\alpha}{\omega} \sin \omega t - \cos \omega t + e^{-\alpha t} \right)$$

$$- \alpha e^{-1} u\left(t - \frac{1}{\alpha} \right) \left\{ \frac{1}{2\alpha^2 \omega^2 + \alpha^4 + \omega^4} \left[(\alpha^2 + \omega^2) e^{-\alpha[t-(1/\alpha)]} \right. \right.$$

$$\left. \left. + (\alpha\omega + \alpha^3) \sin\left(\omega\left(t - \frac{1}{\alpha} \right) \right) - (\alpha^2 + \omega^2) \cos\left(\omega\left(t - \frac{1}{\alpha} \right) \right) \right] \right\}$$

Natural frequency calculations

$$r_1(\omega) := \frac{\omega}{\omega_{11}}$$

Primary system frequency ratio

$$r_2(\omega) := \frac{\omega}{\omega_{22}}$$

Absorber frequency ratio

$$\mu := m\frac{2}{m_1}$$

Mass ratio

$$D(r_1, r_2, \mu) := r_{12} \cdot r_{22} - r_{22} - (1 + \mu) \cdot r_{12} + 1$$

Natural frequencies are values of ω such that D=0.

$$\omega_g := 95 \cdot \frac{rad}{sec}$$

Guess for lower natural frequency

$$\omega_1 := root\left(D\left(r_1\left(\omega_g\right), r_2\left(\omega_g\right), \mu\right), \omega_g\right)$$

Lowest natural frequency

$$\omega_1 = 85.227 \cdot \frac{rad}{sec}$$

$$\omega_g := 145 \cdot \frac{rad}{sec}$$

Guess for higher natural frequency

$$\omega_2 := root\left(D\left(r_1\left(\omega_g\right), r_2\left(\omega_g\right), \mu\right), \omega_g\right)$$

Higher natural frequency

$$\omega_2 = 141.9 \cdot \frac{rad}{sec}$$

Determination of operating range

$$X_1(r_1, r_2, \mu) := \frac{1 - r_{22}}{r_{12} \cdot r_{22} - r_{22} - (1 + \mu) \cdot r_{12} + 1} \cdot \frac{F_0}{k_1}$$

Steady-state amplitude of primary system with absorber

$$TOL := 0.000001$$

Setting tolerance for root procedures

$$\omega_g := 90 \cdot \frac{rad}{sec}$$

Guess for lower end of operating range

$$\omega_1 := root\left(X_{1max} + X_1\left(r_1\left(\omega_g\right), r_2\left(\omega_g\right), \mu\right), \omega_g\right)$$

Lowest operating speed for $X_1 < X_{1max}$

$$\omega_1 = 88.612 \cdot \frac{rad}{sec}$$

$$\omega_g := 135 \cdot \frac{rad}{sec}$$

$$\omega_u := root\left(-X_{1max} + X_1\left(r_1\left(\omega_g\right), r_2\left(\omega_g\right), \mu\right), \omega_g\right)$$

Highest operating speed for $X_1 < X_{1max}$

$$\omega_u = 136.271 \cdot \frac{rad}{sec}$$

Fig. 11-15 (*Continued.*)

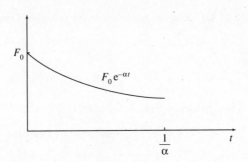

Fig. 11-16

11.11 For what values of m, will the system of Fig. 11-18 have a natural frequency between 60 and 80 rad/s? Use *MAPLE V* to perform the algebra.

The *MAPLE V* worksheet follows (Fig. 11-19). The natural frequencies are the square roots of the eigenvalues of $\mathbf{M}^{-1}\mathbf{K}$. Computer algebra is useful in this problem as the mass matrix contains an unspecified value. Computer algebra is used to develop the characteristic polynomial of $\mathbf{M}^{-1}\mathbf{K}$ in terms of this parameter. Note that since $f(\lambda, m)$ is linear in m, that when $f(\lambda, m) = 0$ is solved for m, only one value of m exists for each λ. A plot of this function between $\lambda = (60\,\text{rad/s})^2 = 3600$ and $\lambda = (80\,\text{rad/s})^2 = 6400$ reveals that the maximum and minimum of the function over this interval are at the ends of the interval (Fig. 11-20). Thus, in order for any of the natural frequencies to be between 60 and 80 rad/s,

$$57.19\;\text{kg} < m < 138.23\;\text{kg}$$

11.12 Use the Rayleigh-Ritz method using trial functions

$$\phi_1(x) = L^3x - 2Lx^3 + x^4 \qquad \phi_2(x) = \tfrac{7}{3}L^4x - \tfrac{10}{3}L^2x^3 + x^5$$

to approximate the lowest natural frequency of a simply supported beam with a concentrated mass m at its midspan. Use computer algebra to help with the manipulations.

The application of *MAPLE V* to approximate the lowest natural frequency of the beam follows in Fig. 11-21. The Rayleigh-Ritz method is applied as illustrated in Chap. 7. The lowest natural frequency is obtained as

$$\omega_1 = 96\sqrt{\frac{42EI}{3968\rho AL^4 + 7875mL^3}}$$

11.13 Use *MAPLE V* to develop the Fourier series representation for the periodic function of Fig. 11-22.

The Fourier series representation for $F(t)$ is

$$F(t) = \frac{a_0}{2} + \sum_{i=1}^{\infty}(a_i\cos\omega_i t + b_i\sin\omega_i t) \qquad \omega_i = \frac{2\pi i}{T}$$

The Fourier coefficients are obtained using *MAPLE V* (Fig. 11-23).

11.14 During operation, a punch press of mass m is subject to an excitation of the form of Fig. 11-22. Set up a spreadsheet to design a vibration isolator of damping ratio ζ for the punch

Solution to Problem 11.10 - Laplace transform solution for response of one-degree-of-freedom system

Setup:
```
>readlib(Heaviside);
 proc(x) ... end
```

Excitation:
```
>F:=t->F0*exp(-alpha*t)*(Heaviside(t)-Heaviside(t-1/alpha));
```

$$F := t \rightarrow F0\, e^{-\alpha\, t}\left(\text{Heaviside}(t) - \text{Heaviside}\left(t - \frac{1}{\alpha}\right)\right)$$

Differential equation:
```
>eq:=(D@@2)(y)(t)+omega^2*y(t)=F(t);
```

$$eq := D^{(2)}(y)(t) + \omega^2\, y(t) = F0\, e^{-\alpha\, t}\left(\text{Heaviside}(t) - \text{Heaviside}\left(t - \frac{1}{\alpha}\right)\right)$$

Solution to differential equation using laplace transform:
```
>dsolve(eq,y(t),laplace);
```

$$y(t) = \frac{y(0)\,\omega^2\cos(\omega t)}{\%1} + \frac{y(0)\,\alpha^2\cos(\omega t)}{\%1} + \frac{D(y)(0)\,\omega\sin(\omega t)}{\%1} + \frac{D(y)(0)\,\alpha^2\sin(\omega t)}{\%1\,\omega}$$

$$+ \frac{F0\,\alpha\sin(\omega t)}{\%1\,\omega} - \frac{F0\cos(\omega t)}{\%1} + \frac{F0\,e^{-\alpha\, t}}{\%1}$$

$$+ \text{invlaplace}\left(\frac{F0\,\text{laplace}\left(-\text{Heaviside}\left(\frac{\alpha t - 1}{\alpha}\right), t, s + \alpha\right)s}{s^3 + s^2\,\alpha + \omega^2\,s + \omega^2\,\alpha}, s, t\right)$$

$$+ \text{invlaplace}\left(\frac{F0\,\text{laplace}\left(-\text{Heaviside}\left(\frac{\alpha t - 1}{\alpha}\right), t, s + \alpha\right)\alpha}{s^3 + s^2\,\alpha + \omega^2\,s + \omega^2\,\alpha}, s, t\right)$$

$$\%1 := \alpha^2 + \omega^2$$

Initial conditions:
```
>subs({y(0)=0,D(y)(0)=0},");
```

$$y(t) = \frac{F0\,\alpha\sin(\omega t)}{(\alpha^2 + \omega^2)\,\omega} - \frac{F0\cos(\omega t)}{\alpha^2 + \omega^2} + \frac{F0\,e^{-\alpha\, t}}{\alpha^2 + \omega^2}$$

$$+ \text{invlaplace}\left(\frac{F0\,\text{laplace}\left(-\text{Heaviside}\left(\frac{\alpha t - 1}{\alpha}\right), t, s + \alpha\right)s}{s^3 + s^2\,\alpha + \omega^2\,s + \omega^2\,\alpha}, s, t\right)$$

$$+ \text{invlaplace}\left(\frac{F0\,\text{laplace}\left(-\text{Heaviside}\left(\frac{\alpha t - 1}{\alpha}\right), t, s + \alpha\right)\alpha}{s^3 + s^2\,\alpha + \omega^2\,s + \omega^2\,\alpha}, s, t\right)$$

Note the following: (1) laplace(Heaviside(t-1/alpha))=exp(-s/alpha)/s;
(2) Replacing s by s+alpha in the above leads to
exp(-s/alpha-1)=exp(-1)exp(-s/alpha)/(s+alpha)

Fig. 11-17

press such that the maximum transmitted force is less than F_{all}. Also use the spreadsheet to determine the maximum displacement of the machine when the isolator is used. Use the spreadsheet to determine the maximum stiffness of an isolator of damping ratio 0.1 for $m = 1000$ kg, $\alpha = 0.2$, $T = 0.1$ s, $F_0 = 20{,}000$ N, and $F_{all} = 3000$ N.

An alternate representation for the Fourier series is

$$F(t) = \frac{a_0}{2} + \sum_{i=1}^{\infty} c_i \sin(\omega_i t + \kappa_i) \qquad c_i = \sqrt{a_i^2 + b_i^2} \qquad \kappa_i = \tan^{-1}\left(\frac{b_i}{a_i}\right)$$

(3) s^3+alpha*s^2+omega^2*s+omega^2*alpha=(s+alpha)*(s^2+omega^2)

Then, the inverse transforms become:

>invlaplace(exp{-s/alpha}/{{s + alpha}^2*{s^2 + omega^2}},s,t);

$$\text{Heaviside}\left(t - \frac{1}{\alpha}\right)\left(2\frac{\alpha\,e^{-\alpha\left(t-\frac{1}{\alpha}\right)}}{\%1} + \frac{\left(t-\frac{1}{\alpha}\right)e^{-\alpha\left(t-\frac{1}{\alpha}\right)}}{\alpha^2+\omega^2} - \frac{\omega\sin\left(\omega\left(t-\frac{1}{\alpha}\right)\right)}{\%1}\right.$$

$$\left.+\frac{\alpha^2\sin\left(\omega\left(t-\frac{1}{\alpha}\right)\right)}{\%1\,\omega} - 2\frac{\alpha\cos\left(\omega\left(t-\frac{1}{\alpha}\right)\right)}{\%1}\right)$$

$$\%1 := 2\,\alpha^2\,\omega^2 + \omega^4 + \alpha^4$$

>invlaplace(exp{-s/alpha}*s/{{s + alpha}^2*{s^2 + omega^2}},s,t);

$$\text{Heaviside}\left(t - \frac{1}{\alpha}\right)\left(\frac{\omega^2\,e^{-\alpha\left(t-\frac{1}{\alpha}\right)}}{\%1} - \frac{\alpha^2\,e^{-\alpha\left(t-\frac{1}{\alpha}\right)}}{\%1} - \frac{\left(t-\frac{1}{\alpha}\right)e^{-\alpha\left(t-\frac{1}{\alpha}\right)}\alpha}{\alpha^2+\omega^2}\right.$$

$$\left.+2\frac{\omega\,\alpha\sin\left(\omega\left(t-\frac{1}{\alpha}\right)\right)}{\%1} - \frac{\omega^2\cos\left(\omega\left(t-\frac{1}{\alpha}\right)\right)}{\%1} + \frac{\alpha^2\cos\left(\omega\left(t-\frac{1}{\alpha}\right)\right)}{\%1}\right)$$

$$\%1 := 2\,\alpha^2\,\omega^2 + \omega^4 + \alpha^4$$

>

Fig. 11-17 (*Continued.*)

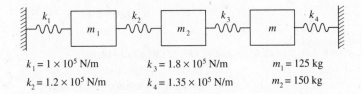

$k_1 = 1 \times 10^5$ N/m $k_3 = 1.8 \times 10^5$ N/m $m_1 = 125$ kg

$k_2 = 1.2 \times 10^5$ N/m $k_4 = 1.35 \times 10^5$ N/m $m_2 = 150$ kg

Fig. 11-18

Defining $r_i = \omega_i/\omega_n$, an upper bound for the maximum displacement is

$$x_{\max} < \frac{1}{m\omega_n^2}\sum_{i=1}^{\infty}\frac{c_i}{\sqrt{(1-r_i^2)^2 + (2\zeta r_i)^2}}$$

Similarly, an upper bound for the transmitted force is

$$F_{\max} < \sum_{i=1}^{\infty} c_i\sqrt{\frac{1+(2\zeta r_i)^2}{(1-r_i^2)^2+(2\zeta r_i)^2}}$$

A spreadsheet for the calculation of the transmitted force and maximum displacement follows in Fig. 11-24. Fourteen terms are taken in the Fourier series evaluation, noting that $c_{14}M_{14}$ is only 0.02 percent of c_1M_1 and $c_{14}T_{14}$ is only 0.15 percent of c_1T_1. The isolator stiffness is entered in the

Solution of Problem 11.11 using MAPLE V

Definitions:
>k1:=1*10^5;

$$k1 := 100000$$

>k2:=1.2*10^5;

$$k2 := 120000.0$$

>k3:=1.8*10^5;

$$k3 := 180000.0$$

>k4:=1.35*10^5;

$$k4 := 135000.00$$

>m1:=125;

$$m1 := 125$$

>m2:=150;

$$m2 := 150$$

>with(linalg);
```
Warning: new definition for    norm
Warning: new definition for    trace
```
[*BlockDiagonal, GramSchmidt, JordanBlock, Wronskian, add, addcol, addrow, adj, adjoint, angle, augment, backsub, band, basis, bezout, blockmatrix, charmat, charpoly, col, coldim, colspace, colspan, companion, concat, cond, copyinto, crossprod, curl, definite, delcols, delrows, det, diag, diverge, dotprod, eigenvals, eigenvects, entermatrix, equal, exponential, extend, ffgausselim, fibonacci, frobenius, gausselim, gaussjord, genmatrix, grad, hadamard, hermite, hessian, hilbert, htranspose, ihermite, indexfunc, innerprod, intbasis, inverse, ismith, iszero, jacobian, jordan, kernel, laplacian, leastsqrs, linsolve, matrix, minor, minpoly, mulcol, mulrow, multiply, norm, normalize, nullspace, orthog, permanent, pivot, potential, randmatrix, randvector, range, rank, ratform, row, rowdim, rowspace, rowspan, rref, scalarmul, singularvals, smith, stack, submatrix, subvector, sumbasis, swapcol, swaprow, sylvester, toeplitz, trace, transpose, vandermonde, vecpotent, vectdim, vector*]

Mass matrix:
>M:=matrix(3,3,[[m1,0,0],[0,m2,0],[0,0,m]]);

$$M := \begin{bmatrix} 125 & 0 & 0 \\ 0 & 150 & 0 \\ 0 & 0 & m \end{bmatrix}$$

Stiffness matrix:
>K:=matrix(3,3,[[k1+k2,-k2,0],[-k2,k2+k3,-k3],[0,-k3,k3+k4]]);

$$K := \begin{bmatrix} 220000.0 & -120000.0 & 0 \\ -120000.0 & 300000.0 & -180000.0 \\ 0 & -180000.0 & 315000.00 \end{bmatrix}$$

>B:=multiply(inverse(M),K);

$$B := \begin{bmatrix} 1760.000000 & -960.0000000 & 0 \\ -800.0000000 & 2000.000000 & -1200.000000 \\ 0 & -180000.0\dfrac{1}{m} & 315000.00\dfrac{1}{m} \end{bmatrix}$$

Fig. 11-19

spreadsheet as a parameter. The transmitted force is calculated for the value entered. If the transmitted force exceeds 3000 N, the isolator stiffness must be lowered. The spreadsheet automatically recalculates when a new stiffness is entered. An isolator stiffness of 1,190,875 N/m leads to a transmitted force of 3000 N.

11.15 A simplified model of a vehicle suspension system is shown in Fig. 11-25. The vehicle traverses a road whose contour is approximately sinusoidal, as shown in Fig. 11-26.

Characteristic polynomial of B, roots are eigenvalues, which are squares

>charpoly(B,lambda);

$$\left(\lambda^3 m - 315000.00\ \lambda^2 - 3760.000000\ \lambda^2 m + .9684000000\ 1 \right.$$
$$\left. - .4867200000\ 10^{12} \right)/m$$

>f:={lambda,m}->lambda^3*m-315000*lambda^2-3760*lambda^2*m+0.9684*10^9*lambda+0.2752*
>10^7*lambda*m-0.4867*10^12;

$$f := (\lambda, m) \rightarrow \lambda^3 m - 315000\ \lambda^2 - 3760\ \lambda^2 m + .9684000000\ 10^9\ \lambda + .2752000000\ 10^7\ m\ \lambda$$
$$- .4867000000\ 10^{12}$$

>g:=lambda->solve(f(lambda,m)=0,m);

$$g := \lambda \rightarrow \text{solve}(\,f(\lambda, m) = 0,\, m\,)$$

>g(lambda);

$$-1.\ \frac{-315000.\ \lambda^2 + .968400000\ 10^9\ \lambda - .4867000000\ 10^{12}}{\lambda^3 - 3760.\ \lambda^2 + .2752000\ 10^7\ \lambda}$$

>plot(g(lambda),lambda = 3600..6400,m = 0..200);

>g(3600);

138.2327410

>g(6400);

57.18886782

>

Fig. 11-19 (*Continued.*)

Develop a spreadsheet program that evaluates the amplitudes of displacement of the cab relative to the wheels, the absolute displacement of the cab, and the absolute acceleration of the cab for vehicle speeds between 0 and 80 m/s.

The frequency of the excitation is

$$\omega = \frac{2\pi v}{d}$$

The equations for the relative displacement, absolute displacement, and absolute acceleration are, respectively,

$$Z = h\,\frac{r^2}{\sqrt{(1 - r^2)^2 + (2\zeta r)^2}}$$

$$X = h\,\sqrt{\frac{1 + (2\zeta r)^2}{(1 - r^2)^2 + (2\zeta r)^2}}$$

$$A = \omega^2 X$$

A spreadsheet program is developed to calculate tables of these amplitudes for varying v. The spreadsheet program is also used to plot these relations (Fig. 11-27).

11.16 Develop a spreadsheet program that uses numerical integration of the convolution integral using piecewise impulse approximations as illustrated in Problems 4.27 and 4.28. Use the spreadsheet program to approximate the response of a 1-degree-of-freedom mass ($m = 100$ kg) − spring ($k = 10,000$ N/m) − dashpot ($c = 150$ N − s/m) system subject to

$$F(t) = 1000e^{-1.5t^2}\text{N}$$

The solution is given in Figs 11-28 and 11-29.

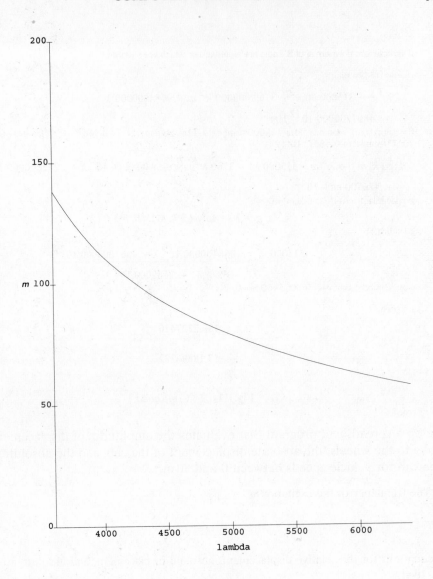

Fig. 11-20

Supplementary Problems

11.17 Use *VIBES* or another dedicated vibrations software package to develop the force spectrum for a system with a damping ratio of 0.2 subject to a triangular pulse.

11.18 Use *VIBES* or another dedicated vibrations software package to approximate the lowest natural frequencies of the system of Fig. 11-30 when 3 degrees of freedom are used to model the beam.

11.19 Use *VIBES* or another dedicated vibrations software package to determine the response of the system of Fig. 11-30 if the machine has a rotating unbalance of magnitude 0.45 kg-m and operates at 200 rad/s.

11.20 A vehicle suspension system is modeled using 1 degree of freedom with $m = 1000$ kg, $k = 1.3 \times 10^7$ N/m, and $\zeta = 0.7$. Use *VIBES* or another dedicated vibrations software package to measure the system's overshoot when the vehicle encounters a 20-cm-deep pothole.

Solution of Problem 11.12. - Rayleigh-Ritz approximation for simply-supported beam with concentrated mass

Trial functions:
>phi1:=x->L^3*x-2*L*x^3+x^4;

$$phi1 := x \rightarrow L^3 x - 2 L x^3 + x^4$$

>phi2:=x->7/3*L^4*x-10/3*L^2*x^3+x^5;

$$phi2 := x \rightarrow \frac{7}{3} L^4 x - \frac{10}{3} L^2 x^3 + x^5$$

Derivatives of trial functions:
>d1:=x->diff(phi1(x),x);

$$d1 := x \rightarrow \text{diff}(\text{phi1}(x), x)$$

>d1(x);

$$L^3 - 6 L x^2 + 4 x^3$$

>d2:=x->diff(phi2(x),x);

$$d2 := x \rightarrow \text{diff}(\text{phi2}(x), x)$$

>d2(x);

$$\frac{7}{3} L^4 - 10 L^2 x^2 + 5 x^4$$

>d1_2:=x->diff(d1(x),x);

$$d1\_2 := x \rightarrow \text{diff}(d1(x), x)$$

>d1_2(x);

$$-12 L x + 12 x^2$$

>d2_2:=x->diff(d2(x),x);

$$d2\_2 := x \rightarrow \text{diff}(d2(x), x)$$

>d2_2(x);

$$-20 L^2 x + 20 x^3$$

Integral evaluations:
>a11:=int(EI*d1_2(x)*d1_2(x),x=0..L);

$$a11 := \frac{24}{5} EI L^5$$

>a12:=int(EI*d1_2(x)*d2_2(x),x=0..L);

$$a12 := 12 EI L^6$$

>a22:=int(EI*d2_2(x)*d2_2(x),x=0..L);

$$a22 := \frac{640}{21} EI L^7$$

>b11:=int(rho*A*phi1(x)*phi1(x),x=0..L)+m*phi1(L/2)*phi1(L/2);

$$b11 := \frac{31}{630} \rho A L^9 + \frac{25}{256} m L^8$$

>b12:=int(rho*A*phi1(x)*phi2(x),x=0..L)+m*phi1(L/2)*phi2(L/2);

$$b12 := \frac{31}{252} \rho A L^{10} + \frac{125}{512} m L^9$$

Fig. 11-21

11.21 Use *VIBES* or another dedicated vibrations software package to simulate the free vibrations of a 1-degree-of-freedom system of mass 10 kg and stiffness 1×10^6 N/m that slides on a surface with friction coefficient $\mu = 0.14$ when it is displaced 3 mm and released from rest.

Use an electronic notepad to perform the calculations required to solve Problems 11.22 through 11.29, which refer to the system of Fig. 11-30. When operating at 200 r/min, the machine has a rotating unbalance of magnitude 0.45 kg-m.

```
>b22:=int(rho*A*phi2(x)*phi2(x),x=0..L)+m*phi2(L/2)*phi2(L/2);
```

$$b22 := \frac{640}{2079}\rho\,A\,L^{11} + \frac{625}{1024}m\,L^{10}$$

```
>with(linalg);
```

[*BlockDiagonal, GramSchmidt, JordanBlock, Wronskian, add, addcol, addrow, adj, adjoint, angle, augment, backsub, band, basis, bezout, blockmatrix, charmat, charpoly, col, coldim, colspace, colspan, companion, concat, cond, copyinto, crossprod, curl, definite, delcols, delrows, det, diag, diverge, dotprod, eigenvals, eigenvects, entermatrix, equal, exponential, extend, ffgausselim, fibonacci, frobenius, gausselim, gaussjord, genmatrix, grad, hadamard, hermite, hessian, hilbert, htranspose, ihermite, indexfunc, innerprod, intbasis, inverse, ismith, iszero, jacobian, jordan, kernel, laplacian, leastsqrs, linsolve, matrix, minor, minpoly, mulcol, mulrow, multiply, norm, normalize, nullspace, orthog, permanent, pivot, potential, randmatrix, randvector, range, rank, ratform, row, rowdim, rowspace, rowspan, rref, scalarmul, singularvals, smith, stack, submatrix, subvector, sumbasis, swapcol, swaprow, sylvester, toeplitz, trace, transpose, vandermonde, vecpotent, vectdim, vector*]

Coefficient matrix:
```
>D:=matrix(2,2,[[a11-omega^2*b11,a12-omega^2*b12],[a12-omega^2*b12,a22-omega^2*b22]]);
```

$$D :=$$

$$\left[\frac{24}{5}EI\,L^5 - \omega^2\left(\frac{31}{630}\rho\,A\,L^9 + \frac{25}{256}m\,L^8\right),\ 12\,EI\,L^6 - \omega^2\left(\frac{31}{252}\rho\,A\,L^{10} + \frac{125}{512}m\,L^9\right)\right]$$

$$\left[12\,EI\,L^6 - \omega^2\left(\frac{31}{252}\rho\,A\,L^{10} + \frac{125}{512}m\,L^9\right),\right.$$

$$\left.\frac{640}{21}EI\,L^7 - \omega^2\left(\frac{640}{2079}\rho\,A\,L^{11} + \frac{625}{1024}m\,L^{10}\right)\right]$$

```
>f:=omega->det(D);
```

$$f := \omega \to \det(D)$$

```
>f(omega);
```

$$\frac{16}{7}EI^2\,L^{12} - \frac{362}{14553}EI\,L^{16}\,\omega^2\,\rho\,A - \frac{125}{2688}EI\,L^{15}\,\omega^2\,m + \frac{31}{2095632}\omega^4\,\rho^2\,A^2\,L^{20}$$

$$+ \frac{125}{4257792}\omega^4\,\rho\,A\,L^{19}\,m$$

```
>solve(f(omega)=0,omega);
```

$$12\,\frac{\sqrt{11}\,\sqrt{EI}}{L^2\,\sqrt{\rho}\,\sqrt{A}},\ -12\,\frac{\sqrt{11}\,\sqrt{EI}}{L^2\,\sqrt{\rho}\,\sqrt{A}},\ 96\,\frac{\sqrt{42}\,L^{3/2}\,\sqrt{3968\,L\,\rho\,A + 7875\,m}\,\sqrt{EI}}{3968\,L^4\,\rho\,A + 7875\,m\,L^3},$$

$$-96\,\frac{\sqrt{42}\,L^{3/2}\,\sqrt{3968\,L\,\rho\,A + 7875\,m}\,\sqrt{EI}}{3968\,L^4\,\rho\,A + 7875\,m\,L^3}$$

```
>
```

Fig. 11-21 (*Continued.*)

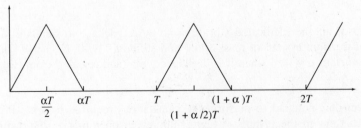

Fig. 11-22

Solution to Problem 11.13 - Fourier coefficients

F(t)= f_1(t) for 0<t<alpha*T/2, f_2(t) for alpha*T/2<t<alpha*T, and 0 for alpha*T<t<T

>f_1:=t->2*F*t/(alpha*T);

$$f\_1 := t \to 2\,\frac{F\,t}{\alpha\,T}$$

>f_2:=t->2*F*(1-t/(alpha*T));

$$f\_2 := t \to 2\,F\left(1 - \frac{t}{\alpha\,T}\right)$$

a_0:
>a_0:=2/T*(int(f_1(t),t=0..alpha*T/2)+int(f_2(t),t=alpha*T/2..alpha*T));

$$a\_0 := F\,\alpha$$

Frequency terms:
>omega:=i->2*pi*i/T;

$$\omega := i \to 2\,\frac{\pi\,i}{T}$$

Cosine coefficients:
>a_i:=2/T*(int(f_1(t)*cos(omega(i)*t),t=0..alpha*T/2)+int(f_2(t)*cos(omega(i)*t),t=alpha*T/2..alpha*T)
>);

$$a\_i := 2\left(\frac{1}{2}\frac{T\left(\cos(\pi\,i\,\alpha)+\pi\,i\,\sin(\pi\,i\,\alpha)\,\alpha\right)F}{\pi^2\,i^2\,\alpha} - \frac{1}{2}\frac{T\,F}{\pi^2\,i^2\,\alpha} - \frac{1}{2}\frac{\cos(2\,\pi\,i\,\alpha)\,T\,F}{\pi^2\,i^2\,\alpha}\right.$$
$$\left. - \frac{1}{2}\frac{T\left(\pi\,i\,\sin(\pi\,i\,\alpha)\,\alpha - \cos(\pi\,i\,\alpha)\right)F}{\pi^2\,i^2\,\alpha}\right)/T$$

>simplify(");

$$-\frac{F\left(-2\cos(\pi\,i\,\alpha)+1+\cos(2\,\pi\,i\,\alpha)\right)}{\pi^2\,i^2\,\alpha}$$

Sine coefficients:
>b_i:=2/T*(int(f_1(t)*sin(omega(i)*t),t=0..alpha*T/2)+int(f_2(t)*sin(omega(i)*t),t=alpha*T/2..alpha*T))
>;

$$b\_i := 2\left(-\frac{1}{2}\frac{T\left(-\sin(\pi\,i\,\alpha)+\pi\,i\,\cos(\pi\,i\,\alpha)\,\alpha\right)F}{\pi^2\,i^2\,\alpha} - \frac{1}{2}\frac{\sin(2\,\pi\,i\,\alpha)\,T\,F}{\pi^2\,i^2\,\alpha}\right.$$
$$\left. + \frac{1}{2}\frac{T\left(\pi\,i\,\cos(\pi\,i\,\alpha)\,\alpha + \sin(\pi\,i\,\alpha)\right)F}{\pi^2\,i^2\,\alpha}\right)/T$$

>simplify(");

$$\frac{F\left(2\sin(\pi\,i\,\alpha)-\sin(2\,\pi\,i\,\alpha)\right)}{\pi^2\,i^2\,\alpha}$$

>

Fig. 11-23

11.22 The beam is to be modeled using 1 degree of freedom with the generalized coordinate chosen as the displacement of the machine. Determine the equivalent stiffness of the beam at this location, the equivalent mass of the beam to approximate the beam's inertia effects, and the system's natural frequency.

11.23 Develop the flexibility matrix for a 4-degree-of-freedom model of the beam.

11.24 Determine approximations to the natural frequencies and normalized mode shapes for a 4-degree-of-freedom model of the beam.

11.25 Approximate the machine's steady-state amplitude using a 4-degree-of-freedom model of the beam.

Solution of Problem 11.14: Design of a vibration isolator for periodic excitation

Parameters			Calculated values		
mass	1000	kg	omega_n	34.50906	rad/s
damping ratio	0.2		F_max	3000.001	N
period	0.1	s	x_max	0.000389	m
alpha	0.2				
F_0	20000	N			
F_all	3000	N			
k	1190875	N/m			

Summation index	Frequency	Frequency ratio	Fourier coefficients			Magnification factor	Transmissibility factor		
i	omega_i	r_i	a_i	b_i	c_i	M_i	T_i	c_i*M_i	c_i*T_i
1	62.83185	1.820735	3130.997	2274.802	3870.125	0.412043116	0.509737846	1594.658	1972.749
2	125.6637	3.64147	1081.733	3329.231	3500.561	0.08099443	0.143102488	283.5259	500.939
3	188.4956	5.462205	-910.784	2803.105	2947.359	0.034580126	0.083091015	101.92	244.899
4	251.3274	7.282941	-1853.58	1346.702	2291.147	0.0191855	0.059091949	43.9568	135.3883
5	314.1593	9.103676	-1621.14	1.99E-13	1621.139	0.012201394	0.046075904	19.78015	74.69544
6	376.9911	10.92441	-823.812	-598.534	1018.287	0.008444277	0.037853387	8.598701	38.54563
7	439.823	12.74515	-167.287	-514.856	541.3516	0.006191213	0.03216465	3.351624	17.41239
8	502.6548	14.56588	67.6083	-208.077	218.7851	0.004733834	0.02798428	1.035692	6.122542
9	565.4867	16.38662	38.65428	-28.084	47.77932	0.0037369	0.024777477	0.178547	1.183851
10	628.3185	18.20735	0	0	0	0.003024914	0.022236972	0	0
11	691.1504	20.02809	25.87601	18.80002	31.9845	0.002498723	0.020173203	0.07992	0.64523
12	753.9822	21.84882	30.04813	92.47864	97.2378	0.002098851	0.018462656	0.204088	1.795268
13	816.8141	23.66956	-48.5033	149.2778	156.9599	0.001787859	0.017021293	0.280622	2.671661
14	879.6459	25.49029	-151.312	109.9349	187.0324	0.001541224	0.015789894	0.288259	2.953222

Fig. 11-24

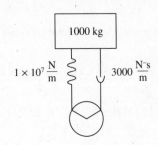

Fig. 11-25

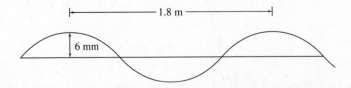

Fig. 11-26

11.26 Use the Rayleigh-Ritz method with the trial functions

$$\phi_1(x) = \tfrac{3}{2}L^2x^2 - \tfrac{5}{2}L^2x^3 + x^4$$

$$\phi_2(x) = \tfrac{7}{2}L^3x^2 - \tfrac{9}{2}L^2x^3 + x^5$$

to approximate the system's lowest natural frequency.

11.27 Use the Rayleigh-Ritz method with the trial functions of Problem 11.26 to approximate the steady-state amplitude of the machine.

Solution of Problem 11.15 - Periodic excitaiton of vehicle suspension system

Parameters: Calculated values

m =   1000   kg   omega_n =   100   rad/s
k =   10000000   N/m   zeta =   0.15
c =   30000   N-s/m
h =   0.006   m
d =   1.8   m

v (m/s)	omega (rad/s)	r	denom	Z (m)	X (m)	omega^2*X (m/s^2)
0	0	0	1	0	0.006	0
1	3.4906585	0.034907	0.998836	7.32E-06	0.006007	0.073197
2	6.98131701	0.069813	0.995346	2.94E-05	0.006029	0.293864
3	10.4719755	0.10472	0.989533	6.65E-05	0.006066	0.665262
4	13.962634	0.139626	0.981399	0.000119	0.006119	1.192947
5	17.4532925	0.174533	0.970951	0.000188	0.006188	1.884964
6	20.943951	0.20944	0.958197	0.000275	0.006274	2.752131
7	24.4346095	0.244346	0.943148	0.00038	0.006379	3.80843
8	27.925268	0.279253	0.925816	0.000505	0.006503	5.071541
9	31.4159265	0.314159	0.906218	0.000653	0.00665	5.563546
10	34.906585	0.349066	0.884375	0.000827	0.006822	8.311854
11	38.3972435	0.383972	0.860312	0.001028	0.00702	10.35042
12	41.887902	0.418879	0.834061	0.001262	0.00725	12.72134
13	45.3785606	0.453786	0.805664	0.001534	0.007516	15.47699
14	48.8692191	0.488692	0.77517	0.001849	0.007823	18.68284
15	52.3598776	0.523599	0.742647	0.002215	0.008178	22.42122
16	55.8505361	0.558505	0.708178	0.002643	0.008591	26.79634
17	59.3411946	0.593412	0.671876	0.003145	0.009071	31.94108
18	62.8318531	0.628319	0.63389	0.003737	0.009632	38.02581
19	66.3225116	0.663225	0.594421	0.00444	0.010292	45.26991
20	69.8131701	0.698132	0.553747	0.005281	0.01107	53.95559
21	73.3038286	0.733038	0.51226	0.006294	0.011993	64.44205
22	76.7944871	0.767945	0.470521	0.00752	0.013086	77.17239
23	80.2851456	0.802851	0.42935	0.009008	0.014374	92.6521
24	83.7758041	0.837758	0.389956	0.010799	0.015865	111.3456
25	87.2664626	0.872665	0.354119	0.012903	0.017514	133.3803
26	90.7571211	0.907571	0.324374	0.015236	0.019171	157.9048
27	94.2477796	0.942478	0.304021	0.01753	0.020509	182.1758
28	97.7384381	0.977384	0.296606	0.019324	0.021081	201.3781
29	101.229097	1.012291	0.304693	0.020179	0.02058	210.89
30	104.719755	1.047198	0.328682	0.020019	0.019134	209.8317
31	108.210414	1.082104	0.366891	0.019149	0.017194	201.3302
32	111.701072	1.117011	0.41672	0.017965	0.015185	189.4659
33	115.191731	1.151917	0.475704	0.016736	0.013345	177.0735
34	118.682389	1.186824	0.541926	0.015595	0.011752	165.5395
35	122.173048	1.22173	0.614016	0.014586	0.010407	155.3435
36	125.663706	1.256637	0.691029	0.013711	0.009279	146.5314
37	129.154365	1.291544	0.772312	0.012959	0.008332	138.9792
38	132.645023	1.32645	0.857407	0.012312	0.007532	132.5154
39	136.135682	1.361357	0.945994	0.011755	0.006851	126.971
40	139.62634	1.396263	1.037838	0.011271	0.006268	122.1969
41	143.116999	1.43117	1.132769	0.010849	0.005764	118.0677
42	146.607657	1.466077	1.230658	0.010479	0.005326	114.4796
43	150.098316	1.500983	1.331409	0.010153	0.004942	111.3479
44	153.588974	1.53589	1.434946	0.009864	0.004604	108.603
45	157.079633	1.570796	1.541211	0.009606	0.004304	106.1882
46	160.570291	1.605703	1.650157	0.009375	0.004036	104.0566
47	164.06095	1.640609	1.761747	0.009167	0.003796	102.1696
48	167.551608	1.675516	1.875951	0.008979	0.00358	100.4949
49	171.042267	1.710423	1.992743	0.008809	0.003384	99.00563
50	174.532925	1.745329	2.112104	0.008653	0.003207	97.67915
51	178.023584	1.780236	2.234017	0.008512	0.003045	96.49627
52	181.514242	1.815142	2.358467	0.008382	0.002897	95.44069
53	185.004901	1.850049	2.485443	0.008263	0.002761	94.49845
54	188.495559	1.884956	2.614934	0.008153	0.002636	93.65755
55	191.986218	1.919862	2.746931	0.008051	0.002521	92.90764
56	195.476876	1.954769	2.881427	0.007957	0.002414	92.23972
57	198.967535	1.989675	3.018416	0.007869	0.002315	91.64595
58	202.458193	2.024582	3.157892	0.007788	0.002223	91.11946
59	205.948852	2.059489	3.29985	0.007712	0.002137	30.65423
60	209.43951	2.094395	3.444286	0.007641	0.002057	90.24493
61	212.930169	2.129302	3.591195	0.007575	0.001983	89.88685
62	216.420827	2.164208	3.740576	0.007513	0.001912	89.57581
63	219.911486	2.199115	3.892423	0.007455	0.001847	89.30807
64	223.402144	2.234021	4.046736	0.0074	0.001785	89.0803
65	226.892803	2.268928	4.203512	0.007348	0.001727	88.88951
66	230.383461	2.303835	4.362748	0.0073	0.001672	88.73301
67	233.87412	2.338741	4.524443	0.007254	0.00162	88.60837
68	237.364778	2.373648	4.688595	0.00721	0.001571	88.51339

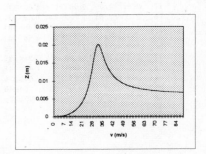

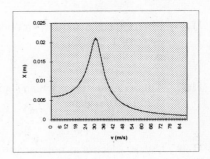

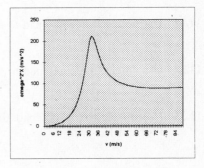

Fig. 11-27

327

69	240.855437	2.408554	4.855202	0.007169	0.001525	88.4461
70	244.346095	2.443461	5.024264	0.00713	0.001481	88.40469
71	247.836754	2.478368	5.195779	0.007093	0.001439	88.3875
72	251.327412	2.513274	5.369745	0.007058	0.001399	88.39304
73	254.818071	2.548181	5.546162	0.007025	0.001362	88.41995
74	258.308729	2.583087	5.725028	0.006993	0.001326	88.46695
75	261.799388	2.617994	5.906344	0.006963	0.001292	88.53289
76	265.290046	2.6529	6.090108	0.006934	0.001259	88.61672
77	268.780705	2.687807	6.276319	0.006906	0.001228	88.71745
78	272.271363	2.722714	6.464977	0.00688	0.001198	88.83419
79	275.762022	2.75762	6.656081	0.006855	0.00117	88.96608
80	279.25268	2.792527	6.849631	0.006831	0.001143	89.11236
81	282.743339	2.827433	7.045626	0.006808	0.001117	89.27232
82	286.233997	2.86234	7.244065	0.006786	0.001092	89.44527
83	289.724656	2.897247	7.444948	0.006765	0.001068	89.63061
84	293.215314	2.932153	7.648276	0.006745	0.001045	89.82775
85	296.705973	2.96706	7.854046	0.006725	0.001023	90.03615
86	300.196631	3.001966	8.06226	0.006707	0.001002	90.25531

Fig. 11-27 (*Continued.*)

11.28 Determine the natural frequency approximations when a three-element finite element model is used for the beam.

11.29 Approximate the machine's steady-state amplitude when a three-element finite element model is used for the beam.

11.30 Use an electronic notepad to solve the appropriate transcendental equation for the three lowest natural frequencies of a uniform fixed-free beam and to plot the corresponding mode shapes.

11.31 Use an electronic notepad to solve Problem 8.25.

11.32 Use an electronic notepad to solve Problem 11.4 if $\zeta = 0.15$ and $\omega = 195$ r/min.

11.33 Use an electronic notepad to numerically evaluate the convolution integral for the system of Problem 11.5 with

$$F(t) = 1000 \tanh (0.1t)$$

Use computer algebra to solve Problems 11.34 through 11.40.

11.34 Derive the elements of the local mass matrix Eq. (9.17) for a beam element.

11.35 Determine the natural frequencies of the system of Fig. 11-31 in terms of the stated parameters.

11.36 Determine the frequency response equation for the block of mass m_1 of the system of Fig. 11-31 if $F(t) = F_0 \sin \omega t$.

11.37 Use the Laplace transform method to determine the response of a 1-degree-of-freedom undamped system subject to the excitation of Fig. 11-32.

Solution of Problem 11.16 - Numerical evaluation of convolution integral
using spreadsheet.

Parameters: Calculated values:
   m =   100 kg   omega_n =   10 rad/s
   k =   10000 N/m   zeta =   0.075
   c =   150 N-s/m   omega_d = 9.971835 rad/s
DELTA t =   0.01 s   r = 0.075212

Time	Force	exp	cos	sin	G_1	sum G_1	G_2	sum G_2	x
0	1000	1	1	0	0	0	0	0	0
0.01	999.9625	1.007528	0.995032	0.099553	10.02056	10.02056	0.500657	0.500657	0.000497
0.02	999.6626	1.015113	0.980178	0.198117	9.992627	20.01319	1.506557	2.007215	0.001973
0.03	999.0629	1.022755	0.955586	0.294713	9.860809	29.874	2.511138	4.518352	0.004399
0.04	998.1642	1.030455	0.921499	0.38838	9.625148	39.49914	3.503384	8.021737	0.007736
0.05	996.9671	1.038212	0.878257	0.478189	9.286883	48.78603	4.472281	12.49402	0.011935
0.06	995.4728	1.046028	0.826289	0.563247	8.84845	57.63448	5.406947	17.90096	0.016941
0.07	993.6825	1.053903	0.766111	0.642709	8.313466	65.94794	6.296767	24.19773	0.022691
0.08	991.598	1.061837	0.698321	0.715784	7.6867	73.63464	7.131525	31.32926	0.029115
0.09	989.221	1.06983	0.623594	0.781749	6.974025	80.60867	7.901528	39.23078	0.036137
0.1	986.5537	1.077884	0.54267	0.839946	6.182362	86.79103	8.597731	47.82852	0.043676
0.11	983.5985	1.085999	0.456355	0.889798	5.319603	92.11063	9.211854	57.04037	0.051646
0.12	980.358	1.094174	0.365506	0.930809	4.394525	96.50516	9.736483	66.77685	0.059959
0.13	976.835	1.102411	0.271025	0.962572	3.416696	99.92185	10.16517	76.94202	0.068524
0.14	973.0328	1.110711	0.173851	0.984772	2.396359	102.3182	10.49252	87.43454	0.077249
0.15	968.9546	1.119072	0.074951	0.997187	1.34432	103.6625	10.71426	98.1488	0.086041
0.16	964.6041	1.127497	-0.02469	0.999695	0.271817	103.9343	10.8273	108.9761	0.094807
0.17	959.9851	1.135985	-0.12409	0.99227	-0.80961	103.1247	10.82977	119.8059	0.103457
0.18	955.1017	1.144537	-0.22226	0.974987	-1.88825	101.2365	10.72107	130.5269	0.111902
0.19	949.958	1.153153	-0.31822	0.948017	-2.95238	98.28411	10.50186	141.0288	0.120056
0.2	944.5586	1.161834	-0.41102	0.911627	-3.99038	94.29372	10.17407	151.2029	0.127838
0.21	938.9083	1.170581	-0.49973	0.86618	-4.9909	89.30282	9.740873	160.9438	0.135169
0.22	933.0117	1.179393	-0.58348	0.812127	-5.94297	83.35985	9.206664	170.1504	0.14198
0.23	926.8741	1.188272	-0.66143	0.750006	-6.83615	76.5237	8.577001	178.7274	0.148203
0.24	920.5007	1.197217	-0.73281	0.680432	-7.66065	68.86305	7.858546	186.586	0.153779
0.25	913.8969	1.20623	-0.79691	0.604098	-8.40746	60.45559	7.058982	193.6449	0.158658
0.26	907.0683	1.215311	-0.85309	0.521762	-9.06844	51.38715	6.186925	199.8319	0.162793
0.27	900.0207	1.22446	-0.9008	0.434242	-9.63643	41.75072	5.251822	205.0837	0.166148
0.28	892.76	1.233678	-0.93955	0.342408	-10.1053	31.6454	4.263837	209.3475	0.168694
0.29	885.2922	1.242965	-0.96897	0.247172	-10.4702	21.17522	3.233726	212.5813	0.170412
0.3	877.6236	1.252323	-0.98876	0.14948	-10.7272	10.44801	2.172714	214.754	0.171287
0.31	869.7604	1.26175	-0.99873	0.050302	-10.8739	-0.42588	1.092353	215.8463	0.171318
0.32	861.7091	1.271249	-0.99878	-0.04937	-10.909	-11.3349	0.004387	215.8507	0.170508
0.33	853.4763	1.280819	-0.9889	-0.14856	-10.8324	-22.1673	-1.07939	214.7713	0.168869
0.34	845.0686	1.290462	-0.9692	-0.24627	-10.6456	-32.8129	-2.14727	212.624	0.166422
0.35	836.4927	1.300174	-0.93987	-0.34154	-10.351	-43.1638	-3.18779	209.4363	0.163195
0.36	827.7555	1.309964	-0.9012	-0.43341	-9.95228	-53.1161	-4.18983	205.2464	0.159223
0.37	818.8638	1.319826	-0.85358	-0.52097	-9.45448	-62.5706	-5.1428	200.1036	0.154547
0.38	809.8247	1.329762	-0.79747	-0.60336	-8.86356	-71.4342	-6.0367	194.0669	0.149216
0.39	800.6451	1.339773	-0.73344	-0.67975	-8.18656	-79.6207	-6.8623	187.2046	0.143283
0.4	791.3321	1.349859	-0.66213	-0.74939	-7.43145	-87.0522	-7.61117	179.5935	0.136807
0.41	781.8929	1.360021	-0.58423	-0.81158	-6.60702	-93.6592	-8.27586	171.3176	0.129851
0.42	772.3345	1.370259	-0.50054	-0.86572	-5.72278	-99.382	-8.84991	162.4677	0.122481
0.43	762.6642	1.380575	-0.41187	-0.91124	-4.78883	-104.171	-9.32798	153.1397	0.114767
0.44	752.889	1.390968	-0.3191	-0.94772	-3.81574	-107.987	-9.70583	143.4339	0.106781
0.45	743.0161	1.40144	-0.22317	-0.97478	-2.81446	-110.801	-9.98043	133.4535	0.098597
0.46	733.0528	1.41199	-0.12502	-0.99215	-1.79608	-112.597	-10.15	123.3035	0.090289
0.47	723.0062	1.42262	-0.02562	-0.99967	-0.77182	-113.369	-10.2138	113.0897	0.081932
0.48	712.8833	1.433329	0.074024	-0.99726	0.247192	-113.122	-10.1725	102.9172	0.073598
0.49	702.6914	1.44412	0.172937	-0.98493	1.250016	-111.872	-10.0279	92.88924	0.06536
0.5	692.4374	1.454991	0.270131	-0.96282	2.226035	-109.646	-9.783	83.10624	0.057289
0.51	682.1283	1.465945	0.364641	-0.93115	3.165078	-106.481	-9.44171	73.66453	0.049451
0.52	671.7712	1.476981	0.455528	-0.89022	4.057534	-102.423	-9.00917	64.65536	0.041911

Fig. 11-28

11.38 Use the Laplace transform method to determine the response of the system of Fig. 11-31 if $F(t)$ is as shown in Fig. 11-32.

11.39 Use the convolution integral to determine the response of a 1-degree-of-freedom system of mass m, natural frequency ω_n, and damping ratio ζ when subject to the excitation of Fig. 11-32.

0.53	661.3729	1.4881	0.54189	-0.84045	4.894462	-97.5286	-8.49141	56.16395	0.034728
0.54	650.9402	1.499303	0.622867	-0.78233	5.667686	-91.8609	-7.89537	48.26858	0.027959
0.55	640.4799	1.51059	0.697656	-0.71643	6.369879	-85.491	-7.22875	41.03983	0.021653
0.56	629.9987	1.521962	0.765513	-0.64342	6.994644	-78.4964	-6.49993	34.53991	0.015857
0.57	619.5031	1.533419	0.825765	-0.56401	7.536568	-70.9598	-5.71788	28.82203	0.010609
0.58	608.9997	1.544963	0.877812	-0.479	7.991275	-62.9685	-4.89199	23.93004	0.005943
0.59	598.4948	1.556594	0.921138	-0.38924	8.35546	-54.6131	-4.03202	19.89802	0.001887
0.6	587.9946	1.568312	0.955312	-0.2956	8.626907	-45.9862	-3.1479	16.75011	-0.00154
0.61	577.5054	1.580119	0.979994	-0.19903	8.8045	-37.1817	-2.24968	14.50043	-0.00432
0.62	567.0332	1.592014	0.994939	-0.10048	8.888212	-28.2935	-1.34736	13.15307	-0.00645
0.63	556.5838	1.603999	1	-0.00093	8.879085	-19.4144	-0.45079	12.70228	-0.00793
0.64	546.1632	1.616074	0.995124	0.098629	8.779197	-10.6352	0.430459	13.13274	-0.00876
0.65	535.7769	1.628241	0.980362	0.197207	8.591618	-2.04356	1.287169	14.41991	-0.00895
0.66	525.4304	1.640498	0.955859	0.293825	8.32035	6.276788	2.110618	16.53053	-0.00853
0.67	515.1292	1.652848	0.92186	0.387524	7.970258	14.24705	2.892651	19.42318	-0.00751
0.68	504.8785	1.665291	0.878701	0.477373	7.546998	21.79404	3.625771	23.04895	-0.00593
0.69	494.6833	1.677828	0.826812	0.562479	7.056925	28.85097	4.303216	27.35217	-0.00382
0.7	484.5486	1.690459	0.766708	0.641997	6.507007	35.35798	4.919016	32.27118	-0.00121
0.71	474.4792	1.703185	0.698986	0.715135	5.904724	41.2627	5.468054	37.73924	0.001842
0.72	464.4797	1.716007	0.62432	0.781169	5.257966	46.52066	5.946099	43.68534	0.005299
0.73	454.5545	1.728925	0.54345	0.839441	4.574929	51.09559	6.34984	50.03518	0.009107
0.74	444.708	1.741941	0.457181	0.889373	3.864011	54.9596	6.676902	56.71208	0.013213
0.75	434.9442	1.755055	0.36637	0.930469	3.133703	58.09331	6.925845	63.63792	0.017564
0.76	425.2672	1.768267	0.271919	0.96232	2.392488	60.4858	7.096164	70.73409	0.022102
0.77	415.6809	1.781579	0.174766	0.98461	1.648737	62.13453	7.188266	77.92235	0.026771
0.78	406.1887	1.794991	0.075877	0.997117	0.910615	63.04515	7.203439	85.12579	0.031512
0.79	396.7942	1.808504	-0.02377	0.999718	0.185985	63.23113	7.143814	92.26961	0.036268
0.8	387.5007	1.822119	-0.12317	0.992385	-0.51767	62.71346	7.012316	99.28192	0.040983
0.81	378.3114	1.835836	-0.22136	0.975193	-1.19335	61.52011	6.812604	106.0945	0.0456
0.82	369.2293	1.849657	-0.31734	0.948312	-1.83455	59.68556	6.549007	112.6435	0.050067
0.83	360.257	1.863581	-0.41017	0.912009	-2.43541	57.25015	6.226449	118.87	0.054333
0.84	351.3974	1.877611	-0.49893	0.866644	-2.99068	54.25947	5.850377	124.7204	0.05835
0.85	342.6528	1.891746	-0.58273	0.812669	-3.49582	50.76364	5.426675	130.147	0.062072
0.86	334.0256	1.905987	-0.66073	0.75062	-3.94701	46.81664	4.961579	135.1086	0.065459
0.87	325.518	1.920336	-0.73218	0.681113	-4.34117	42.47547	4.461595	139.5702	0.068473
0.88	317.1319	1.934792	-0.79635	0.604838	-4.67598	37.79949	3.933408	143.5036	0.071082
0.89	308.8692	1.949358	-0.85261	0.522555	-4.94989	32.8496	3.383798	146.8874	0.073257
0.9	300.7315	1.964033	-0.90039	0.435079	-5.1621	27.6875	2.819552	149.707	0.074976
0.91	292.7204	1.978819	-0.93923	0.343281	-5.31254	22.37496	2.247387	151.9544	0.07622
0.92	284.8372	1.993716	-0.96874	0.248072	-5.40186	16.9731	1.67387	153.6282	0.076976
0.93	277.0832	2.008725	-0.98863	0.150398	-5.43139	11.54171	1.10534	154.7336	0.077236
0.94	269.4594	2.023847	-0.99869	0.05123	-5.40309	6.138616	0.547846	155.2814	0.076997
0.95	261.9668	2.039083	-0.99883	-0.04845	-5.31953	0.819088	0.007082	155.2885	0.076262
0.96	254.6061	2.054433	-0.98904	-0.14764	-5.1838	-4.36471	-0.51167	154.7768	0.075037
0.97	247.3781	2.069899	-0.96943	-0.24537	-4.99948	-9.36418	-1.00359	153.7732	0.073336
0.98	240.2831	2.085482	-0.94019	-0.34066	-4.77056	-14.1347	-1.46434	152.3089	0.071174
0.99	233.3216	2.101182	-0.9016	-0.43257	-4.50142	-18.6362	-1.89014	150.4188	0.068573
1	226.4938	2.117	-0.85406	-0.52018	-4.19668	-22.8328	-2.27775	148.141	0.065559

Fig. 11-28 (*Continued.*)

11.40 Determine the Fourier coefficients for the excitation of Fig. 11-33.

11.41 Develop a spreadsheet program to solve Problem 11.14 if the machine is subject to the excitation shown in Fig. 11-33 with $T = 0.1$ s, $\alpha = 0.3$, and $F_0 = 21,000$ N.

11.42 Develop a spreadsheet program that determines the frequency response of a system when a damped vibration absorber is added. Use the program to analyze the system of Problem 11.9 when an optimally damped absorber of mass ratio 0.15 is added to the machine.

11.43 Develop a spreadsheet program similar to the program of Problem 11.16 to determine the response of the system due to an excitation of the form of Problem 11.15.

Numerical Evaluation of Convolution Integral

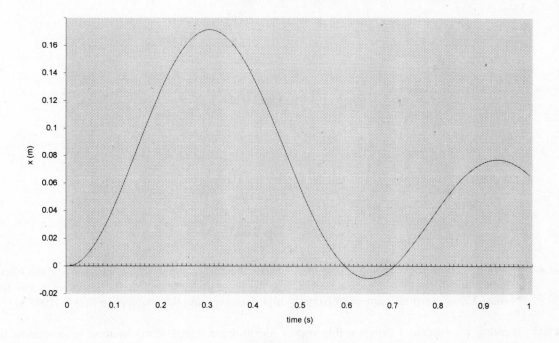

Fig. 11-29

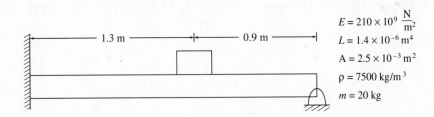

$E = 210 \times 10^9 \dfrac{N}{m^2}$

$L = 1.4 \times 10^{-6} \, m^4$

$A = 2.5 \times 10^{-3} \, m^2$

$\rho = 7500 \, kg/m^3$

$m = 20 \, kg$

Fig. 11-30

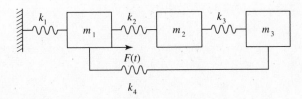

Fig. 11-31

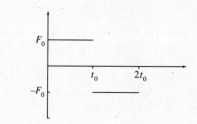

Fig. 11-32

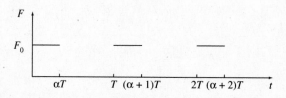

Fig. 11-33

11.44 Develop a spreadsheet program that uses Euler's method to numerically approximate the forced response of a 1-degree-of-freedom system. Use the program to determine the response of a system of mass 150 kg, natural frequency 210 rad/s, and damping ratio 0.05 subject to $F(t) = 1200t^{3/2}e^{-1.5t}$.

11.45 Develop a spreadsheet program that uses a fourth-order Runge-Kutta method to determine the forced response of a 1-degree-of-freedom system. Use the program to determine the response of the system of Problem 11.44.

REFERENCE

Kelly, S. G. *Fundamentals of Mechanical Vibrations*, McGraw-Hill, New York, 1993.

Appendix

SAMPLE Screens From
The Companion *Schaum's Electronic Tutor*

This book has a companion *Schaum's Electronic Tutor* which uses Mathcad[®] and is designed to help you learn the subject matter more readily. The *Electronic Tutor* uses the LIVE-MATH environment of Mathcad technical calculation software to give you on-screen access to approximately 100 representative solved problems from this book, together with summaries of key theoretical points and electronic cross-referencing and hyperlinking. The following pages reproduce a representative sample of screens from the *Electronic Tutor* and will help you understand the powerful capabilities of this electronic learning tool. Compare these screens with the associated solved problems from this book (the corresponding page numbers are listed at the start of each problem) to see how one complements the other.

In the companion *Schaum's Electronic Tutor*, you'll find all related text, diagrams, and equations for a particular solved problem together on your computer screen. As you can see on the following pages, all the math appears in familiar notation, including units. The format differences you may notice between the printed *Schaum's Outline* and the *Electronic Tutor* are designed to encourage your interaction with the material or show you alternate ways to solve challenging problems.

As you view the following pages, keep in mind that every number, formula, and graph shown *is completely interactive when viewed on the computer screen*. You can change the starting parameters of a problem and watch as new output graphs are calculated before your eyes; you can change any equation and immediately see the effect of the numerical calculations on the solution. Every equation, graph, and number you see is available for experimentation. Each adapted solved problem becomes a "live" worksheet you can modify to solve dozens of related problems. The companion *Electronic Tutor* thus will help you to learn and retain the material taught in this book and you can also use it as a working problem-solving tool.

The Mathcad icon shown on the right is printed throughout this *Schaum's Outline* to indicate which problems are found in the *Electronic Tutor*.

For more information about the companion *Electronic Tutor,* including system requirements, please see the back cover.

[®] Mathcad is a registered trademark of MathSoft, Inc.

Underdamped System Response and Overshoot
(Schaum's Mechanical Vibrations, Solved Problems 2.21, 2.22, and 2.23, pp. 51-52)

Statement Overshoot for an underdamped system is defined as the maximum displacement of the system at the end of its first half cycle when the system is subject to an initial displacement with zero initial velocity.

(a) What is the minimum damping ratio of a system such that it is subject to no more than 5% overshoot?

(b) A suspension system is being designed for a vehicle of mass m when empty. It is estimated that the maximum added mass from passengers, cargo and fuel is m_p . When the vehicle is empty its static deflection is to be Δ. What is the minimum value of the damping coefficient such that the vehicle is subject to no more than 5% overshoot, empty of full?

(c) The suspension system for the vehicle of part (b) is designed using the minimum damping coefficient determined in part (b). What is the overshoot when the vehicle encounters a bump of height **h** when loaded with passengers, fuel, and cargo of total mass m_1?

System Parameters

$mass := 2000 \cdot kg$ Empty mass of vehicle

$\Delta := 3.1 \cdot 10^{-3} \cdot m$ Static deflection when empty

$m_p := 1000 \cdot kg$ Maximum mass of passengers, cargo, and fuel

$m_1 := 220 \cdot kg$ Mass of passengers, cargo, and fuel for part (c)

$h := 5.5 \cdot 10^{-2} \cdot m$ Height of bump

$g := 9.81 \cdot \dfrac{m}{sec^2}$ Acceleration due to gravity

Solution

(a) The free vibration response of an underdamped system is

$$x = A \cdot \exp\left(-\zeta \cdot \omega_n \cdot t\right) \cdot \sin\left(\omega_d \cdot t + \phi_d\right)$$

where ζ is the system's damping ratio, ω_n is its natural frequency, ω_d is its damped natural frequency, and A and ϕ_d are constants of integration. If x_0 is the system's initial displacement, then

$$x_0 = A \cdot \sin\left(\phi_d\right)$$

The displacement from equilibrium at the end of the first half cycle is $x_h = -x(T_d/2)$

$$x_h = -A \cdot \exp\left(-\zeta \cdot \omega_n \cdot \frac{T_d}{2}\right) \cdot \sin\left(\omega_d \frac{T_d}{2} + \phi_d\right)$$

Noting that

$$T_d = \frac{2 \cdot \pi}{\omega_d} \qquad\qquad \sin\left(\pi + \phi_d\right) = -\sin\left(\phi_d\right)$$

leads to

$$x_h = x_0 \cdot \exp\left(-\frac{\zeta \cdot \pi}{\sqrt{1 - \zeta^2}}\right)$$

For 5% overshoot, $x(T_d/2) = -0.05 x_0$

$$0.05 = \exp\left(-\frac{\zeta \cdot \pi}{\sqrt{1 - \zeta^2}}\right)$$

which is solved for ζ as

$$\zeta := \sqrt{\frac{\ln(0.05)^2}{\pi^2 + \ln(0.05)^2}} \qquad\qquad \zeta = 0.69$$

(b)   The required suspension stiffness is

$$k := \frac{mass \cdot g}{\Delta}$$
$$k = 6.329 \cdot 10^6 \cdot \frac{newton}{m}$$

The results of part (a) show that the damping ratio must be smaller than 0.69 to limit the overshoot to 5%. The stiffness of the system is fixed, but its mass varies, depending on the passengers, fuel, and cargo. If the damping coefficient is also fixed, the damping ratio is larger for a larger mass ($\zeta = c/[2(mk)^{1/2}]$). Hence, in order for the overshoot to be limited to 5% for all possible loading conditions, the damping ratio must be limited to 0.69 when the vehicle is fully loaded.

$$c := 2 \cdot \zeta \cdot \sqrt{(mass + m_p) \cdot k}$$
$$c = 1.902 \cdot 10^5 \cdot kg \cdot sec^{-1}$$

(c)   The natural frequency of the system with this loading is

$$\omega_n := \sqrt{\frac{k}{(mass + m_1)}}$$
$$\omega_n = 53.394 \cdot \frac{rad}{sec}$$

The system's damping ratio is

$$\zeta_1 := \frac{c}{2 \cdot \sqrt{(mass + m_1) \cdot k}}$$
$$\zeta_1 = 0.802$$

From part (a) the overshoot is

$$x_h := h \cdot exp\left(-\frac{\zeta_1 \cdot \pi}{\sqrt{1 - \zeta_1^2}}\right)$$
$$x_h = 8.072 \cdot 10^{-4} \cdot m$$

Extension   Plot the overshoot as a function of the total mass where the mass varies from the mass of the empty vehicle to the mass when the vehicle is fully loaded.

$i := 0, 1 .. 20$

For evaluation and plotting the mass ranges from **mass** to **mass** + m_p in increments of $m_p/20$.

$$ms_i := mass + \frac{i}{20} \cdot m_p$$

$$\omega_i := \sqrt{\frac{k}{ms_i}}$$

The natural frequency varies with the mass

$$\zeta_i := \frac{c}{2 \cdot \sqrt{ms_i \cdot k}}$$

The damping ratio varies with the mass.

$$x_i := h \cdot \exp\left[-\frac{\zeta_i \cdot \pi}{\sqrt{1 - \left(\zeta_i\right)^2}} \right]$$

The overshoot

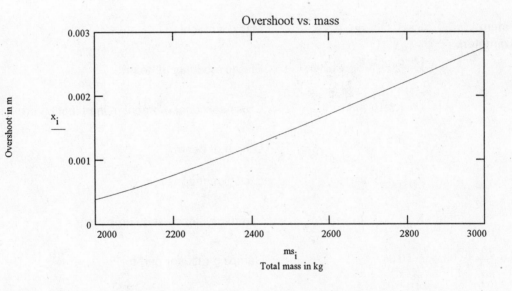

Overshoot vs. mass

Further study

(1) Plot the overshoot as a function of damping ratio for $0 < \zeta < 1$.

(2) Plot the overshoot as a function of the damped period. Assume the stiffness is fixed and the damping ratio varies between 0 and 1.

Undamped and Damped Response to Harmonic Excitation

(Schaum's Mechanical Vibrations, Solved Problems 3.4 and 3.8, pp. 74, 77)

Statement   A machine of mass **m** is placed at the end of a cantilever beam of length **L**, elastic modulus **E**, and cross-sectional moment of inertia **I**. As it operates the machine produces a harmonic force of magnitude F_0. At what operating speed will the machine's steady-state amplitude be less than X_{max} if

(a)   the system is modeled as an undamped 1 degree-of-freedom system?

(b)   the beam is modeled as a 1degree-of-freedom system with a viscous damping ratio ζ?

System Parameters

$mass := 45 \cdot kg$ Mass of machine

$E := 200 \cdot 10^9 \cdot newton \cdot m^{-2}$ Elastic modulus of beam

$I := 1.6 \cdot 10^{-5} \cdot m^4$ Cross-sectional moment of inertia of beam

$L := 1.6 \cdot m$ Length of beam

$F_0 := 125 \cdot newton$ Excitation magnitude

$X_{max} := 2 \cdot 10^{-4} \cdot m$ Maximum steady-state amplitude

$\zeta := 0.08$ Damping ratio for part b

Solution

(a)   The system is modeled as a one-degree-of-freedom system. The equivalent stiffness is the stiffness of a fixed-free beam at its end is

$$k := \frac{3 \cdot E \cdot I}{L^3}$$ $k = 2.344 \cdot 10^6 \cdot \dfrac{newton}{m}$

If the inertia of the beam is neglected, the natural frequency of vibration of the machine is

$$\omega_n := \sqrt{\frac{k}{mass}} \qquad\qquad \omega_n = 228.218 \cdot \frac{rad}{sec}$$

In order for the steady-state amplitude to be limited to X_{max}, the maximum magnification factor is

$$M_{max} := \frac{mass \cdot \omega_n^2 \cdot X_{max}}{F_0} \qquad M_{max} = 3.75$$

The magnification factor is related to the frequency ratio and damping ratio by

$$M(r, \zeta) := \frac{1}{\sqrt{\left(1 - r^2\right)^2 + (2 \cdot \zeta \cdot r)^2}}$$

For an undamped system $\zeta = 0$. There are two values of r such that $M(r,0) = M_{max}$. Finding first the value of **r < 1**

$r := 0.8$ **Guess for r < 1**

$r_1 := root\left(M(r,0) - M_{max}, r\right) \qquad r_1 = 0.856$ **Solving for r**

$\omega_1 := r_1 \cdot \omega_n \qquad\qquad \omega_1 = 195.434 \cdot \frac{rad}{sec} \qquad$ **X < X_{max} for $\omega < \omega_1$**

Now for r > 1

$r := 1.2$ **Guess for r > 1**

$r_2 := root\left(M(r,0) - M_{max}, r\right) \qquad r_2 = 1.125$ **Solving for r**

$\omega_2 := r_2 \cdot \omega_n \qquad\qquad \omega_2 = 256.851 \cdot \frac{rad}{sec} \qquad$ **X < X_{max} for $\omega > \omega_2$**

(b)   There is viscous damping ratio ζ

$r := .8$ Guess for **r < 1**

$r_1 := \text{root}\left(M(r, \zeta) - M_{max}, r\right)$ $r_1 = 0.879$ Solving for **r**

$\omega_1 := r_1 \cdot \omega_n$ $\omega_1 = 195.434 \cdot \dfrac{rad}{sec}$ **X < X$_{max}$ for $\omega < \omega_1$**

$r := 1.2$ Guess for **r > 1**

$r_2 := \text{root}\left(M(r, \zeta) - M_{max}, r\right)$ $r_2 = 1.096$ Solving for **r**

$\omega_2 := r_2 \cdot \omega_n$ $\omega_2 = 250.099 \cdot \dfrac{rad}{sec}$ **X < X$_{max}$ for $\omega > \omega_2$**

Alternate method of solution which can be used if your version of MATHCAD has symbolic capabilities:

First load the symbolic processor. The magnification factor **M** is written symbolically as a function of **r** and ζ. The [ctrl]= is used to type the equal sign.

$$M = \frac{1}{\sqrt{\left(1 - r^2\right)^2 + (2 \cdot \zeta \cdot r)^2}}$$

$$\begin{bmatrix} \sqrt{1 - 2 \cdot \zeta^2 + \sqrt{-4 \cdot \zeta^2 + 4 \cdot \zeta^4 + \dfrac{1}{M^2}}} \\[2em] -\sqrt{1 - 2 \cdot \zeta^2 + \sqrt{-4 \cdot \zeta^2 + 4 \cdot \zeta^4 + \dfrac{1}{M^2}}} \\[2em] \sqrt{1 - 2 \cdot \zeta^2 - \sqrt{-4 \cdot \zeta^2 + 4 \cdot \zeta^4 + \dfrac{1}{M^2}}} \\[2em] -\sqrt{1 - 2 \cdot \zeta^2 - \sqrt{-4 \cdot \zeta^2 + 4 \cdot \zeta^4 + \dfrac{1}{M^2}}} \end{bmatrix}$$

Now select the variable **r** in the equation for **M**. Choose "Solve for Variable" from the "Symbolic" menu. Four solutions for r in terms of **M** and ζ are shown to the left, in an array. Note that the second and fourth solutions are negatives of the first and third solutions and of no interest.

The first and third solutions are set equal to functions of **M** and ζ by copying them to the clipboard and pasting them to the right hand side of a function. The third solution is the smaller value of **r** and set equal to r_1 while the first solution is set equal to r_2

$$r_1(M,\zeta) := \sqrt{1 - 2\cdot\zeta^2 - \sqrt{-4\cdot\zeta^2 + 4\cdot\zeta^4 + \frac{1}{M^2}}}$$

$$r_2(M,\zeta) := \sqrt{1 - 2\cdot\zeta^2 + \sqrt{-4\cdot\zeta^2 + 4\cdot\zeta^4 + \frac{1}{M^2}}}$$

The solution to the problem is obtained by evaluating the functions for **M = M_{max}** and ζ **= 0** and ζ **= 0.08**.

$\zeta := 0$ $r_1(M_{max}, \zeta) = 0.856$ $r_2(M_{max}, \zeta) = 1.125$

$\zeta := 0.08$ $r_1(M_{max}, \zeta) = 0.879$ $r_2(M_{max}, \zeta) = 1.096$

The values obtained above are identical to those previously attained using the root function. The upper bound on the speed for $\omega < \omega_n$ and the lower bound on the speed for $\omega > \omega_n$ are obtained as before. This method is useful if these speeds are to be determined for a range of ζ.

Further study

(1)   Make a plot of the upper bound on the operating range for **r < 1** vs ζ for $0 < \zeta < 0.7$

(2)   Plot the lower bound on the operating range for **r > 1** as a function of the length of the beam for 1 m < **L** < 4 m assuming ζ = 0.08 and all other parameters as given.

Natural Frequencies and Mode Shapes for a 3 DOF System

(Schaum's Mechanical Vibrations, Solved Problems 5.30, 5.31, and 5.38, pp. 160, 164)

Statement Determine the natural frequencies and normalized mode shapes of the following system.

System Parameters

Assume

$$mass := 1 \cdot kg$$

$$k := \frac{1 \cdot newton}{m}$$

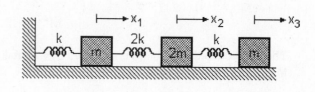

Solution The mass and stiffness matrices for this 3 degree-of-freedom system are

$$M := \begin{pmatrix} mass & 0 \cdot kg & 0 \cdot kg \\ 0 \cdot kg & 2 \cdot mass & 0 \cdot kg \\ 0 \cdot kg & 0 \cdot kg & 2 \cdot mass \end{pmatrix} \qquad K := \begin{bmatrix} 3 \cdot k & -2 \cdot k & 0 \cdot \dfrac{newton}{m} \\ -2 \cdot k & 3 \cdot k & -k \\ 0 \cdot \dfrac{newton}{m} & -k & k \end{bmatrix}$$

The inverse of the mass matrix is calculated as

$$M_{inv} := \frac{1}{\left(4 \cdot mass^3\right)} \cdot \begin{bmatrix} 4 \cdot mass^2 & 0 \cdot kg^2 & 0 \cdot kg^2 \\ 0 \cdot kg^2 & 2 \cdot mass^2 & 0 \cdot kg^2 \\ 0 \cdot kg^2 & 0 \cdot kg^2 & 2 \cdot mass^2 \end{bmatrix} \qquad M_{inv} = \begin{pmatrix} 1 & 0 & 0 \\ 0 & 0.5 & 0 \\ 0 & 0 & 0.5 \end{pmatrix} \cdot kg^{-1}$$

The natural frequencies are the square roots of the eigenvalues of $\mathbf{M_{inv}K}$. To this end

$$D := M_{inv} \cdot K \qquad\qquad D = \begin{pmatrix} 3 & -2 & 0 \\ -1 & 1.5 & -0.5 \\ 0 & -0.5 & 0.5 \end{pmatrix} \cdot sec^{-2}$$

$$w2 := eigenvals(D) \qquad\qquad w2 = \begin{pmatrix} 3.871 \\ 1 \\ 0.129 \end{pmatrix} \cdot sec^{-2}$$

Thus the natural frequencies are

$$\omega_1 := \sqrt{w2_2} \qquad\qquad \omega_2 := \sqrt{w2_1} \qquad\qquad \omega_3 := \sqrt{w2_0}$$

$$\omega_1 = 0.359 \cdot \frac{rad}{sec} \qquad\qquad \omega_2 = 1 \cdot \frac{rad}{sec} \qquad\qquad \omega_3 = 1.967 \cdot \frac{rad}{sec}$$

The mode shapes are the eigenvectors of $\mathbf{M_{inv}K}$.

$$E := eigenvecs(D) \qquad\qquad E = \begin{pmatrix} 0.915 & 0.577 & 0.383 \\ -0.399 & 0.577 & 0.55 \\ 0.059 & -0.577 & 0.742 \end{pmatrix}$$

The eigenvectors for the modes are **u**, **v**, and **w** respectively where

$$u := E^{<2>} \qquad\qquad v := E^{<1>} \qquad\qquad w := E^{<0>}$$

$$u = \begin{pmatrix} 0.383 \\ 0.55 \\ 0.742 \end{pmatrix} \qquad\qquad v = \begin{pmatrix} 0.577 \\ 0.577 \\ -0.577 \end{pmatrix} \qquad\qquad w = \begin{pmatrix} 0.915 \\ -0.399 \\ 0.059 \end{pmatrix}$$

The mode shapes are normalized with respect to the mass matrix

$$c_1 := u^T \cdot M \cdot u \qquad\qquad c_2 := v^T \cdot M \cdot v \qquad\qquad c_3 := w^T \cdot M \cdot w$$

$$c_1 = 1.853 \cdot kg \qquad\qquad c_2 = 1.667 \cdot kg \qquad\qquad c_3 = 1.162 \cdot kg$$

The normalized mode shapes are

$$u_n := \frac{1}{\sqrt{c_{1_{0,0}}}} \cdot u \qquad v_n := \frac{1}{\sqrt{c_{2_{0,0}}}} \cdot v \qquad w_n := \frac{1}{\sqrt{c_{3_{0,0}}}} \cdot w$$

$$u_n = \begin{pmatrix} 0.282 \\ 0.404 \\ 0.545 \end{pmatrix} \cdot kg^{-0.5} \qquad v_n = \begin{pmatrix} 0.447 \\ 0.447 \\ -0.447 \end{pmatrix} \cdot kg^{-0.5} \qquad w_n = \begin{pmatrix} 0.849 \\ -0.37 \\ 0.055 \end{pmatrix} \cdot kg^{-0.5}$$

The previous solution follows the methods used in Schaum's Outline in Mechanical Vibrations. However, **Mathcad** has a feature which determines eigenvalues and eigenvectors for a generalized eigenvalue problem of the form **Kx = λMx**. The natural frequencies are the square roots of the eigenvalues of the generalized eigenvalue problems and the mode shapes are the normalized eigenvectors. However to use this feature the matrix **K** must be dimensionless. To this end

$$M1 := \begin{pmatrix} 1 & 0 & 0 \\ 0 & 2 & 0 \\ 0 & 0 & 2 \end{pmatrix} \qquad\qquad K1 := \begin{pmatrix} 3 & -2 & 0 \\ -2 & 3 & -1 \\ 0 & -1 & 1 \end{pmatrix}$$

$$w3 := genvals(K1, M1) \qquad w3 = \begin{pmatrix} 3.871 \\ 1 \\ 0.129 \end{pmatrix}$$

$$E1 := genvecs(K1, M1) \qquad E1 = \begin{pmatrix} 0.915 & -0.577 & -0.383 \\ -0.399 & -0.577 & -0.55 \\ 0.059 & 0.577 & -0.742 \end{pmatrix}$$

Note that the eigenvectors returned using **genvecs** are not normalized with respect to the mass matrix.

Extension   Investigate mode shape orthogonality using the normalized eigenvectors.

$$u_n^T \cdot M \cdot u_n = 1 \qquad\qquad u_n^T \cdot M \cdot v_n = 0 \qquad\qquad u_n^T \cdot M \cdot w_n = 0$$

$$v_n^T \cdot M \cdot u_n = 0 \qquad\qquad v_n^T \cdot M \cdot v_n = 1 \qquad\qquad v_n^T \cdot M \cdot w_n = 0$$

$$w_n^T \cdot M \cdot u_n = 0 \qquad\qquad w_n^T \cdot M \cdot v_n = 0 \qquad\qquad w_n^T \cdot M \cdot w_n = 1$$

Further study

(1)   Show numerically that the mode shapes are also orthogonal with respect to the stiffness matrix.

(2)   Show numerically that $u_n^T K u_n = \omega_1^2$

Undamped Absorber Design

(Schaum's Mechanical Vibrations, Solved Problems 8.27 and 8.28, p. 254)

Statement   A machine of mass **m** is attached to a spring of stiffness k_1. During operation the machine is subject to a harmonic excitation of magnitude F_0 and frequency ω.

(a) Determine the stiffness and mass of an undamped absorber of minimum mass such that the steady-state amplitude of the machine is zero and the steady-state amplitude of the absorber mass is less than X_{2max} when the machine operates at ω.

(b) What are the natural frequencies of the system with the absorber in place?

System Parameters

$$m_1 := 200 \cdot kg \qquad k_1 := 4 \cdot 10^5 \cdot \frac{newton}{m} \qquad F_0 := 500 \cdot newton$$

$$\omega := 50 \cdot \frac{rad}{sec} \qquad X_{2max} := 0.002 \cdot m$$

The natural frequency of the primary system is

$$\omega_{11} := \sqrt{\frac{k_1}{m_1}} \qquad\qquad \omega_{11} = 44.721 \cdot \frac{rad}{sec}$$

Solution

(a)   The steady-state amplitude of the primary mass is zero when the absorber is tuned to the excitation frequency. When this occurs the steady-state amplitude of the absorber mass is

$$X_2 = \frac{F_0}{k_2}$$

which is rearranged to yield

$$k_2 = \frac{F_0}{X_2}$$

The minimum absorber stiffness such that $X_2 < X_{2max}$ is

$$k_2 := \frac{F_0}{X_{2max}} \qquad\qquad k_2 = 2.5 \cdot 10^5 \cdot \frac{newton}{m}$$

In order for $X_1 = 0$, the natural frequency of the absorber $\omega_{22} = (k_2/m_2)^{1/2}$ must be equal to the excitation frequency ω. Thus since the ratio of k_2 to m_2 is fixed, the absorber with the minimum possible mass corresponds to using the minimum allowable stiffness. This leads to.

$$m_2 := \frac{k_2}{\omega^2} \qquad\qquad m_2 = 100 \cdot kg$$

(b)   The mass ratio is

$$\mu := \frac{m_2}{m_1} \qquad\qquad \mu = 0.5$$

The natural frequencies of the absorber are the frequencies such that the denominator of Eq.(8.12) is zero. This leads to the following

$$\omega_{22} := \omega \qquad\qquad f(\omega) := \omega^4 - \left[(1 + \mu) \cdot \omega_{22}^2 + \omega_{11}^2\right] \cdot \omega^2 + \omega_{11}^2 \cdot \omega_{22}^2$$

The natural frequencies are the values of ω such that $f(\omega) = 0$. When the absorber is added, the system has two natural frequencies, $\omega_1 < \omega_{11}$, $\omega_2 > \omega_{11}$

$$\omega := 50 \cdot \frac{rad}{sec} \qquad\qquad \text{guess value}$$

$$\omega_1 := root(f(\omega), \omega) \qquad\qquad \omega_1 = 32.679 \cdot \frac{rad}{sec}$$

$$\omega := 80 \cdot \frac{rad}{sec} \qquad\qquad \text{guess value}$$

$$\omega_2 := root(f(\omega), \omega) \qquad\qquad \omega_2 = 68.426 \cdot \frac{rad}{sec}$$

Extension   With the absorber in place, for what values of ω will the steady-state amplitude of the absorber be less than X_{1max} ?

$$X_{1max} := 2 \cdot 10^{-3} \cdot m$$

The steady-state amplitude of the absorber in terms of ω is expressed by

$$r_1(\omega) := \frac{\omega}{\omega_{11}} \qquad\qquad r_2(\omega) := \frac{\omega}{\omega_{22}}$$

$$X_1(\omega) := \frac{F_0}{k_2} \cdot \frac{1 - r_2(\omega)^2}{\left[r_1(\omega)^2 \cdot r_2(\omega)^2 - r_2(\omega)^2 - (1+\mu) \cdot r_1(\omega)^2 + 1 \right]}$$

Analysis shows that the denominator above is negative for $\omega_1 < \omega < \omega_2$. The numerator is positive for $\omega < \omega_{22}$ and negative for $\omega > \omega_{22}$. Thus the appropriate range of ω is such that $\omega_a < \omega < \omega_b$ where $X_1(\omega_a) = -X_{1max}$ and $X_1(\omega_b) = X_{1max}$.

$$TOL := 0.000001$$

The default value of **TOL** = 0.001 is not tight enough in this problem. Try using the method below with the default value and notice the difference in answers. The smaller the value of **TOL**, the more accurate the answer.

$$\omega := 42 \cdot \frac{rad}{sec}$$

$$\omega_a := root\left(X_{1max} + X_1(\omega), \omega\right) \qquad \omega_a = 40.442 \cdot \frac{rad}{sec}$$

$$\omega := 54 \cdot \frac{rad}{sec}$$

$$\omega_b := root\left(X_{1max} - X_1(\omega), \omega\right) \qquad \omega_b = 61.238 \cdot \frac{rad}{sec}$$

Further study

(1)   Study the design of an undamped absorber if the goal is to reduce the steady-state amplitude of the primary mass to 2 mm for all speeds between 35 rad/sec and 65 rad/sec.

(2)   Repeat the extension of this problem if the excitation is a frequency squared excitation caused by a rotating imbalance, that is, $F_0 = 0.2\omega^2$.

Index